PROTEINS OF IRON METABOLISM

A Grune & Stratton Rapid Manuscript Production

PROTEINS OF IRON METABOLISM

Edited by

Elmer B. Brown, M.D.

*Washington University School of Medicine
St. Louis, Missouri*

Philip Aisen, M.D.

*Albert Einstein College of Medicine
Bronx, New York*

Jack Fielding, M.D.

*St. Mary's Hospital
London, England*

Robert R. Crichton, Ph.D.

*University of Louvain
Louvain-la-Neuve, Belgium*

GRUNE & STRATTON
A Subsidiary of Harcourt Brace Jovanovich, Publishers
New York   San Francisco   London

© 1977 by Grune & Stratton, Inc.
All rights reserved. No part of this publication
may be reproduced or transmitted in any form or
by any means, electronic or mechanical, including
photocopy, recording, or any information storage
and retrieval system, without permission in
writing from the publisher.

Grune & Stratton, Inc.
111 Fifth Avenue
New York, New York 10003

Distributed in the United Kingdom by
Academic Press, Inc. (London) Ltd.
24/28 Oval Road, London NW 1

Library of Congress Catalog Number 77-79947
International Standard Book Number 0-8089-1050-7

Printed in the United States of America

CONTENTS

Preface	xi
Contributors	xiii
Plenary Lecture - Chemistry and Biochemistry of Iron, *T. Spiro*	xxiii

SECTION I: FERRITIN STRUCTURE AND FUNCTION

1. Structure and Ion-Binding Properties of Ferritin and Apoferritin   3
   A. Treffry, S. H. Banyard, R. J. Hoare, and P. M. Harrison

2. Ferritin: Comparative Structural Studies, Iron Deposition and Mobilisation   13
   R. R. Crichton, D. Collet-Cassart, Y. Ponce-Ortiz, M. Wauters, F. Roman, and E. Paques

3. Ultrastructural Studies of Disrupted Ferritin   23
   W. H. Massover

4. The Restrictive Nature of Apoferritin Channels as Measured by Passive Diffusion   31
   M. E. May and W. W. Fish

5. Binding of Protons and Other Monvalent Ions by Ferritin and Apoferritin   39
   E. Breslow and S. T. Silk

SECTION II: ISOFERRITINS

6. Synthesis of Ferritin Subunits by Free and Membrane Bound Polysomes   49
   T. G. Adelman and J. W. Drysdale

7. Alterations in Serum and Tissue Isoferritins in Disease States:
   I. Duodenal Ferritin Content and Isoferritin Composition   57
   J. W. Halliday, U. Mack, and L. W. Powell
   Alterations in Serum and Tissue Isoferritins in Disease States:
   II: Hemochromatosis and Malignant Diseases   61
   L. W. Powell, J. W. Halliday, L. B. McKeering and R. Tweedale

8. Assembly of Ferritin Interspecies and Intertissue Hybrids   65
   Y. Niitsu, S. Ohtsuka, N. Watanabe, J. Koseki, U, Kohgo, and I. Urushizaki

9. Ferritin and Apoferritin from Human Liver: Aspects of Heterogeneity   71
   D. J. Lavoie, D. M. Marcus, K. Ishikawa, and I. Listowsky

10. Biochemical and Immunological Properties of Human Isoferritins   79
    M. Worwood, M. Wagstaff, B. M. Jones, S. Dawkins, and A. Jacobs

SECTION III: FERRITIN METABOLISM

11. Assembly of the Ferritin Molecule in Rat Hepatoma Cells in Vitro   91
    S. S. C. Lee and G. W. Richter

12. Red Cell Ferritin and Iron Storage During Animal Development   99
    E. C. Theil and J. E. Brown

13. Induction of Ferritin Formation in Hepatocytes   107
    H. Kief, R. R. Crichton, H. Bähr, K. Engelbart, and R. Lattrell

14. Studies on the Carbohydrate Components of Ferritin   115
    M. A. Cynkin and M. Knowlton

15. Sites of Ferritin Synthesis and Nature of Subunit Product   121
    M. C. Linder, J. Zahringer, B. S. Baliga, R. L. Drake, B. Barres, and H. Munro

SECTION IV: TRANSFERRIN STRUCTURE

16. Structure and Evolution of Serum Transferrin   133
    R. T. A. MacGillivray, E. Mendez, and K. Brew

17. Comparative Structural and Conformational Studies of Polypeptide Chain, Carbohydrate Moiety and Binding Sites of Human Serotransferrin and Lactotransferrin   143
    G. Spik and J. Mazurier

18. Iron-Binding Fragments Obtained by Proteolysis of Bovine Transferrin and Lactoferrin   153
    J. H. Brock, F. R. Arzabe, N. E. Richardson, and E. V. Deverson

19. Chemical Modification of Basic Amino Acid Residues in the Transferrins   161
    T. B. Rogers, R. A. Gold, and R. E. Feeney

20. The Iron Binding Property of Ovotransferrin   169
    J. Williams and R. W. Evans

21. Differences between Ovotransferrin and Human Serum Transferrin in Structural and Metal-Binding Cooperativity   179
    J. W. Donovan

22. Vanadyl(IV) Labelled Transferrin: An Overview N. D. Chasteen, R. C. Campbell, L. K. White, and J. D. Casey	187
23. Different Physical Properties and Similar Functional Properties of the Two Sites of Human Transferrin D. C. Harris	197
24. Transferrin Iron-Binding: Observation on Nonrandom (Heterogeneic) Binding E. J. Zapolski and J. V. Princiotto	205
25. Histidyl Residues of Transferrin and Conalbumin as NMR Reporter Groups for the Binding of Hydrogen Ions, Metal Ions and Anions R. C. Woodworth, R. J. P. Williams, and B. M. Alsaadi	211

SECTION V: TRANSFERRIN FUNCTION

26. Spectrophotometric and Differential Scanning Calorimetric Measurements of Zn(II), Al(III) and Ga(III) Binding to Ovo- and Human Serum Transferrin Y. Tomimatsu and J. W. Donovan	221
27. Iron Release from Transferrin Mediated by Organic Phosphate Compounds E. H. Morgan	227
28. Evidence for the Direct Involvement of ATP in the Iron Uptake by Reticulocytes A. Egyed	237
29. ATP-Induced Release of Fe(III) from Transferrin F. J. Carver and E. Frieden	245
30. A Study of Plasma Transferrin in Normal and Iron Deficient Rats H. Huebers, E. Huebers, S. Linck, and W. Rummel	251
31. Variables in the Rat Iron-Transferrin Reticulocyte System S. Okada and E. B. Brown	261
32. Plasma Iron Kinetics in Man K. Skårberg, A. Christensen, G. Marsaglia, and C. Finch	267

SECTION VI: CELLULAR AND SUBCELLULAR IRON METABOLISM

33. Factors Influencing the Rate of Iron Release from Fe^{3+}-Transferrin-CO_3^{2-} G. A. Graham and G. W. Bates	273

34. Studies on Transferrin Receptors of Erythroid Cells 281
    P. Aisen, A. Leibman, H-Y. Y. Hu, and A. I. Skoultchi

35. The Isolation of Transferrin Receptors from Reticulocyte Membranes 291
    B. Ecarot-Charrier, V. Grey, A. Wilczynska, and H. M. Schulman

36. On the Binding Sites of Iron-Transferrin on Rat Reticulocytes 299
    C. van der Heul, M. J. Kroos, and H. G. van Eijk

37. New Evidence for the Internalization of Functional Transferrin in Rabbit
    Reticulocytes 305
    J. Martinez-Medellin, H. M. Schulman, E. de Miguel, and L. Benavides

38. Ferritin as a Cytosol Transport Protein: Does Transferrin Enter
    Reticulocytes? 311
    J. Fielding and B. E. Speyer

39. The Role of Mitochondria in the Control of Iron Delivery to Hemoglobin
    Molecules 319
    P. Pǒnka, J. Neuwirt, J. Borová, and O. Fuchs

40. Iron Mobilisation from Isolated Rat Hepatocytes 327
    E. Baker, F. R. Vicary, and E. R. Huehns

41. The Incorporation of Iron into Isolated Rat Hepatocytes 335
    D. Grohlich, C. G. D. Morley, R. J. Miller, and A. Bezkorovainy

42. Mitochondrial Iron Uptake and Heme Synthesis in Copper Deficiency 341
    D. M. Williams, A. J. Barbuto, C. L. Atkin, and G. R. Lee

43. Mitochondrial "Non-Heme Non-FeS Iron" and its Significance in the
    Cellular Metabolism of Iron 349
    T. Flatmark and A. Tangerås

44. A General Model of Intracellular Iron Metabolism 359
    B. F. Trump and I. K. Berezesky

45. Membrane Receptors for Microbial Iron Transport Compounds
    (Siderophores) 365
    J. B. Neilands and R. R. Wayne

46. Siderophore Transport In *Bacillus megaterium* 371
    B. R. Byers, J. E. L. Arceneaux, A. H. Haydon, and J. E. Aswell

SECTION VII: OTHER ASPECTS OF IRON METABOLISM

47. Brain Iron in the Rat: A Possible Basis for Irreversible Depletion of
    Brain Non-Heme Iron Following a Brief Period of Iron Deficiency 381
    P. R. Dallman and R. A. Spirito

48. Iron Stores, Serum Ferritin and Iron Absorption — 387
   R. W. Charlton, D. Derman, B. Skikne, S. R. Lynch, M. H. Sayers, J. D. Torrance, and T. H. Bothwell

49. A New Iron Binding Protein with Wide Tissue Distribution — 393
   S. Pollack and F. D. Lasky

50. Iron Binding Compounds in Particulate Fraction of Intestinal Mucosa — 397
   Y. Yoshino, S. Yamakawa, and Y. Hirai

51. Regulation of Iron Absorption by Control of Heme Biosynthesis in the Intestinal Mucosa — 403
   J. Hegenauer, L. Ripley, and P. Saltman

52. The Induction of Diabetic Changes in Rats by Intraperitoneal Injection of Fe^{3+}-NTA
   M. Awai, M. Narasaki, and S. Seno

53. Pathogenesis of Impaired Iron Release in Inflammation — 417
   C. Hershko and A. M. Konijn

54. Iron Status and Host Defense — 427
   A. M. Ganzoni and M. Puschmann

55. The Inhibitory Effect of Serum on the Immunoradiometric Assay of Ferritin — 433
   D. Lipschitz and J. Cook

56. The Development of New Iron Chelating Drugs for the Treatment of Patients with Thalassemia — 439
   A. Cerami, R. W. Grady, C. M. Peterson, R. L. Jones, and J. H. Graziano

PREFACE

Meetings to discuss the biochemistry of ferritin and transferrin have previously been held in London and Louvain-la-Neuve in 1973 and 1975. The present volume records the proceedings of the Third International Meeting which was held from April 10-13, 1977 in the pleasant surroundings of Arden House, New York, the conference center of Columbia University. Over one hundred participants enjoyed three days of exceptional spring weather, to most of which they remained oblivious in the intensive program of formal presentations and discussions, both in and out of the conference room. To all of those, both speakers and discussants, who made the meeting so successful, we are most indebted.

The influence of the two previous meetings showed itself remarkably in the work which had been carried out in the interim and reported at this third meeting. The problem areas had been more sharply defined and had clearly influenced the direction and quality of the research undertaken since then. It is in this happy experience that we find the justification for such a gathering of biochemists, biologists and clinicians.

Once again, we have chosen to publish contributions in the form of camera-ready typescripts: what is lost in elegance is gained in rapidity of publication.

We are particularly indebted to Fay Steinfeld and Doris McCall for their help beyond any reasonable call of duty with the organization of the meeting. The conference was made possible by a grant from the National Institutes of Health (AM 19446). We are also grateful to the CIBA-GEIGY Corporation, to Mallinckrodt Inc. and Mead Johnson Laboratories for their generous assistance.

The Organizing Committee
Philip Aisen,
Elmer Brown,
Robert Crichton,
Jack Fielding

CONTRIBUTORS

Thomas G. Adelman, Ph.D.
Tufts University School of Medicine
Boston, Massachusetts

Philip Aisen, M.D.
Albert Einstein College of Medicine
Bronx, New York

Basim M. Alsaadi. B.Sc., M.Sc.
University of Oxford
Oxford, England

J. E. L. Arceneaux, Ph.D.
University of Mississippi Medical Center
Jackson, Mississippi

Fanny R. Arzabe
Fundacion F Cuenca Villoro
Zaragoza, Spain

J. E. Aswell, Ph.D.
Virginia Polytechnic Institute
Blacksburg, Virginia

Curtis L. Atkin, Ph.D.
University of Utah College of Medicine
Salt Lake City, Utah

Michiyasu Awai, M.D., Ph.D.
Okayama University Medical School
Okayama, Japan

Hermann Bähr, D.V.M.
Hoechst AG
Frankfurt, Germany

Erica Baker, Ph.D.
University College Hospital Medical School
London, England

B. S. Baliga, Ph.D.
Massachusetts Institute of Technology
Cambridge, Massachusetts

Stephen H. Banyard, D. Phil.
University of Sheffield
Sheffield, England

A. J. Barbuto, B.S.
University of Utah College of Medicine
Salt Lake City, Utah

B. Barres, B.S.
Massachusetts Institute of Technology
Cambridge, Massachusetts

George W. Bates, Ph.D.
Texas A&M University
College Station, Texas

L. Benavides
Universidad Nacional Autonoma de Mexico
Mexico D.F., Mexico

Irene K. Berezesky, B.A.
University of Maryland School of Medicine
Baltimore, Maryland

Anatoly Bezkorovainy, Ph.D.
Rush-Presbyterian-St. Luke's Medical Center
Chicago, Illinois

Jitka Borová
Charles University
Prague, Czechoslovakia

Thomas H. Bothwell, M.D.
University of the Witwatersrand
Johannesburg, Republic of South Africa

Esther Breslow, Ph.D.
Cornell University Medical College
New York, New York

Keith Brew, Ph.D.
University of Miami School of Medicine
Miami, Florida

J. H. Brock, Ph.D.
Fundacion F Cuenca Villoro
Zaragoza, Spain

Elmer B. Brown, M.D.
Washington University School of Medicine
St. Louis, Missouri

J. Edward Brown, Ph.D.
North Carolina State University
Raleigh, North Carolina

B. Rowe Byers, Ph.D.
University of Mississippi Medical Center
Jackson, Mississippi

Robert C. Campbell, Ph.D.
University of New Hampshire
Durham, New Hampshire

Franklin J. Carver, Ph.D.
Florida State University
Tallahassee, Florida

J. David Casey
University of New Hampshire
Durham, New Hampshire

Anthony Cerami, Ph.D.
Rockefeller University
New York, New York

R. W. Charlton, M.D.
University of the Witwatersrand
Johannesburg, Republic of South Africa

N. Dennis Chasteen, Ph.D.
University of New Hampshire
Durham, New Hampshire

Alan Christensen
University of Washington
Seattle, Washington

Daniel Collet-Cassart
University of Louvain
Louvain-la-Neuve, Belgium

James Cook, M.D.
University of Kansas Medical Center
Kansas City, Kansas

Robert R. Crichton, Ph.D.
University of Louvain
Louvain-la-Neuve, Belgium

Morris A. Cynkin, Ph.D.
Tufts University School of Medicine
Boston, Massachusetts

Peter R. Dallman, M.D.
University of California
San Francisco, California

Contributors

Sara Dawkins, B.Sc.
Welsh National School of Medicine
Cardiff, Wales

D. Derman, M.B.
University of the Witwatersrand
Johannesburg, Republic of South Africa

E. V. Deverson
Institute of Animal Physiology
Cambridge, England

John W. Donovan, Ph.D.
Western Regional Research Center, USDA
Berkeley, California

R. L. Drake, Ph.D.
Massachusetts Institute of Technology
Cambridge, Massachusetts

James W. Drysdale, Ph.D.
Tufts University School of Medicine
Boston, Massachusetts

B. Ecarot-Charrier, Ph.D.
Lady Davis Institute for Medical Research
Jewish General Hospital
Montreal, Canada

Andrew Egyed, Ph.D.
National Institute of Hematology and Blood Transfusion
Budapest, Hungary

Klaus Engelbart, M.D.
Hoechst AG
Frankfurt, Germany

Robert W. Evans, Ph.D.
University of Bristol
Bristol, England

R. E. Feeney, Ph.D.
University of California
Davis, California

Jack Fielding, F.R.C.P., F.R.C.Path., D.P.H.
St. Mary's Hospital
London, England

Clement A. Finch, M.D.
University of Washington School of Medicine
Seattle, Washington

Wayne W. Fish, Ph.D.
Medical University of South Carolina
Charleston, South Carolina

Torgeir Flatmark, Ph.D., M.D.
University of Bergen
Bergen, Norway

Earl Frieden, Ph.D.
Florida State University
Tallahassee, Florida

Ota Fuchs
Charles University
Prague, Czechoslovakia

Andreas M. Ganzoni, M.D.
University of Ulm
Ulm, Germany

R. A. Gold, M.S.
University of Monterrey School of Medicine
Monterrey, Mexico

Robert W. Grady, Ph.D.
Rockefeller University
New York, New York

Gary A. Graham, Ph.D.
Texas A&M University
College Station, Texas

Joseph H. Graziano, Ph.D.
New York Hospital
Cornell Medical Center
New York, New York

V. Grey, Ph.D.
Lady Davis Institute for Medical
  Research
Jewish General Hospital
Montreal, Canada

Dietmar Grohlich, M.A.
Rush-Presbyterian-St. Luke's Medical
  Center
Chicago, Illinois

June W. Halliday, Ph.D.
University of Queensland
Queensland, Australia

Daniel C. Harris, Ph.D.
University of California
Davis, California

Pauline M. Harrison, D. Phil.
University of Sheffield
Sheffield, England

A. H. Haydon, Ph.D.
University of Mississippi Medical
  Center
Jackson, Mississippi

Jack Hegenauer, Ph.D.
University of California at San Diego
La Jolla, California

Chaim Hershko, M.D.
Hadassah University Hospital
Jerusalem, Israel

Yukihiko Hirai, B.S.
Nippon Medical School
Tokyo, Japan

Richard J. Hoare, D. Phil.
University of Sheffield
Sheffield, England

Hsiang-Yun Yang Hu, Ph.D.
Albert Einstein College of Medicine
Bronx, New York

E. Huebers
University of the Saarland
Homburg/Saar, West Germany

Helmut Huebers, M.D.
University of the Saarland
Homburg/Saar, West Germany

E. R. Huehns, M.D.
University College Hospital Medical
  School
London, England

Kunitsugu Ishikawa, M.D.
Albert Einstein College of Medicine
Bronx, New York

Allan Jacobs, M.D.
Welsh National School of Medicine
Cardiff, Wales

Brian M. Jones, F.I.M.L.T.
Welsh National School of Medicine
Cardiff, Wales

Robert L. Jones, M.D.
Rockefeller University
New York, New York

Heiner Kief, M.D.
Hoechst AG
Frankfurt, Germany

Contributors

Margaret H. Knowlton
Tufts University School of Medicine
Boston, Massachusetts

Utaka Kohgo, M.D.
Sapporo Medical College
Sapporo, Japan

Abraham M. Konijn, Ph.D.
Hadassah University Hospital
Jerusalem, Israel

Junichi Koseki
Sapporo Medical College
Sapporo, Japan

M. J. Kroos
Medical Faculty, Erasmus University
Rotterdam, The Netherlands

Fred D. Lasky, Ph.D.
Albert Einstein College of Medicine
Bronx, New York

Rudolf Lattrell, Ph.D.
Hoechst AG
Frankfurt, Germany

Daniel J. Lavoie, Ph.D.
Albert Einstein College of Medicine
Bronx, New York

G. R. Lee, M.D.
University of Utah College of Medicine
Salt Lake City, Utah

S.S.C. Lee, M.D., Ph.D.
University of Rochester
Rochester, New York

Adela Leibman, M.S.
Albert Einstein College of Medicine
Bronx, New York

S. Linck
University of the Saarland
Homburg/Saar, West Germany

Maria C. Linder, Ph.D.
California State University
Fullerton, California

David Lipschitz, M.D., Ph.D.
University of Kansas Medical Center
Kansas City, Kansas

Irving Listowsky, Ph.D.
Albert Einstein College of Medicine
Bronx, New York

S. R. Lynch, M.D.
University of the Witwatersrand
Johannesburg, Republic of South Africa

Ross. T. A. MacGillivray, Ph.D.
University of Miami School of
   Medicine
Miami, Florida

U. Mack, B.Sc.
University of Queensland
Queensland, Australia

Donald M. Marcus, M.D.
Albert Einstein College of Medicine
Bronx, New York

George Marsaglia
University of Washington School of
   Medicine
Seattle, Washington

J. Martinez-Medellin Ph.D.
Universidad Nacional Autonoma de
   Mexico
Mexico D.F., Mexico

William H. Massover, M.D., Ph.D.
Brown University
Providence, Rhode Island

Michael E. May, Ph.D.
Medical University of South Carolina
Charleston, South Carolina

Joēl Mazurier, Ph.D.
Université des Sciences et Techniques de Lille I
Villeneuve d'Ascq, France

L. B. McKerring, M.B., B.S.
University of Queensland
Queensland, Australia

Enrique Mendez, Ph.D.
University of Miami School of Medicine
Miami, Florida

E. de Miguel
Lady Davis Institute for Medical Research
Jewish General Hospital
Montreal, Canada

Robin J. Miller, M.D.
Rush-Presbyterian-St. Luke's Medical Center
Chicago, Illinois

Evan H. Morgan, M.B., B.S., Ph.D., D.Sc.
University of Western Australia
Nedlands, Western Australia

Colin G. D. Morley, Ph.D.
Rush-Presbyterian-St. Luke's Medical Center
Chicago, Illinois

Hamish N. Munro, D.Sc., M.B.
Massachusetts Institute of Technology
Cambridge, Massachusetts

Mikio Narasaki, M.D.
Okayama University Medical School
Okayama, Japan

J. B. Neilands, Ph.D.
University of California
Berkeley, California

Jan Neuwirt M.D., Ph.D.
Charles University
Prague, Czechoslovakia

Yoshiro Niitsu, M.D.
Sapporo Medical College
Sapporo, Japan

Shinobu Ohtsuka
Sapporo Medical College
Sapporo, Japan

Shigeru Okada, M.D., Ph.D.
Washington University School of Medicine
St. Louis, Missouri

Eric Paques
University of Louvain
Louvain-la-Neuve, Belgium

Charles M. Peterson, M.D.
Rockefeller University
New York, New York

Simeon Pollack, M.D.
Albert Einstein College of Medicine
Bronx, New York

Yezid Ponce-Ortiz
University of Louvain
Louvain-la-Neuve, Belgium

Contributors

Přemysl Poňka, M.D., Ph.D.
Charles University
Prague, Czechoslovakia

Lawrie W. Powell, M.D., Ph.D.
University of Queensland
Queensland, Australia

J. V. Princiotto, M.D., Ph.D.
Georgetown University Schools of
 Medicine and Dentistry
Washington, D.C.

M. Puschmann
University of Ulm
Ulm, Germany

N. E. Richardson
Institute of Animal Physiology
Cambridge, England

Goetz W. Richter, M.D.
University of Rochester
Rochester, New York

Larry Ripley, M.S.
University of California at San Diego
La Jolla, California

T. B. Rogers, Ph.D.
University of California
Davis, California

Francoise Roman
University of Louvain
Louvain-la-Neuve, Belgium

W. Rummel, M.D.
University of the Saarland
Homburg/Saar, West Germany

Paul Saltman, Ph.D.
University of California at San Diego
La Jolla, California

M. H. Sayers, M.B.
University of the Witwatersrand
Johannesburg, Republic of South Africa

H. M. Schulman, Ph.D.
Lady Davis Institute for Medical
 Research
Jewish General Hospital
Montreal, Canada

Satimaru Seno, M.D., Ph.D.
Okayama University Medical School
Okayama, Japan

Susan T. Silk, Ph.D.
Cornell University Medical College
New York, New York

Karl Skårberg, M.D.
Karolinska Sjukhuset
Stockholm, Sweden

B. Skikne, M.B.
University of the Witwatersrand
Johannesburg, Republic of South Africa

Arthur I. Skoultchi, Ph.D.
Albert Einstein College of Medicine
Bronx, New York

Barbara E. Speyer, Ph.D.
St. Mary's Hospital
London, England

Geneviève Spik, Ph.D.
Université des Sciences et Techniques
 de Lille I
Villeneuve d'Ascq, France

Robert A. Spirito, M.S.
University of California
San Francisco, California

Thomas G. Spiro, Ph.D.
Princeton University
Princeton, New Jersey

Arild Tangerås
University of Bergen
Bergen, Norway

Elizabeth C. Theil, Ph.D.
North Carolina State University
Raleigh, North Carolina

Yoshio Tomimatsu, Ph.D.
Western Regional Research Center, USDA
Berkeley, California

J. D. Torrance, Ph.D.
University of the Witwatersrand
Johannesburg, Republic of South Africa

Amyra Treffry, Ph.D.
University of Sheffield
Sheffield, England

Benjamin F. Trump, M.D.
University of Maryland School of Medicine
Baltimore, Maryland

R. Tweedale, B.Sc.
University of Queensland
Queensland, Australia

Ichiro Urushizaki, M.D.
Sapporo Medical College
Sapporo, Japan

C. van der Heul, M.D.
Medical Faculty, Erasmus University
Rotterdam, The Netherlands

H. G. van Eijk, Ph.D.
Medical Faculty, Erasmus University
Rotterdam, The Netherlands

F. R. Vicary
University College Hospital Medical School
London, England

Michael Wagstaff, M.Sc.
Welsh National School of Medicine
Cardiff, Wales

Naoki Watanabe
Sapporo Medical College
Sapporo, Japan

Mireille Wauters
University of Louvain
Louvain-la-Neuve, Belgium

R. R. Wayne
University of California
Berkeley, California

Lawrence K. White, Ph.D.
University of New Hampshire
Durham, New Hampshire

A. Wilczynska, M.Sc.
Lady Davis Institute for Medical Research
Jewish General Hospital
Montreal, Canada

Darryl M. Williams, M.D.
University of Utah College of Medicine
Salt Lake City, Utah

John Williams, Ph.D.
University of Bristol
Bristol, England

Contributors

Robert J. P. Williams, M.A., D. Phil.
University of Oxford
Oxford, England

Robert C. Woodworth, B.S., Ph.D.
University of Vermont
Burlington, Vermont

Mark Worwood, Ph.D.
Welsh National School of Medicine
Cardiff, Wales

Sanae Yamakawa, B.S.
Nippon Medical School
Tokyo, Japan

Minoru Yokota, M.D., Ph.D.
Sapporo Medical College
Sapporo, Japan

Yoshio Yoshino, M.D.
Nippon Medical School
Tokyo, Japan

J. Zähringer, M.D.
Massachusetts Institute of Technology
Cambridge, Massachusetts

E. J. Zapolski, M.S.
Georgetown University Schools of
    Medicine and Dentistry
Washington, D.C.

CHEMISTRY AND BIOCHEMISTRY OF IRON

Thomas G. Spiro

Princeton University

I. INTRODUCTION

Among the metallic elements, iron easily qualifies for preeminent interest in biochemistry and medicine. Not only is it the most abundant metal, but its biochemical functions are extremely diverse, ranging from the activation of oxygen, nitrogen and hydrogen to the control of electron flow through numerous bioenergetic pathways. The human body contains 4-5 grams of iron. Zinc is second, among the metals, with about 2 grams. While zinc is a constituent of many key enzymes, its chemistry is limited to that of a Lewis acid. Other metals are present in much smaller (80, 20, 9, 6 and 1 mg for Cu, Mn, Mo, Cr and Co, respectively (1) or genuinely "trace" quantities. Because of the importance of iron, much more is known about its biochemistry than that of other metals, and outstanding issues are more sharply focused. It is most likely that all the essential metals have specific transport and storage systems, but our knowledge of them is currently very limited. In contrast, enough is known about iron transport and storage to support an entire research specialty, represented at this conference, and to permit lively speculation on the blanks in our knowledge.

II. OXIDATION AND SPIN STATES

In common with other third-row transition metals, the valence electrons of iron belong to the $3\underline{d}$ subshell. Because the $3\underline{d}$ electrons are of comparable energy, the transition metals can exist in a variety of oxidation states. In different chemical settings, iron is found in oxidation states as high as Fe(VI) and as low as Fe(-II). In an aqueous environment, however, only the oxidation states Fe(III) and Fe(II) are stable - although Fe(IV) is implicated in intermediate products of the peroxidase reaction (2). These forms of iron have 5 and 6 valence electrons, respectively.

In the free (gas phase) ions, Fe^{2+} and Fe^{3+}, the five $3\underline{d}$ orbitals are of equal energy. The valence electrons distribute themselves among these orbitals with maximum unpairing of their spins, i.e. five unpaired electrons for Fe^{3+} and four for Fe^{2+}. This situation carries over to the normal "high-spin" complexes of Fe(III) and Fe(II). If the ligands bind very strongly, however, then they may destabilize some of the $\underline{d}$ orbitals sufficiently that electrons are paired up in the remaining ones. In most such cases the complexes are basically octahedral, with two high- and three low-lying orbitals. Placement of the valence electrons in the latter produces "low-spin" Fe(III) and Fe(II) complexes, with one and zero unpaired electrons respectively.

A change in spin state has marked consequences not only for the electronic properties of the iron ion, (magnetism, absorption spectra, etc.) but also for its stereochemistry. The vacant orbitals in the low-spin complexes are those pointing at the ligands, which are therefore unencumbered in forming strong, short bonds. In high-spin complexes all the orbitals are partially filled and iron-ligand distances are consequently longer, in effect an iron ion is bigger when high-spin than when low-spin.

This size effect has important consequences for the chemistry of heme proteins. The porphyrin ring has a central cavity just big enough to accommodate a low-spin iron ion. When two strong axial ligands (e.g. imidazole and O_2) are bound, the iron sits in the heme plane; but when one of them is weak (e.g. water) or absent (as in deoxyhemoglobin) then the iron pops out of the plane (toward the strong axial ligand) and becomes high-spin. The displacement of the iron atom upon deoxygenation of hemoglobin is believed to be a key facet of both reversibility and cooperativity in oxygen binding (3,4).

III. COMPLEXATION AND HYDROLYSIS

A. Iron (III)

The Fe^{3+} ion is small and highly charged. It shows a marked preference for small anions such as F^-, CN^-, O^{2-}, OH^- and also RO^- i.e. alkoxide, phenolate and carboxylate. Its affinity for alkoxide is such that alcoholic OH groups are readily deprotonated at neutral pH upon coordination to Fe^{3+}. The binding of Fe^{3+} to hydroxy-acids and sugars is strong. Its affinity for hydroxide is demonstrated by the fact that the aquo-Fe^{3+} ion, $(H_2O)_6Fe^{3+}$, is a stronger acid than acetic acid; its pK_a is about 3.

The product of hydrolysis is not simply $(H_2O)_4FeOH^{2+}$, because OH^- has a strong tendency to bridge polyvalent metal ions, forming polynuclear complexes. At low degrees of hydrolysis the main product (5) is $Fe_2(OH)_2{}^{2+}$ (coordinated water molecules omitted for clarity). At higher degrees of hydrolysis, large polycations, of approximate formula $[Fe(OH)^{+0.5}_{2.5}]_n$, with n ranging up to 1000, are formed (6). These are essentially amorphous particles of hydrous ferric oxide, with a positively charged surface. Their reactions with acid or other depolymerizing agents are slow, and become slower with time, probably as a result of dehydration (7). They are stable in solution for long periods of time, but precipitate if the pH is raised to neutrality. This is prevented, however, if complexing agents (e.g. acetate [8,9] or fructose [10]) are present which can coat the particles. A sufficient excess of complexing agent can prevent polymerization (9) or can reduce the size and increase the reactivity (10) of the polymers.

It is also possible to limit polymerization at the dimer stage via chelating agents which bind tightly to Fe^{3+}, but leave one or two water molecules coordinated (6). These include EDTA (ethylenediamine tetraacetate), and NTA (nitrilotriacetate), for example, and also porphyrins; in each of these cases the main species present at neutral pH is $(LFe)_2O$, where L is the chelating agent. (It is not possible to tell, in the absence of structure data, whether the formula should be $(LFe)_2O$ or $(LFe)_2(OH)_2$, which differ only by a water molecule. Both linear oxo and bent hydroxy bridges have been seen in crystal structures (10), the former being more common). The dimers are labile, and react rapidly with acid.

The solubility product constant of $Fe(OH)_3$ is about 10^{-39}. This means that the equilibrium concentration of aqueous Fe^{3+} cannot exceed about $10^{-18}M$ at pH 7, although it can rise to $1\underline{M}$ at pH 1. This does not mean, of course, that significant quantities of ferric iron cannot be found in neutral solution, but only that its chemistry depends completely on the nature of the complexed species and their hydrolysis products. Their reactions are apt to be slow, either because of polymerization, or because of the need to displace tightly bound chelating agents.

B. Iron (II)

The Fe^{2+} ion is less polarizing than Fe^{3+}. It is a much weaker aquoacid, with a pK_a around 7, and the solubility product of $Fe(OH)_2$ is around 10^{-15}, allowing a $(H_2O)_6Fe^{2+}$ concentration near 0.1 M at pH 7. Its affinity for anionic oxygen-containing ligands is less pronounced. Both Fe^{2+} and Fe^{3+} appear to have comparable affinities for thiolate and sulfide ligands, as judged from the extensive chemistry of iron-sulfur proteins and chemical analogs which has recently been developed (12).

Ferrous iron has special affinity for chelating ligands containing the α-diimine structure $\overset{\nearrow\,-\,\searrow}{N\quad N}$ including α,α bipyridyl, o-phenanthrolene and porphyrin. These form low-spin complexes with Fe^{2+}, in which the paired iron electrons are stabilized by delocalization into low-lying π* orbitals of the ligands.

C. Stereochemistry

Low-spin complexes, whether Fe(II) or (III), are six coordinate and octahedral, with bond angles close to 90° at the iron atom. The spatial disposition of the filled and unfilled d orbitals enforce this geometry. In high-spin complexes the orbitals are all partially filled and there is no directional preference. The stereochemistry is imposed by the ligands. Small ligands, such as water, favor six-coordination, while larger ones favor five, or four coordination.

Chloride forms tetrahedral $FeCl_4^-$ and $FeCl_4^{2-}$, and tetrahedral geometry is maintained in iron-sulfur chemistry (12). For chelating ligands, constraints of ring size often determine the coordination geometry. The out-of-plane structure of five-coordinate high-spin iron heme and its relevance to oxygenation, has already been mentioned. In the case of Fe^{3+}-EDTA chelate, the six binding groups cannot stretch far enough to occupy regular octahedral coordination sites; they leave room for a bound water molecule, and the complex is seven-coordinate (13).

IV. OXIDATION-REDUCTION

A. Potentials

The relative stability of Fe^{3+} and Fe^{2+} in a given ligand environment is measured by the redox potential. Some examples are given below, along with reduction potentials for O_2 and H^+. For reactions which consume protons as well as electrons the potential is strongly pH dependent. While the reduction of aquo-Fe^{3+} does not consume a proton, the potential, 0.77 v, is nevertheless limited to strongly acid solution, because hydrolysis becomes important at pH values higher than 1.

	E_o' (volt)	
	pH 0	pH 7
$O_2 + 4H^+ + 4e^- = 2H_2O$	1.23	0.82
$Fe(phen)_3^{3+} + e^- = Fe(phen)_3^{2+}$ (phen = o-phenanthrolene)	1.12	
$Fe(H_2O)_6^{3+} + e^- = Fe(H_2O)_6^{2+}$	0.77	
$Fe(CN)_6^{3-} + e^- = Fe(CN)_6^{4-}$		0.36
Fe(III) cytc + e^- = Fe(II) cytc (cytc = cytochrome c)		0.25
Fe(III) Hb + e^- = Fe(II)Hb (Hb = hemoglobin)		0.14
Fe(III) HRP + e^- = Fe(II) HRP (HRP = horseradish peroxidase)		-0.27
$2H^+ + 2e^- = H_2$	0.0	-0.41

Relative to water, o-phenanthrolene stabilizes Fe^{2+}, raising the reduction potential to 1.12. Most other ligands stabilize Fe^{3+}, including porphyrin, which, although strongly delocalizing, like o-phenanthrolene, is a dianionic ligand. The value of the potential varies widely for iron porphyrins, however, depending on the nature of the axial ligands and the chemical setting. Imidazole shows a preference for Fe^{3+}, and horseradish peroxidase (HRP), which is thought to have imidazole and water as axial ligands, has extremely low redox potential; a nearby anionic side-chain may also contribute to the preference for Fe^{3+} (14). But hemoglobin (Hb) which has these same axial ligands, has a reduction potential nearly half a volt higher (a fortunate circumstance since Fe[II]Hb is the physiologically active form). There is resonance Raman evidence that the Fe^{3+} form of Hb is destabilized by protein induced strain on the porphyrin ring (4). In the case of cytochrome c, which also has an imidazole axial ligand, the high redox potential is thought to result from the sixth ligand being methionine, which favors (low-spin) Fe^{2+}.

B. Mechanisms

The rates of redox reactions are not unconnected to the potentials. The greater the potential difference between two half-reactions, the greater the driving force for electron transfer. Oxygen is certainly capable of oxidizing Fe^{2+}, but the reaction is not especially rapid in acid solution, where the Fe^{3+} reduction potential is relatively high. The rate accelerates markedly as the pH is raised, however (15), reflecting the stabilization of Fe^{3+} by hydroxide, and the resultant drop in the reduction potential. In general Fe^{2+} oxidation is accelerated by ligands favoring Fe^{3+}. For example, transferrin, which avidly binds Fe^{3+}, increases the rate of Fe^{2+} oxidation (16), even though there is no evidence for specific binding of Fe^{2+} to transferrin (17).

There may, however, be mechanistic barriers to redox reactions. While electron transfer to and from simple iron complexes is mechanistically straightforward, this is less likely for iron buried in a protein site, or polymerized in an oxide particle. Here steric considerations and specific binding sites are likely to be important. On the other side, the oxidant or reductant may be mechanistically complex.

This is certainly the case for molecular oxygen, which requires four electrons for reduction to water. Four electrons can be supplied in a single step only by the terminal oxidases, the complex multi-metal ion proteins evolved by nature for this purpose. In other cases, reduction of oxygen leads to complex intermediates or products. The initial step in the reduction by Fe^{2+} is most likely the formation of a transient complex.

$$Fe^{2+} + O_2 \rightarrow Fe^{3+}\text{-}O_2^-$$

with partial electron transfer to form bound superoxide (18). In acid solution, this complex is subject to attack by

$$Fe^{3+}-O_2^- + H^+ \rightarrow Fe^{3+} + HO_2$$

protons to produce HO_2 which could then obtain another electron from a second Fe^{2+} to produce peroxide. Alternatively, the second Fe^{2+} could directly attack the bound superoxide

$$Fe^{3+}-O_2^- + Fe^{2+} \rightarrow Fe^{3+}-O_2^{2-}-Fe^{3+}$$

producing a peroxo-bridged species, for which there is precedent in cobalt chemistry (18), (and perhaps also in oxy-hemerythrin [19]). If the breakdown of the superoxide or peroxide complexes is rate-limiting, then the reaction can be accelerated by agents which rapidly exchange electrons with both Fe^{2+} and O_2. This is presumably the basis for the catalytic effect of cupric ion (20) and of ceruloplasmin (21).

Whether such catalysis is physiologically required to account for iron turnover in the plasma (21) is difficult to assess, in view of the likelihood that iron redox and transfer reactions take place at specific sites rather than in the circulating plasma. However, ceruloplasmin, or some other oxidase, would have the biological virtue of reducing O_2 to H_2O in a single step, thereby preventing the release of superoxide and peroxide (also hydroxyl radicals, for which Fenton's reagent, $Fe^{2+} + H_2O_2$, is a commonly used source) which are damaging to biological tissue. It is possible, of course, that nature's supply of superoxide dismutase and catalase are sufficient to eliminate these noxious byproducts at the rates normally needed.

V. BIOLOGICAL IMPLICATIONS

It seems evident that nature has selected iron in large part because of its ability to vary the Fe^{3+}/Fe^{2+} redox potential over the entire range from oxygen to hydrogen. This virtuosity is demonstrated in the mitochondrial respiratory chain, where every electron transfer stage involves either iron-heme or iron-sulfur complexes (the oxygen terminus, cytochrome oxidase, also uses copper).

To maintain an adequate supply of this multi-purpose element, a substantial transport and storage system is needed. Although iron is an abundant element in the earth's crust, most of it is not readily assimilated biologically. We live in an aerobic world, in which Fe(III) is the stable oxidation state, and insoluble ferric oxide is the stable chemical form. Microbes have evolved selective Fe^{3+} chelating agents, the siderochromes (22), which they excrete into their environment to scavenge iron, while plants exude organic acids for the same purpose. Humans rely instead on gastric juice, whose acidity mobilizes a modest proportion of the iron in the diet (23). It is difficult to guess what happens to iron when the stomach's contents pass to the alkaline milieu of the intestine. Competing hydrolysis and complexation reactions must determine the absorbability of the Fe^{3+} ions, and there is evidently some reduction to Fe^{2+} (23). Dietary supplements of iron pass through this same complicated reaction vessel. Simple ferric salts are ineffective because of the immobile polymers which they form on neutralization, but the smaller, labile polymers found in ferric fructose solutions (10) are apparently absorbed (24). Ferrous salts have long been known to be effective in iron uptake.

Once across the intestinal wall, iron is transported by transferrin and stored in ferritin. Both of these represent highly stable environments for Fe^{3+}. The transferrin binding constant for Fe^{3+} is comparable to those of the siderochromes, and ferritin is essentially a protein-coated ferric oxide particle. How is iron delivered to and from these sites? Small, highly specific Fe^{3+} chelators are one possibility, although their concentrations and binding constants would have to be carefully regulated. Reduction to Fe^{2+} followed by reoxidation is another, and perhaps more plausible alternative. Most studies have dwelt on in vitro measurements with chemical model systems. While these provide a necessary foundation, the real challenge is to uncover the specific in vivo mechanisms that are operative.

REFERENCES

1. Worwood, M. in "Iron in Biochemistry and Medicine" (Jacobs, A. and Worwood, M. eds.), p. 345, Academic Press, N.Y., 1974.

2. Rakhit, G., Spiro, T.G. and Uyeda, M., Biochem. and Biophys. Res. Commun. 71, 803 (1976).

3. Perutz, M.F., Feisht, A.R., Simon, S.R. and Roberts, G.K., Biochem. 13, 2174 (1974).

4. Spiro, T.G. and Burke, M.J., J. Am. Chem. Soc. 98, 5982 (1976).

5. Hedstrom, B.O.A. Arkiv. Kemi 6, 1 (1953).

6. Spiro, T.G. and Saltman, P., Structure and Bonding 6, 116 (1969).

7. Sommer, B.A., Margerum, D.W., Renner, J., Saltman, P. and Spiro, T.G., Bioinorg. Chem. 2, 295 (1973).

8. Spiro, T.G., Pape, L. and Saltman, P., J. Am. Chem. Soc. 89, 5555 (1967).

9. Spiro, T.G., Bates, G. and Saltman, P., J. Am. Chem. Soc. 89, 5559 (1967).

10. Bates, G., Hegenauer, J., Renner, J., Saltman P. and Spiro, T.G., Bioinorg. Chem. 2, 311 (1973).

11. Thich, J.A., Ou, C.C., Powers, D., Vasiliou, B., Mastropalo, D., Potenza, J.A. and Schugar, H.J., J. Am. Chem. Soc. 98, 1425 (1976).

12. Holm, R.H., Endeavor 34, 38 (1975).

13. Lund, M.D., Hamor, M.J., Hamor, T.A., and Hoard, J.L., Inorg. Chem. 3, 34 (1964).

14. Williams, R.J.P. in "Iron in Biochemistry and Medicine" (Jacobs, A and Worwood, M. eds.) p. 198, Academic Press, N.Y., 1974.

15. Goto, K., Tamura, H. and Nagayama, M., Inorg. Chem. 9, 963 (1970).

16. Bates, G.W., Workman, E.F. and Schlabach, M.R., Biochem. and Biophys. Res. Commun. 50, 84 (1973).

17. Gaber, B.P. and Aisen, P., Biochim. Biophys. Acta 221, 228 (1970).

18. Vaska, L., Accts. Chem. Res. 9, 175 (1976).

19. Kurtz, D.M., Shriver, D.F. and Klotz, I.M., J. Am. Chem. Soc., 98, 5035 (1976).

20. Kurimura, Y., and Murakami, K., Bull. Chem. Soc. Jap. 42, 2715 (1969).

21. Osaki, S., Johnson, D.A. and Frieden, E., J. Biol. Chem. 241, 2746 (1966).

22. Neilands, J.B., Structure and Bonding 11, 145 (1972).

23. Turnbull, A. in "Iron in Biochemistry and Medicine" (Jacobs, A. and Worwood, M. eds.) p. 370. Academic Press, N.Y., 1974.

24. Bates, G.W., Boyer, J., Hegenauer, J.C. and Saltman, P., Am. J. Clin. Nutr. 25, 983 (1972).

PROTEINS OF IRON METABOLISM

Section I

FERRITIN STRUCTURE AND FUNCTION

STRUCTURE AND ION-BINDING PROPERTIES OF FERRITIN & APOFERRITIN

Amyra Treffry, Stephen H. Banyard,
Richard J. Hoare & Pauline M. Harrison
Department of Biochemistry, University of Sheffield

STRUCTURE OF HORSE SPLEEN APOFERRITIN

Horse spleen apoferritin is a protein of molecular weight 440,000 composed of 24 subunits, arranged as a hollow shell (12-13nm outside and about 8nm inside diameter). It may accommodate up to 4500 Fe atoms as microcrystalline hydrous ferric oxide-phosphate, which is structurally unrelated to apoferritin at the atomic level(1). A molecular model of horse spleen apoferritin based on our X-ray analysis at 600pm resolution(2) is shown in Fig. 1. The analysis was made on cubic

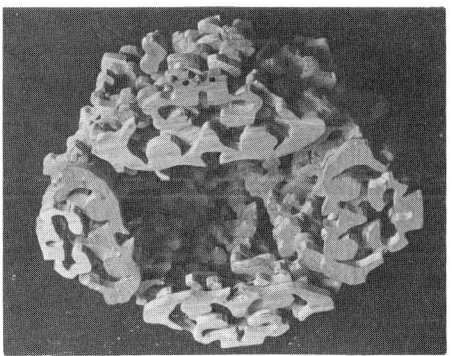

Fig. 1. Balsa wood model of half an apoferritin molecule based on the 0.6nm resolution electron density map of Hoare et al.(2). Four of the six channels which pass through the shell to the central cavity are visible.

crystals in which molecular and crystal 432 symmetry relating the structurally equivalent subunits are coincident. Subunits meet near, but cannot cross, symmetry axes, but between axes their boundaries are not clearly defined in our model. A striking feature is the presence of channels along 4-fold axes, providing a means of access for iron and small molecules (not

more than about 1.3nm across) to the central cavity. The high
resolution analysis in progress should elucidate the nature of
the side chains around the channels. Our static picture cannot
tell us the rates at which molecules or ions enter the molecule
but we find the following at internal sites: mercuribenzoate,
UO_2^{2+}, Tb^{3+}, CH_3Hg^+, $(Nb_6Cl_{12})^{2+}$. We note that several of
these are cations and that a roughly spherical complex of
diameter 1nm, $(Nb_6Cl_{12})^{2+}$, has entered the molecule. Binding
of UO_2^{2+} and Tb^{3+} probably involves carboxyls; mercurials
might be expected to bind to thiols. Covalently bound mercuribenzoate is found at two sites, only one of which is shared by
CH_3Hg^+, so possibly only this one site locates cysteine.

The inner protein surface is not smooth, but contains
eight shallow pockets centred on the 3-fold axes. The comple-
mentary space of the iron-core is thus lobed, explaining the
low angle diffraction patterns of ferritin crystals(1) and the
apparent sub-unit structure observed by electron microscopists
(3). Protein conformations in ferritin and apoferritin
crystals are identical to 250pm resolution.

STRUCTURE-FUNCTION RELATIONSHIPS IN FERRITIN

Remarkably little is known about how ferritin acquires or
releases its iron in vivo. In vitro the most successful method
of forming ferritin from apoferritin is to present the latter
with Fe(II) and an oxidant (4,5). The stable form of iron at
physiological pH, however, is Fe(III). Iron injected into rats
as ferric ammonium citrate(6) is accumulated by liver ferritin,
and its distribution amongst molecules of different Fe-contents
resembles that found when Fe(II) is given in vitro. This does
not prove, however, that the pathway involves Fe(II). We show
below that ferritin molecules can bind limited amounts of
Fe(III), although we were unable to demonstrate any binding to
apoferritin.

Ferritin molecules contain phosphate associated with their
iron-cores, which may explain differences in their titration
behaviour(7). Ferritin phosphate may be in dynamic equilibrium
with the 1mM or greater intracellular concentration of
inorganic phosphate(8) and our in vitro experiments support
this conclusion. Moreover they suggest that ferritin's Fe and
P_i are added separately. This could mean that Fe is acquired
at a membrane site, perhaps the site at which transferrin
releases its Fe.

It is tempting to suppose that one or more of the Tb^{3+} or
UO_2^{2+} sites we have observed on apoferritin's inner surface
are also binding sites for iron and represent the nucleation

sites we postulated (4) as part of the mechanism of iron
accumulation. We know that apoferritin also binds Zn(II) (9)
and that zinc inhibits iron incorporation both in vitro (9)
and in vivo (10). We compare below the effects of zinc and
terbium on Fe-uptake into ferritin and apoferritin and the
competition between these two ions for binding sites on the
protein.

Fe(III)-BINDING BY FERRITIN

We examined binding from solutions of Fe(III)-chelates in
two ways. Either MSE dialysis cells were used and apoferritin
or ferritin fractions of different iron contents were dialysed
against buffered solutions of ^{59}Fe-labelled Fe(III)-chelates
for periods up to 7 days, or unfractionated native ferritin
was incubated under similar conditions and, after washing to
remove excess reagents, was then fractionated by sucrose
density gradient centrifugation. Binding of Fe(III) by
ferritin (but not apoferritin) was found in the presence of
oxalate, nitrilotriacetate or citrate at neutral or, better, at
alkaline pH. The number of Fe atoms bound per molecule was
dependent on concentrations of protein and Fe-chelate, buffer
and pH. Results for two experiments are shown in Fig. 2 below
expressed as ions bound per molecule (a) and per Fe atom (b).

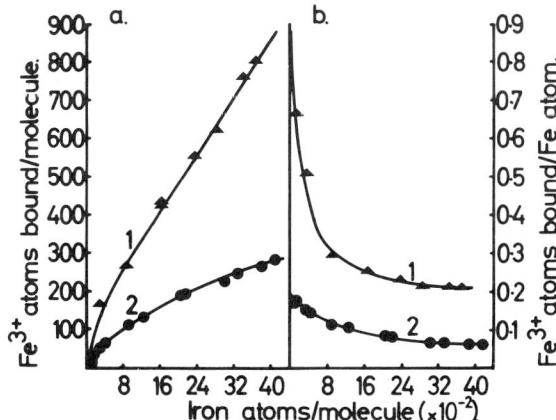

Fig. 2. ^{59}Fe(III)-binding by native horse spleen ferritin
a. Equilibrium dialysis (7 days) of native ferritin fractions.
Protein (150µg/ml) dialysed against 0.5mM ^{59}Fe(NO$_3$)$_3$/10mM Na
citrate in MSE Dianorm 2 x 1 ml cells at 25°C. Buffer: 20mM

glycine/NaOH, pH 9.4, containing 0.1M $NaNO_3$, 0.02% NaN_3.
b. Unfractionated ferritin incubated 7 days with 0.5mM ^{59}Fe $(NO_3)_3$/10mM Na citrate, buffer as in a, then washed by ultrafiltration and fractionated by density gradient centrifugation.

The difference between the two sets of results, 1 and 2, represents the loosely bound Fe removed by washing in ultrafiltration cells. It can be seen that Fe(III)-binding is dependent on molecular Fe-content. The distribution is somewhat different from those obtained when small amounts of Fe(II) plus oxident were presented or initial rates of Fe(II) oxidation were examined (these showed maxima for fractions containing about 1000 Fe/molecule (11)), but in all cases relatively high specific binding was given by iron-poor fractions, which we have attributed to the relatively high surface/volume of their small iron-core particles (11). The iron-poor molecules did not seem to be capable of building up stable iron-cores under the conditions used in these experiments. The Fe(III) was judged to be <u>inside</u> the molecule in representative washed samples since their isoelectric focussing patterns were unaltered. Whether the relatively small amount of Fe(III) bound at pH 9.4 in these experiments has any physiological significance is an open question. It does indicate that, however inefficiently, iron can be bound by the micelles without obligatory passage through an oxidation site.

Although the experimental evidence indicates clearly that a small amount of Fe(III) can be taken up by ferritin molecules the nett binding of ^{59}Fe(III) from ferric citrate may have been compensated for, or partly compensated for, by loss of ^{56}Fe(III) from the iron-cores. In those samples in which we measured total Fe as well as ^{59}Fe, we found a nett Fe increase in the ferritin compartment.

PHOSPHATE BINDING AND RELEASE

Ferritin reconstituted in the absence of phosphate from Fe(II) and apoferritin (4) was found to bind P_i on incubation at pH 7.0 with ^{32}P-phosphate and the product was indistinguishable from native ferritin. When we reconstituted ferritin in the presence of phosphate, however, we obtained a product differing from the above in its molar extinction at 420nm. Its ^{32}P-phosphate was less readily exchangeable by cold anions and its iron was released more rapidly. We also compared the release of ^{32}P-phosphate from these reconstituted ferritins, and from native ferritin which had been incubated with the labelled anion, with release of cold P_i from native ferritin as a function of Fe- release with thioglycollate at pH 4.3. The results are shown in Fig. 3. It can be seen that P_i is

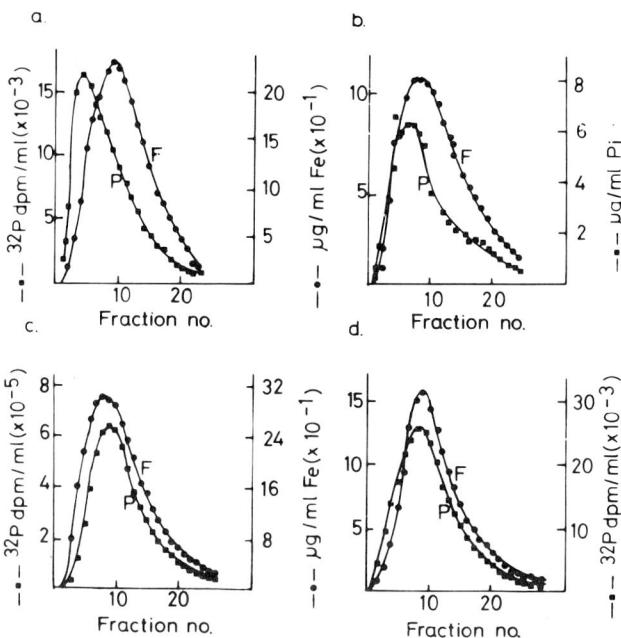

Fig. 3. Release of phosphate and iron from horse spleen ferritin by treatment with thioglycollic acid/sodium acetate at pH 4.25. A continuous ultrafiltration method was used with a cell volume of 5ml. Iron was measured in successive 1ml fractions as its bipyridyl complex and phosphate was measured either colorimetrically or as ^{32}P.
a. native ferritin which had been incubated with ^{32}P-phosphate.
b. native ferritin. c. ferritin reconstituted from apoferritin and Fe(II) in the presence of ^{32}P-phosphate. d. reconstituted ferritin incubated with ^{32}P-phosphate after reconstitution.
P, phosphate released   F, iron released as Fe(II)

removed well ahead of Fe in a, slightly ahead in b and d and with or slightly behind Fe in c. When we relate this to the micellar nature of ferritin's iron-core, we conclude that in a the P_i is largely on surface sites, in b and d it is both surface and internal (in intercrystallite spaces or imperfections) while in c the P_i is distributed throughout the iron-core.

BINDING OF Zn(II) AND Tb(III) AND THEIR EFFECTS ON Fe-UPTAKE

A. Inhibition by Tb(III) and Zn(II) of Ferritin Fe-Uptake

We have studied the effects of adding $TbCl_3$ or $ZnSO_4$ on Fe-uptake into ferritin and apoferritin. The system used was similar to that described previously (4,9). Iron, to concn. 150μM, was added as $Fe(NH_4)_2(SO_4)_2$ to solutions of apoferritin or a ferritin fraction containing about 2000 Fe atoms/molecule (about 1-2μM) in 20mM imidazole buffer containing excess KIO_3/$Na_2S_2O_3$ as oxidant. The formation of the red-brown hydrous ferric oxide micelles inside ferritin was followed by recording optical densities at 310 or 420nm. $ZnSO_4$ or $TbCl_3$ was added before $Fe(NH_4)_2(SO_4)_2$. Marked inhibition was observed even when the concn. of Zn(II) or Tb(III) was no more than that of Fe(II). Inhibition with both metal ions was greater with apoferritin than with the ferritin fraction. These results support our previous conclusion that Fe-uptake into apoferritin is a two-stage process involving an initial slow step, in which iron-core crystallites are nucleating on the apoferritin inner surface, followed by a more rapid growth phase, when Fe is deposited directly on these nuclei. Both phases are inhibited by extraneous metal ion, and we conclude tentatively that both Zn and Tb ions compete with Fe ions for binding sites on apoferritin and on iron-core crystallites.

B. Zn(II)-binding by Apoferritin and Ferritin and Competition by Tb(III)

Both Zn(II) and Tb(III) were shown to bind to apoferritin in a previous study in which binding was measured as release of protons at acid pH (9). Two classes of site were found containing 1-2 and 3-4 metal ions, the former class having the higher affinity. We have now reinvestigated $^{65}Zn(II)$-binding to apoferritin by equilibrium dialysis in MSE equilibrium dialysis cells and have studied the effect of $TbCl_3$ on this binding. Results such as those shown in Fig. 4 indicate the presence of two classes of site containing 1-2 and 2-4 Zn(II) ions and that Tb(III) competes for the 1-2 high affinity sites and possibly also for the 2-4 low affinity sites.

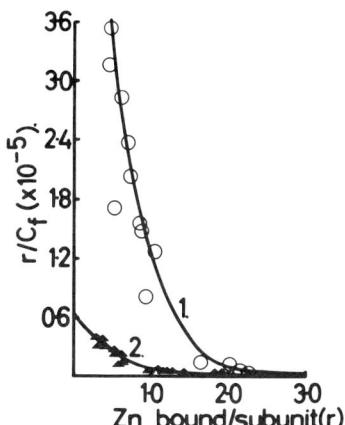

Fig. 4. ^{65}Zn(II)-binding by horse spleen apoferritin. Binding in MSE Dianorm equilibrium dialysis 2 x 1ml cells at 25°C. Apoferritin (1mg/ml) dialysed against 0.04-2.0mM ^{65}ZnSO$_4$. Buffer: 20mM imidazole, pH 7.0, containing 0.1M NaNO$_3$, 0.02% NaN$_3$. For competition of Tb(III), 0.196 μmoles TbCl$_3$ was added to each compartment (curve 2).

Binding of Zn(II) to a ferritin fraction containing 2400 Fe atoms/molecule in equilibrium dialysis cells indicated about 2 high affinity and a much larger number of low affinity sites, some or all of which may represent binding to the iron-core. In a preliminary experiment we also found that Tb(III) inhibits binding of ^{59}Fe(III) citrate to ferritin.

C. The Effects of Chemical Modification on Binding and Inhibition by Zn(II) and Tb(III)

Apoferritin was treated with bromoacetate as described by Silk and Breslow (7) under conditions in which two histidines and one cysteine are carboxymethylated (7) and we also observed the incorporation of three ^{14}C-acetate groups/subunit. We found that this modification affected neither Zn(II)-binding nor the competition of Tb(III) for Zn(II)-binding sites. Iron uptake rates were unaffected or even enhanced.

Carboxyl groups in apoferritin and a ferritin fraction (2400 Fe atoms/molecule) were modified with glycineamide as described by Wetz and Crichton (12). These authors found 11 carboxyls/subunit modified in apoferritin and 7 in full ferritin. In agreement with Wetz and Crichton we found carboxyl modification of apoferritin reduced its ability to oxidize Fe(II) almost to zero. We also found Zn(II)-binding was now practically nil. Apoferritin derived from the modified ferritin fraction was capable of accumulating iron, but at a lower rate than that of unmodified apoferritin, which was also subject to inhibition by both zinc and terbium. It also gave fewer Zn(II) ions bound in equilibrium dialysis experiments and Tb(III) competition for these sites was again observed.

D. Tb(III) Sites in Apoferritin Crystals

Apoferritin was crystallized for X-ray analysis in approx. 40mM $CdSO_4$. In a three dimensional electron density difference map, calculated with data obtained from crystals bathed in 37mM $CdSO_4$ plus 100mM $TbCl_3$ compared with data from crystals in 37mM $CdSO_4$ alone, we found one major Tb(III) site tucked in a hollow off the inner surface, a minor site also on the inner surface (both these sites were within 0.1nm of sites previously found for UO_2^{2+}), another minor site close to the 3-fold axis and near the outer surface and a negative peak near this last site, which may have been due to displaced Cd(II). Two further small peaks (on inner and outer surfaces) may possibly be Tb-binding sites of very low occupancy. We have not examined Zn(II)-binding because of its relatively low atomic number and the presence of cadmium in the crystals.

E. Conclusions

Our results indicate competition between Zn(II)- and Tb(III)-binding, competition between these ions and Fe-uptake and elimination of Zn(II)-binding and Fe-uptake by carboxyl modification. We conclude that one or more of the sites at which Fe(II), Zn(II) and Tb(III) are bound involves one or more carboxyl group. Since carboxyl groups are protected from modification in full ferritin (12) it seems highly probable that one or more of the inner Tb(III) sites observed by X-ray analysis is also an Fe-oxidation site postulated as a nucleation centre for ferritin iron-core formation.

REFERENCES

1. Harrison, P.M., Hoare, R.J., Hoy, T.G., Macara, I.G., in "Iron in Biochemistry and Medicine" (Jacobs, A. & Worwood, M. eds.) pp.73-114, Academic Press, London, 1974.

2. Hoare, R.J., Harrison, P.M., Hoy, T.G., Nature 255, 653 (1975).

3. Farrant, J.L., Biochim.Biophys.Acta 13, 569 (1954).

4. Macara, I.G., Hoy, T.G., Harrison, P.M., Biochem.J. 126, 151 (1972).

5. Bryce, C.F.A., Crichton, R.R., Biochem.J. 133, 301 (1973).

6. Hoy, T.G., Harrison, P.M., Brit.J.Haematol. 33, 497 (1976).

7. Silk, S.T., Breslow, E., J.Biol.Chem. 251, 6963 (1976).

8. Van Kreel, B.K., Pijnenburg, A.M.C.M., Van Eijk, H.G., Leijnse, B., Biochim.Biophys.Acta 273, 243 (1972).

9. Macara, I.G., Hoy, T.G., Harrison, P.M., Biochem.J. 135, 785 (1973).

10. Coleman, C.B., Matrone, G., Biochim.Biophys.Acta 177, 106 (1969).

11. Harrison, P.M., Hoy, T.G., Macara, I.G., Hoare, R.J., Biochem.J. 143, 445 (1974).

12. Wetz, K., Crichton, R.R., Eur.J.Biochem. 61, 545 (1976).

We thank the Medical and Science Research Councils for support and Mr. A. Doyle, Mrs. J. Brentnall and Mrs. J. Sowerby for expert technical assistance.

FERRITIN : COMPARATIVE STRUCTURAL STUDIES, IRON DEPOSITION AND MOBILISATION

ROBERT R. CRICHTON, DANIEL COLLET-CASSART, YEZID PONCE-ORTIZ, MIREILLE WAUTERS, FRANCOISE ROMAN, ERIC PAQUES

Unité de Biochimie, Université de Louvain
Louvain-la-Neuve, Belgium

I. STRUCTURAL STUDIES ON HORSE SPLEEN APOFERRITIN

It is now generally accepted that horse spleen apoferritin consists of 24 chemically identical polypeptide chains (hereafter referred to as subunits) each of molecular weight 18,500 (1). The presence of components of lower molecular weight than the subunit (12,000 and 6,000) was originally attributed by Björk and Fish to proteolysis (2) and this view was reinforced by the observation that amino acid compositions, tryptic peptide patterns, N- and C-terminal residues as well as the molecular weights of the fragments were consistent with this interpretation (3-5). More recently we have observed that addition of inhibitors of proteolysis such as phenylmethane sulphonyl fluoride, to the homogenising medium, give ferritin preparations which no longer contain the lower molecular weight components, as judged by SDS-gel electrophoresis (Collet-Cassart D. and Crichton R.R. in preparation). It has also been reported that commercial preparations of ferritin exhibit a proteolytic activity(6).

The sequence determination of horse spleen apoferritin is being carried out by the following strategy : (i) isolation of tryptic peptides resulting from cleavage of the maleylated protein, (ii) isolation of peptides from CNBr cleavage (iii) isolation of peptic peptides. Sequence analysis on small peptides (up to 15-20 residues) was carried out by dansyl-Edman (7) and of longer peptides by automated Edman degradation (8). Previously, an N-terminal pentapeptide and C-terminal octapeptide were described (9,10).

Hydrolysis of maleylated apoferritin with trypsin yields 11-12 arginyl peptides ; the cleavage is complete as judged by SDS-acrylamide gel electrophoresis whereas trypsin digestion of non-maleylated apoferritin gives a very poor yield of tryptic peptides. Six of the arginyl peptides were isolated by gel filtration followed by preparative electrophoresis and chromatography on paper, and their sequence determined by dansyl-Edman. Two large arginyl peptides were isolated by gel filtration ; the smaller of these was purified by chromatography in 7M guanidine hydrochloride on Sephadex G75, desalted and the N-terminal 20 residues sequenced. In all, together with the N-terminal pentapeptide (9) 75 residues were determined in eight arginyl peptides. The other arginyl peptide is in the process of purification.

CNBr cleavage gives four major peptides (11). However, their isolation presents considerable problems because of their strong tendency to associate, even in denaturing conditions. We have, however, isolated two of these peptides (molecular weights 6,000 and 3,000) by gel filtration and have determined 30 residues of the larger and 23 residues of the smaller using the liquid phase sequencer. The smallest CNBr peptide (MW 1000) has recently been obtained, and work is in progress to obtain the largest fragment (MW 8,500).

Peptic cleavage produces a large number of peptides, of which we have isolated those containing at least one arginine residue by cation exchange chromatography in pyridine acetate buffers and by preparative paper chromatography. Nine peptides were obtained and their amino acid sequence determined manually. This has given us a number of overlaps between the arginyl peptides.

To date some 75 % of the amino acid sequence of horse spleen apoferritin has been determined and work is continuing to obtain peptides from the N-terminal part of the protein, which is the most difficult region to characterise.

II. COMPARATIVE STUDIES ON MAMMALIAN AND PLANT FERRITINS.

We have isolated ferritin from human and horse liver, spleen and heart. In the isolation of all these proteins (except that of horse spleen) the techniques used in general for ferritin isolation do not yield homogeneous preparations. By far the most difficult case has been that of human liver, where amino acid analysis clearly shows that even preparations which give only a single band on SDS-electrophoresis are still contaminated (glycine values typically of 14-16 residues/subunit and leucine values of 16-18 residues/subunit). The method finally used in all cases involves two or three cycles of ultracentrifugation after preliminary heat denaturation, $(NH_4)_2SO_4$ precipitation and in some cases CM-cellulose and Sepharose 6B chromatography. The subunit molecular weight of the human ferritins are in the normal range (18-19,000); however horse liver ferritin was found to give consistently two bands (figure 1b) corresponding to molecular weights of 18,000 and 21,000. In contrast horse heart and horse spleen ferritin contain only the component of molecular weight 18,000-19,000. The isoelectric points of the human and horse ferritin were also determined (Huebers H

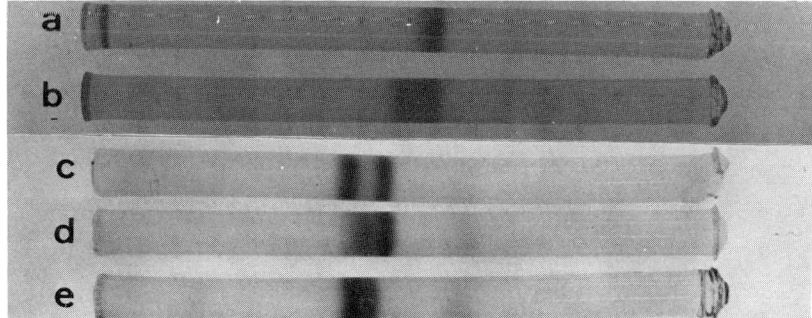

Figure 1: SDS-polyacrylamide gels of
    a) horse spleen ferritin
    b) horse liver ferritin
    c) pea ferritin + lentil ferritin
    d) pea ferritin + horse spleen ferritin
    e) lentil ferritin + horse spleen ferritin

and Crichton R.R. unpublished data) on columns of Ampholine stabilised by sucrose. The human ferritins had pI values consistently higher than those of the horse proteins (5.0-5.5). The horse liver ferritin gave two sharp bands on focussing which merged into one single peak when the column was eluted. Amino acid composition of the human and horse proteins are given in table 1, and are all remarkably similar, though each ferritin is distincly different from each other, confirming previous results (12,13).

Table 1: Amino acid compositions of mammalian and plant ferritins
Ho=horse   S=spleen   P=pea
Hu=human   L=liver   L=lentil
Ph=phyto   H=heart

Amino Acid	HoS	HoL	HoH	HuS	HuL	HuH	PhP	PhL
Cys	2.9	2.6	nd	1.7	1.5	nd	nd	nd
Asx	17.3	17.9	18.4	19.3	19.2	19.0	30.3	20.9
Thr	5.5	5.6	7.7	6.1	6.2	7.6	4.0	5.0
Ser	9.0	8.9	11.2	7.7	9.3	7.8	14.7	9.7
Glx	23.9	25.3	19.9	22.3	23.9	24.5	28.9	27.6
Pro	2.8	3.1	6.4	3.8	2.9	5.7	3.1	6.0
Gly	9.9	10.2	10.2	10.8	10.1	11.0	11.0	9.6
Ala	14.0	13.4	12.7	13.8	13.7	15.3	14.7	15.4
Val	6.9	6.9	7.4	6.0	6.3	7.2	12.8	17.4
Met	2.8	2.5	4.8	2.7	2.9	1.9	3.0	3.9
Ile	3.5	3.6	4.9	3.8	2.5	4.7	7.4	6.9
Leu	25.0	24.2	20.3	23.3	23.4	24.0	15.8	17.4
Tyr	5.0	4.1	4.8	4.4	6.0	5.2	6.0	6.5
Phe	7.3	7.7	7.3	7.0	6.9	6.0	8.4	9.5
Trp	2.1	2.1	nd	2.2	2.2	nd	1.6	1.7
His	5.8	6.5	8.4	6.9	5.4	5.0	5.0	9.7
Lys	8.7	8.8	13.5	10.4	10.4	8.7	8.9	13.3
Arg	9.5	9.0	6.7	9.3	8.4	7.6	6.7	8.9

Ferritin was also isolated from dried peas and lentils by a method described elsewhere (14) using ultracentrifugation and DEAE cellulose chromatography. The phytoferritins were judged homogeneous on the basis of electrophoresis, both in non denaturing conditions and in SDS. Subunit molecular weights were estimated from SDS gels (fig.1c) to be 20,300 and 21,400 respectively for pea and lentil apoferritins. The isoelectric points (Huebers H. and Crichton R.R. unpublished data) were higher than for mammalian ferritins. The amino acid composition of the two phytoferritins are given in table 1. What is perhaps the most striking are the similarities with the mammalian ferritins, and for pea ferritin, in view of the much higher pI, we must assume that the increase in Glx and Asx is mostly amide. We also compute from a neutron low angle scattering study that the internal volume of the phytoferritins is greater than for mammalian ferritins, and hence that they can store more iron. They still consist, however, of 24 polypeptide chains (15). The comparative studies are now being extended to a comparison of the primary structure of these different ferritins.

III. IRON OXIDATION AND DEPOSITION

We have recently developed a novel mechanism for ferritin iron oxidation and deposition (16-17). We consider that there are four steps in the process (fig.2) : (i) fixation of Fe^{2+} at specific binding sites on adjacent polypeptide chains. We assume that Fe^{2+} has a higher affinity for these sites than does Fe^{3+}. (ii) once the iron is bound, a dioxygen molecule is fixed between the two iron atoms (iii) Reduction of the dioxygen gives an end-on peroxo-complex in which the oxygen ligand is coordinated to two iron atoms : the formal valence state of iron is now ferric.(iv) Hydrolysis of the peroxo complex : this is accompanied by migration of the ferric oxyhydroxide to the interior of the protein shell and occupation of the binding sites by incoming Fe^{2+}. We assume that there are heteronucleation sites in the interior of the protein to which the first iron atoms migrate. Thereafter the growing micelle of FeO.OH acts as a heteronucleation site.

The evidence for this model is now briefly summarised. Step (i) involves a catalytic site composed of two adjacent subunits. There are two arguments which support this idea. Firstly, kinetic studies using stopped-flow techniques in sodium borate-cacodylate buffer show that the initial velocity of iron deposition varies in a sigmoidal fashion with the concentration of iron added. Li-

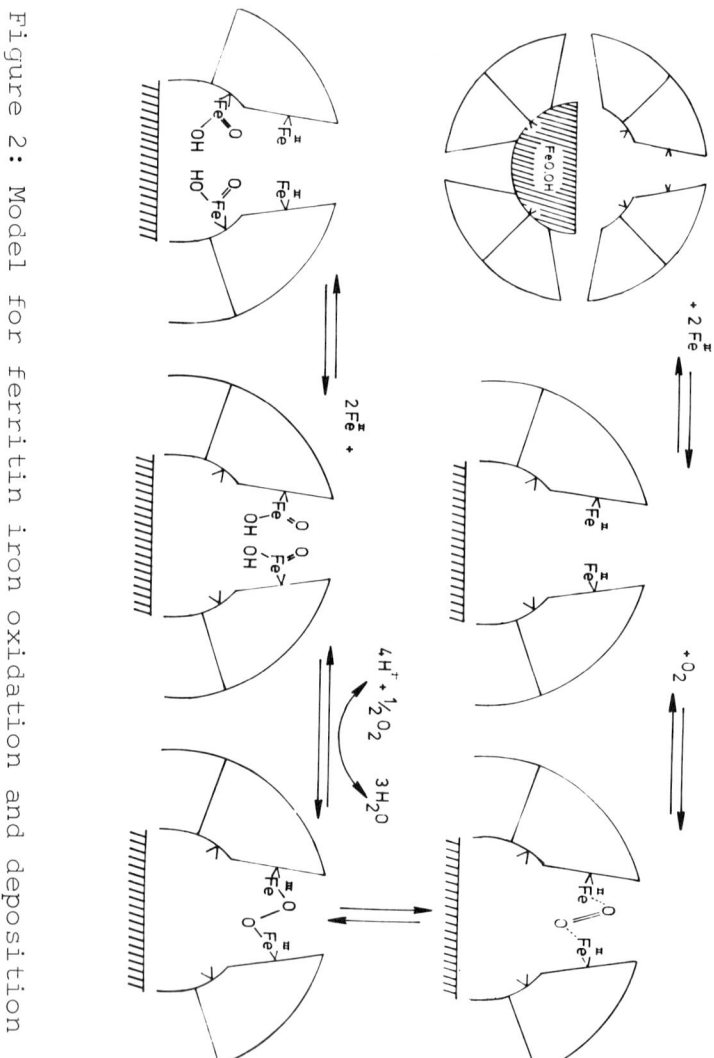

Figure 2: Model for ferritin iron oxidation and deposition

nearisation of this curve can be achieved by plotting $1/v$ against $1/(Fe\ II)^2$ suggesting that ferritin formation under these conditions is second order with respect to iron (19). Further, 80-85% inhibition of iron deposition is observed when the apoferritin is preincubated with CrIII. For a 100-fold range in CrIII concentration the amount of CrIII fixed per subunit corresponds to 0.5 atoms/subunit. This again suggests that two subunits are involved in the catalytic site (18), since we have no reason to suppose that the apoferritin subunit has more than one iron binding site.

The second step, involving binding of molecular oxygen as electron acceptor is supported by the observations that (i) FeIII is not assimilated by ferritin, or if so at a rate which is at least 10^3 times less rapid than FeII, and (ii) that other biological electron acceptors, such as NAD^+ and $NADP^+$ are ineffective in ferritin formation (Crichton R.R. unpublished data).

Logically if we have a catalytic site involving two iron atoms, we must expect a formal peroxo-intermediate as the next step of ferritin formation (19). Support for the peroxo-intermediate comes from the following observations. We have been unable (18) to detect superoxide as an intermediate in ferritin iron deposition. This suggests either that a two electron oxidation is involved, or else that any superoxide formed remains bound to the metal, and thus is not detected by the analytical methods used (18). The second argument for a peroxointermediate is the release of iron II from ferritin following incubation with 2,2'-bipyridyl and triphenylphosphine. Preincubation of ferritin under anaerobic conditions with a quantity of 2,2'-bipyridyl (17) or of triphenylphosphine (Crutzen A., Bönemann K. and Crichton R.R., unpublished data) which is stoichiometric with the number of subunits present, results in a "burst" of Fe(II) liberation when an excess of bipyridyl is added, with concomittant formation of bipyridyl-N-oxide (17) and triphenylphosphine oxide (Crutzen A., Bönemann K. and Crichton R.R. unpublished data). A parallel with this situation is found in a number of metalloorganic complexes of Fe(II) in which the iron is bound by oxygen bridges to aluminium alkoxy residues. These complexes dimerise to Fe(III)-peroxo derivatives on oxygenation and are reduced to the Fe(II) complex on addition of bipyridyl, with formation of bipyridyl-N-oxide (Teyssié P., personnal communication).

The fourth step, namely hydrolysis of the peroxo-complex, and its displacement from the catalytic sites is dictated by the fact that the bulk of the iron in ferritin is present in the form of FeO.OH, by the fact that the iron must migrate towards the interior of the protein shell, where it interacts with the protein (20), and by the tendency of iron III to hydrolyse at physiological pH. Support for the idea that there is an equilibrium between the ferric oxyhydroxide in the interior of ferritin and the peroxo-complex comes from detailed studies of iron release from ferritin by bipyridyl (17). We have observed that bipyridyl can mobilise several hundred iron atoms from ferritin in the absence of any other reducing agent with concomittant formation of bipyridyl-N-oxide, whereas the same phenomenon is not observed with ferritin ferric oxyhydroxide cores (17). We propose that in the absence of Fe(II), the iron in ferritin exists in a dynamic equilibrium between the peroxo-complex on the catalytic sites and the ferric oxyhydroxide in the interior of the protein. This interpretation is supported by the fact that bipyridyl mobilises more than 24 iron atoms/apoferritin molecule (17). Further experiments to characterise the peroxo-intermediate are in progress.

IV MOBILISATION OF IRON FROM FERRITIN

We have previously shown that ferritin iron can be mobilised by the system FMN/NADH in presence of bipyridyl(18,21,22). Three phases can be distinguished : a lag phase corresponding to consumption of O_2 by the $FMNH_2$ generated, a phase of iron mobilisation and a plateau, which is a function of the total iron mobilised. We studied changes in these three parameters at a series of concentrations of FMN, NADH and of ferritin. The velocity of iron mobilisation varied in a linear fashion with ferritin concentration. When the concentrations of FMN and bipyridyl are maintained at 5mM and the ferritin at 0.29 mM we varied the NADH concentration from 0.5mM to 10mM and found that the lag phase increased as the NADH concentration decreased, and that there was a sigmoidal variation of the amount of iron released (with a maximum at 40%) and of the velocity of iron mobilisation as a function of NADH concentration. With constant NADH, bipyridyl and ferritin concentrations of 5mM, 5mM and 0.145mM respectively, we found that on varying the FMN concentration from 0.1mM to 5mM, the lag phase increased as the FMN concentration decreased, the velocity of iron mobilisation varied in a sigmoidal fashion as a function of FMN concentration while the amount of iron mobilised remained constant at around 40%. We have compared these results with those obtained in the system FMN/NADH without ferritin or bipyridyl where we have followed the formation of the semi-quinone form of FMN, FMNH which has an absorption maximum at 590nm. The sigmoidal aspects of the kinetics of iron release is paralled in the kinetics of FMNH formation, and thus suggests that FMNH may be involved in iron reduction. Further, we found that incubation of FMN and NADH leads, as expected, to formation of both the semi-quinone FMNH and the fully reduced $FMNH_2$. The dimeric form $(FMNH)_2$ absorbs at 780nm. Incubation of ferritin in the presence of FMN/NADH results in a decrease in the absorption at 590nm; this also suggests that FMNH is consumed during ferritin iron reduction. In contrast we found a decrease in the absorption at 780nm on incubation of FMN/NADH with apoferritin, suggesting that FMNH also has an affinity for apoferritin.

From these studies we conclude that FMNH as well as $FMNH_2$ can reduce ferritin iron (we know from studies with $FMNH_2$ obtained by catalytic reduction that $FMNH_2$ can reduce ferritin iron(18)). This supports our previous proposal(21-23) that FMN is the coenzyme for ferritin iron reduction and that NADH, or NADPH is the source of electrons

REFERENCES
(1) Bryce C.F.A.& Crichton R.R.,J.Biol.Chem.246, 4198 (1971).
(2) Björk I. and Fish W.W., Biochemistry 10,2844 (1971)
(3) Collet-Cassart D. & Crichton R.R., Hoppe Seyler's Z. Physiol.Chem. 355, 15 (1974).

(4) Collet-Cassart D. & Crichton R.R. in "Proteins of iron storage and transport in biochemistry & medicine" (ed.R.Crichton) p.185, North Holland Amsterdam, 1975.
(5) Ishitani K., Niitsu Y. & Listowsky I., J. Biol.Chem. $\underline{250}$, 3142 (1975).
(6) Freedman M.L., Cohen H.S., Rosman J. & Forte F., Brit.J.Haematol. $\underline{32}$, 579 (1976)
(7) Gray W.R. in "Methods in Enzymology" vol XXV p.121 (Hirs & Timasheff eds.) Academic Press N.Y. (1972).
(8) Edman P. & Begg G., Eur.J.Biochem. $\underline{1}$, 80 (1967)
(9) Suran A.P., Arch.Biochem.Biophys. $\underline{113}$, 1 (1966).
(10) Mainwaring W.I.P. & Hofmann T., Arch.Biochem. Biophys. $\underline{125}$, 975 (1968).
(11) Bryce C.F.A. & Crichton R.R., J.Chromatog. $\underline{63}$, 267 (1971).
(12) Crichton R.R., Millar J.A., Cumming R.L.C. & Bryce C.F.A., Biochem.J. $\underline{131}$, 51 (1973).
(13) Huebers H., Huebers E., Rummel W. & Crichton R.R., Eur.J.Biochem. $\underline{66}$, 447 (1976).
(14) Ponce-Ortiz Y. & Crichton R.R., Biochem.Soc. Trans., in press (1977).
(15) Crichton R.R., Ponce-Ortiz Y., Koch M.H.J., Parfait R. & Stuhrmann H.B., in preparation.
(16) Crichton R.R., in Proc. 16th Int. Symp. Haematol. in press (1977).
(17) Crichton R.R. & Roman F., Biochem.Soc.Trans., in press (1977).
(18) Wauters M., Crichton R.R. & Michelson A.M., Biochimie, in press (1977).
(19) Vaska L., Accts.Chem.Res. $\underline{9}$, 175 (1976).
(20) Stuhrmann H.B., Haas J., Ibel K., Koch M.H.J. & Crichton R.R., J.Mol.Biol. $\underline{100}$, 399 (1976).
(21) Crichton R.R., Wauters M. & Roman F. in "Proteins of iron storage and transport in biochemistry and medicine (Ed. R.R. Crichton) p. 287, North Holland, Amsterdam (1975).
(22) Crichton R.R., Wauters M. & Roman F., Biochem.Soc.Trans. $\underline{3}$, 946 (1975).
(23) Crichton R.R. in "Iron Metabolism and its Disorders" (Ed. H. Kief) p. 81 Excerpta Medica, Amsterdam, 1976.

ULTRASTRUCTURAL STUDIES OF DISRUPTED FERRITIN

William H. Massover

Division of Biology and Medicine, Brown University,
Providence, Rhode Island 02912, U.S.A.

I. INTRODUCTION

The metalloprotein, ferritin, presently is believed to be a remarkably stable assemblage of polymerized macromolecular subunits (1, 2). Rather drastic chemical and physical conditions that produce the depolymerization or destruction of many other multimeric proteins do not seem to perturb ferritin. Free monomeric subunits have not been detected even at very high dilutions (3, 4). The stability of the polymer must be due mainly to strong bonding interactions between the individual protein subunits. Several studies recently have proposed that the protein also is attached to the ferric mineral (5 - 8); this interaction could be the primary reason why the polymerized subunits in iron-rich ferritin are even more stable to certain probes than are those in the iron-free apoprotein, apoferritin (2). Although the polymeric protein shell often is considered to be of importance with regard to the binding, uptake, and release of iron by ferritin (e.g., 1, 2, 9, 10), the nature of the protein-mineral interactions remains poorly defined.

Both the protein-protein and protein-mineral interactions mostly have been investigated by indirectly examining entire populations of ferritin. The heterogeneity of many ferritin populations (11), as well as the existence of a subunit-polymer equilibrium in all populations (e.g., 12), suggests that these interactions also should be investigated in individual molecules. In order to gain such information, the present study has utilized high resolution electron microscopy to examine single members in negatively stained populations of naturally and experimentally disrupted ferritin.

II. MATERIALS AND METHODS

Purified horse spleen ferritin (Calbiochem) was further purified by gel chromatography with Sephadex G-150 or G-200 (Pharmacia Fine Chemicals). Samples were experimentally treated as described in the text; control samples were untreated or were diluted to an equivalent extent with pure distilled water. All samples were deposited upon carbon film coated grids that were made hydrophilic by exposure to glow discharge, and then were negatively stained with 2% sodium silicotungstate (pH 7.0). The air-dried specimens were examined at 80 kV with a JEOL JEM-100B electron microscope. Polyacrylamide (5%) disc gel electrophoresis was carried out with a Tris-glycine buffer (pH 9.0).

III. RESULTS AND DISCUSSION

A. Disrupted Particles in Untreated Samples

Several previously published electron microscope studies have detected some gaps in the profile of the protein shells of horse spleen ferritin (e.g., 5, 12). The observation of particles with incomplete shells becomes more and more frequent during long-term storage at 4°C (Figs. 1 and 2). The size of the gaps in the shells must indicate the absence of one or more subunits. On occasion, one can find small bits of protein which slightly protrude from a shell, as if they are about to move away (Fig. 2). Both very small and larger fragments of the naturally disrupted shells are found in some preparations (Fig. 1), particularly when the highly concentrated commercial samples are diluted sufficiently.

The uncrystallizable fraction of horse spleen ferritin usually has been considered to be composed of particles with broken shells (e.g., 12). A satisfying rationalization for the presence of these incomplete particles has remained elusive (e.g., 12, 13), especially in view of the very high stability of the polymeric shells. Since the small bits of detached protein can be found even in ferritin samples that elute near the void volume on gel filtration (Fig. 1), these fragments probably represent dissociated pieces of formerly complete shells. The progressive continuation of this process during storage suggests that ferritin is subject to some type of aging. The observed ultrastructural changes in the stored samples could be the basis for the time-dependent changes in hydrogen ion titration sites recently reported by Silk and Breslow (14). If these structural disruptions in

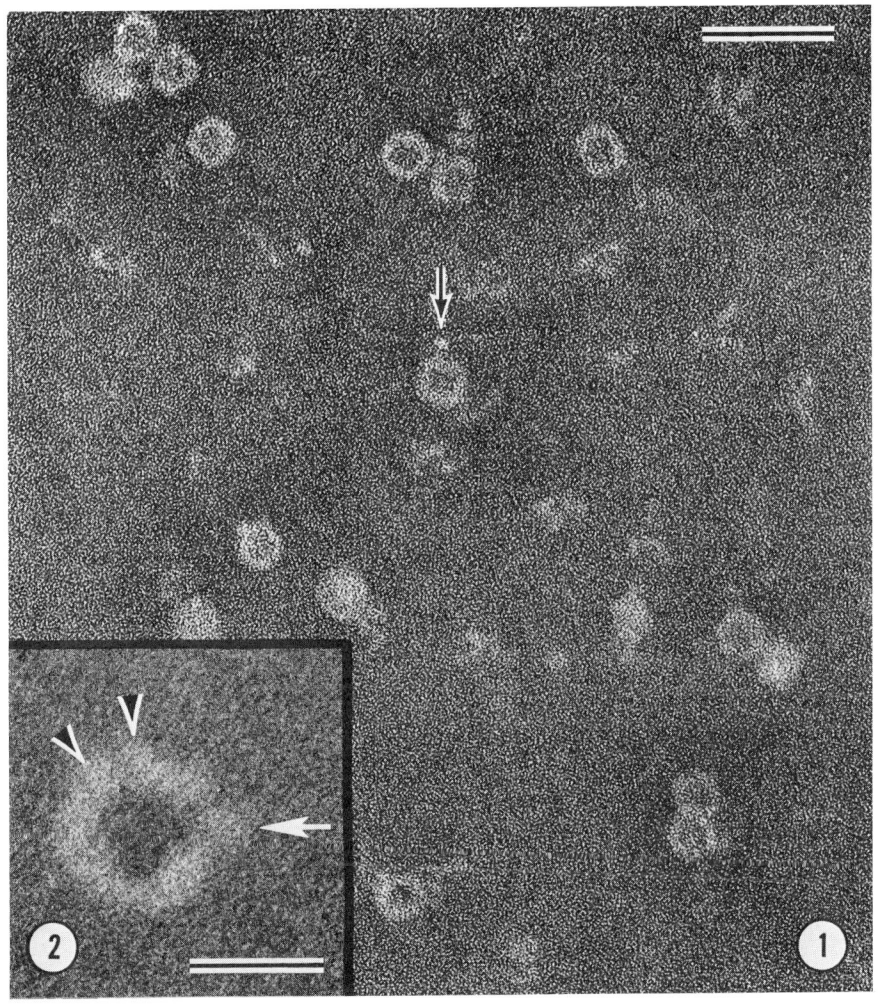

Fig. 1: Horse spleen ferritin, negatively stained after dilution with pure distilled water. Most particles in this field appear to be intact, but a variety of protein fragments are present. Arrow indicates a subunit-sized bit of protein. Bar represents 300 Å.

Fig. 2: Negatively stained horse spleen ferritin, as in Fig. 1. A small blob of protein (arrow) is slightly detached from this shell, leaving a gap filled with negative stain. Two small blobs with a similar size can be recognized in this shell (arrowheads). Bar represents 100 Å.

the protein shell also occur in vivo, this process might produce the protein depletion in ferritin that is thought to give rise to the formation of hemosiderin (e.g., 13).

B. Disrupted Particles Produced by Physical Treatments

Freezing and thawing (Fig. 3), boiling and cooling, or drying and rehydrating all will produce some disruption of the protein shells in ferritin (8). The smallest gaps seen in otherwise intact shells probably represent the removal of one or several individual subunits. Some of the mineral cores in disrupted samples only have a few very small bits of protein attached to them (Fig. 4). Dispersed crystallites, each of which would comprise only a portion of a single core in full ferritin, can remain intimately apposed to fragments of their disrupted shells (Fig. 5). The individual particles shown in Figs. 4 and 5 provide direct evidence for the postulated attachment of the protein shell to the mineral core. In view of the very small size of some bits of protein that remain apposed to the core mineral (e.g., Fig. 4), it seems likely that protein-mineral interactions take place at the level of individual subunits or groups of a few subunits.

C. Disrupted Particles Produced by Chemical Treatments

Acetone, which has been used to precipitate ferritin from its aqueous solution (15), causes a considerable disruption of the protein shells (16). All varieties of incomplete shells and dispersed fragments, as well as naked cores, can be found in resolubilized samples after acetone treatment (Fig. 6). Polyacrylamide disc gel electrophoresis of these samples shows a pronounced smearing of the protein bands, in contrast to the results from control samples. This smearing probably represents the size and/or charge heterogeneity in the population that is induced by this chemical treatment. Since drying produces a similar ultrastructural disruption of the protein shells (8), one might suspect that dehydration is the actual damaging agent (e.g., see also 17).

IV. CONCLUDING REMARKS

The disruptive effects of the physical and chemical treatments used in the present study all produce the removal of some single subunits and groups of subunits from the

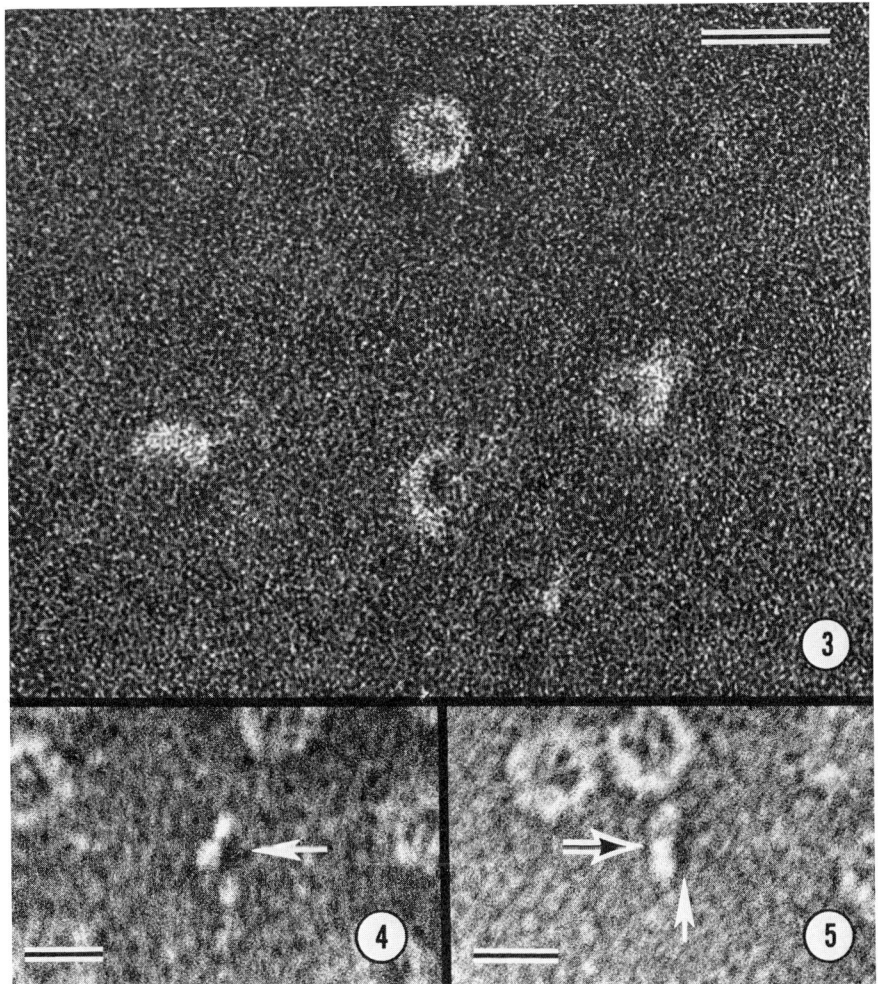

Fig. 3: Horse spleen ferritin, negatively stained after three rounds of freezing and thawing. Note intact and disrupted shells, and dispersed shell fragments. Bar is 200 Å.

Fig. 4: Three small bits of protein (white) remain attached to the core at the center (black object at arrow). Sample prepared as in Fig. 3. Bar represents 100 Å.

Fig. 5: A fragment of a disrupted protein shell (white object at large arrow) remains closely apposed to a small core crystallite (black object at small arrow). The sample was prepared as in Fig. 3. Bar represents 100 Å.

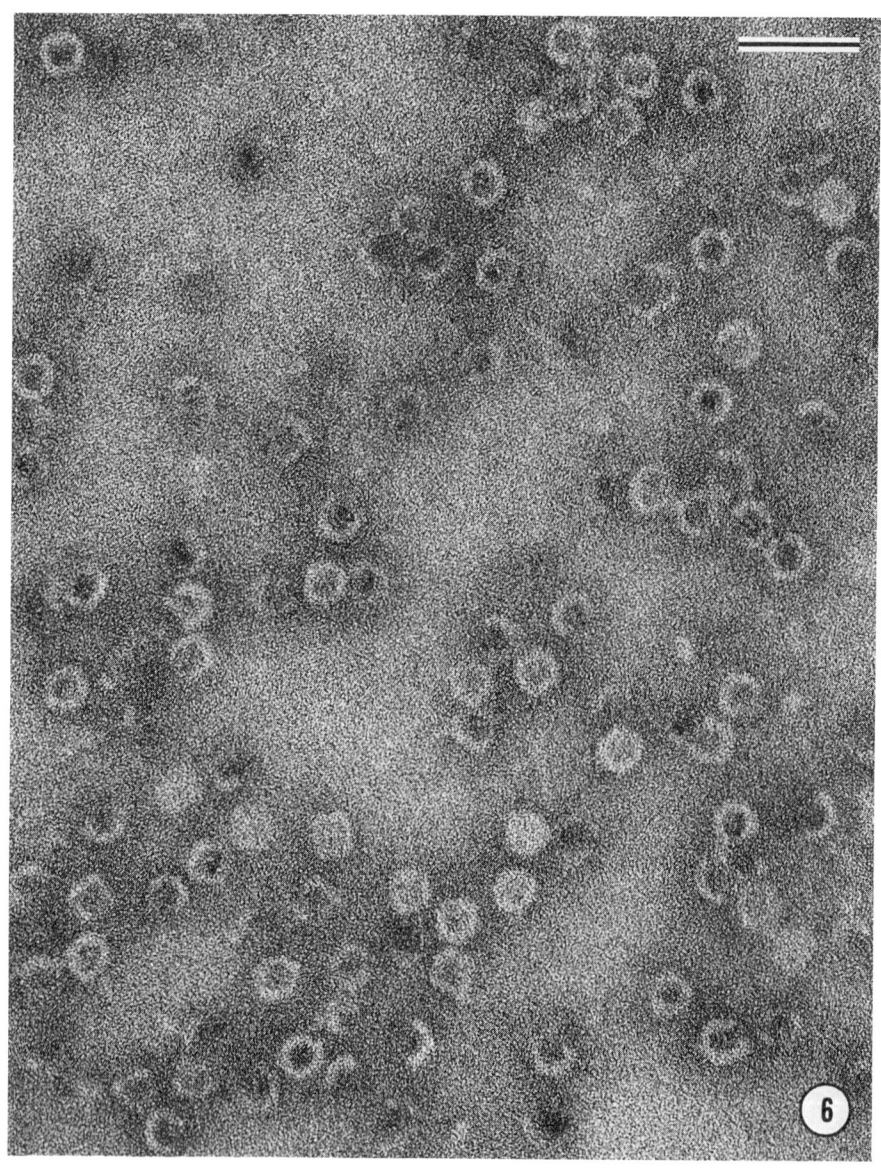

Fig. 6: Horse spleen ferritin, negatively stained after precipitation with acetone and resolubilization with pure distilled water. A variety of incomplete shells and shell fragments are present amongst the intact particles. The bar represents 300 Å.

protein shell of ferritin. This partial disruption seems to be quite different from the complete dissociation into single subunits which is produced by treatment with agents such as 67% acetic acid (18) or sodium dodecyl sulfate (19). The fact that only some particles in the total population are visibly disrupted by the agents used is quite compatible with either a great stability of the polymer, or a very rapid reassembly by the dissociated subunits and fragments. In this latter regard, one should note that dissociated subunits are known to be able to reassociate into complete shells (18, 19).

The larger size of some dispersed fragments from disrupted ferritin shells indicates that they must be composed of several associated subunits. Crichton has described some chromatographic evidence for subunit dimers and tetramers in populations of single subunits which are permitted to reassociate (2, 20). Further structural and biochemical work is needed to determine whether the observed multi-subunit associations also are found as discrete entities within the complete shells of ferritin.

Several analytical studies of ferritin populations (3, 4) have provided data leading to the widely accepted conclusion that individual protein subunits do not spontaneously dissociate from complete shells even at very high dilutions. The present observations of some subunit-sized particles in ferritin populations treated only with negative stain appear to be in direct conflict with this proposal. If the negative stain alone was causing the present observations, one would not expect to see the many fully intact shells in the same preparations. Moreover, standard polyacrylamide gel electrophoresis of untreated samples sometimes shows a rapidly migrating iron-free band at the position expected for free subunits (unpublished results). Since apparent subunits are observed even in preparations of ferritin that has been fractionated by gel filtration, it now seems likely that there is indeed some spontaneous subunit dissociation from the shells. The possibility of a very rapid reassociation after any subunit dissociation should be critically reconsidered by further biophysical and biochemical studies.

V. REFERENCES

1. Harrison, P.M., Sem. Hematol. 14, 55 (1977).
2. Crichton, R.R., Struct. Bond. 17, 67 (1973).
3. Jaenicke, R., Bartmann, P., Biochem. Biophys. Res. Comm. 49, 884 (1972).

4. Crichton, R.R., Eason, R., Barclay, A., Bryce, C.F.A., Biochem. J. 131, 855 (1973).
5. Massover, W.H., Cowley, J.M., in "Proteins of Iron Storage and Transport in Biochemistry and Medicine" (Crichton, R.R., ed.), p. 237. North-Holland, Amsterdam, 1975.
6. Stuhrmann, H.B., Haas, J., Ibel, K., Koch, M.H.J., Crichton, R.R., in "Proteins of Iron Storage and Transport in Biochemistry and Medicine" (Crichton, R.R., ed.), p. 261. North-Holland, Amsterdam, 1975.
7. Stuhrmann, H.B., Haas, J., Ibel, K., Koch, M.H.J., Crichton, R.R., J. Mol. Biol. 100, 399 (1976).
8. Massover, W.H., J. Mol. Biol. in press (1977).
9. Stefanini, S., Chiancone, E., Antonini, E., Finazzi-Agro, A., FEBS Lett. 69, 90 (1976).
10. Stefanini, S., Chiancone, E., Vecchini, P., Antonini, E., Mol. Cell. Biochem. 13, 55 (1976).
11. Drysdale, J.W., Adelman, T.G., Arosio, P., Casareale, D., Fitzpatrick, P., Hazard, J.T., Yokota, M., Sem. Hematol. 14, 71 (1977).
12. Pape, L., Multani, J.S., Stitt, C., Saltman, P., Biochem. 7, 606 (1968).
13. Harrison, P.M., Hoare, R.J., Hoy, T.G., Macara, I.G., in "Iron in Biochemistry and Medicine" (Jacobs, A., Worwood, M., eds.), p. 73. Academic Press, London, 1974.
14. Silk, S.T., Breslow, E., J. Biol. Chem. 251, 6963 (1976).
15. Makino, Y., Konno, K., J. Biochem. 65, 471 (1969).
16. Massover, W.H., in "Proceedings of the 35th. Annual Meeting, Electron Microscopy Society of America, Boston" (Bailey, G.W., ed.), in press. Claitor's Publ. Div., Baton Rouge, La., 1977.
17. Listowsky, I., Betheil, J.J., England, S., Biochem. 6, 1341 (1967).
18. Harrison, P.M., Gregory, D.W., Nature. 220, 578 (1968).
19. Smith-Johannsen, H., Drysdale, J.W., Biochim. Biophys. Acta. 194, 43 (1969).
20. Crichton, R.R., Biochem. J. 130, 35P (1972).

Acknowledgments: I thank Lourdes V. Concepcion and Frederick J. Szymanski for technical assistance. This study was supported in part by Grant CA-18530 from the National Cancer Institute (NIH, USPHS, DHEW).

THE RESTRICTIVE NATURE OF APOFERRITIN

CHANNELS AS MEASURED BY PASSIVE DIFFUSION

Michael E. May
Wayne W. Fish

Medical University of South Carolina
Charleston, South Carolina

I. INTRODUCTION

The results of in vitro studies on the uptake of iron by ferritin or the mobilization of iron from within ferritin have resulted in the proposal of four different hypotheses for the molecular mechanisms of the two processes: that iron is mobilized into or out of the protein shell via specific oxidation-reduction sites on the protein (1); that iron is mobilized into and out of ferritin by the active participation of chelators and reductants in transporting iron to and releasing it from the surface of the hydrated ferric oxide crystallite (2); that iron is accumulated into and mobilized from the protein shell by both pathways (3); and that the hydrated ferric oxide crystallite initially forms in solution and the protein shell subsequently associates about the polynuclear Fe(III) aggregate (4). The fundamental rationale for the first three of the hypotheses with respect to iron accumulation and for all four of the hypotheses with respect to iron mobilization requires a definite directionality, i.e. Fe(III) precipitated inside the protein shell, Fe(II) released outside the protein shell.

The existence of radial channels in the protein shell which are of defined dimensions and which are sufficiently large enough to allow mononuclear iron ions or chelates to enter the protein shell but small enough to retain the ferric polynuclear crystallite within the shell is uniformly assumed in the discussions of iron binding and release by ferritin (2,5,6). Additionally, it would seem that the vectorial nature of iron incorporation and mobilization must be conferred by the protein, and these radial channels in apoferritin, which result from the unique arrangement of the apoferritin subunits in forming the protein shell, would appear to play a major role in the iron storage and mobilization processes. The most recent x-ray crystallographic information on apoferritin structure indicates that there are six channels in the protein shell and that these channels are of 9-15 Å in diameter (7). These crystallographic results have stimulated limited conjectural (8) and experimental (9) considerations of the restrictive nature of the apoferritin channels to molecules other than iron.

Theoretical (10) and semiempirical (11) treatments of the simple diffusion of small molecules through channels of dimensions similar to those of the diffusing molecule indicate that a severe steric restriction is placed on the rate of simple diffusion of these molecules through the channels. Certainly the molecular dimensions of the reducing agents, iron chelators, and amino acid side chain modifying reagents employed in ferritin structure-function studies possess dimensions which are of the same order of magnitude as the radial channels in apoferritin. For this reason and to test the feasibility of the passive diffusion of reductants and iron chelators through apoferritin channels as an important feature in iron storage and/or mobilization, we have attempted to quantitate diffusion rates into the apoferritin shell of a number of small, innocuous organic molecules.

II. MATERIALS AND METHODS

Porcine ferritin was prepared from fresh pig spleen (12). Apoferritin was prepared from porcine ferritin or from commerical equine ferritin (Miles Laboratories, 2 x crystallized) according to the procedure routinely employed in our laboratory (12,13). The quality of the apoferritin preparations was evaluated by their circular dichroic properties and any preparation which did not meet established criteria (13) was not used. Apoferritin concentrations were determine by the differential refractometric method of Babul and Stellwagen (14).

The following organic compounds were employed for the measurements of diffusion rates into apoferritin: ^{14}C-methanol, uniformly labeled ^{14}C-glucose, ^{14}C-methylammonium chloride, and 1,2-^{14}C-sodium acetate. The hydrated radii assumed for the diffusants and which were based on published hydrodynamic properties were: methanol, 1.47 Å (15); glucose, 3.6 Å (11); methylammonium ion, 3.5 Å (16); and acetate ion, 4.5 Å (16).

Rates of permeation through the apoferritin shell by these small molecules were determined by incubating the protein and diffusant together at 10°C, removing aliquots at various time intervals, separating the protein (together with its accumulated diffusant) from the free diffusant by rapid gel chromatography[1] and determining the amount of apoferritin-associated radioactivity. A more detailed account of the methodology is given elsewhere (17). The expected equilibrium values for the permeation processes were calculated from the protein concentration, the diffusant concentration, the volume external to the protein shell, and the volume enclosed by the protein shell. Data were then plotted as the fraction of equilibrium, F, <u>versus</u> time. Treatment of the data in terms of various mathematical models for the diffusion process are not dealt with in this report but have been discussed (17).

III. RESULTS

Fig. 1 illustrates the behavior of the system when observed with respect to the rate of "uptake" and "release" of one of the diffusants employed. As can be seen, the rate of entry of glucose into porcine apoferritin is quite slow; only about half the anticipated equilibrium amount of glucose had diffused into the protein after a week. To support the argument that the radioactivity associated with the protein had indeed diffused inside the protein shell, apoferritin which had been incubated with ^{14}C-glucose for over two weeks was separated from excess diffusant and then monitored with respect to the flux of diffusant from within the protein out into the unlabeled medium. As can be seen, this process of diffusant escape was also quite slow.

Fig. 2 illustrates that indeed the flux of diffusant into apoferritin was still proceeding after these extremely long periods of time. In this instance, after about 270 hr of incubation of apoferritin in ^{14}C-methanol at 10°C, the

---

[1] The time between application of the sample to the gel chromatography column and elution of the apoferritin was about four minutes. Protein and diffusant radioactivity associated with the protein eluted in superimposable peaks and returned to baseline before the "free" diffusant radioactivity zone began to emerge.

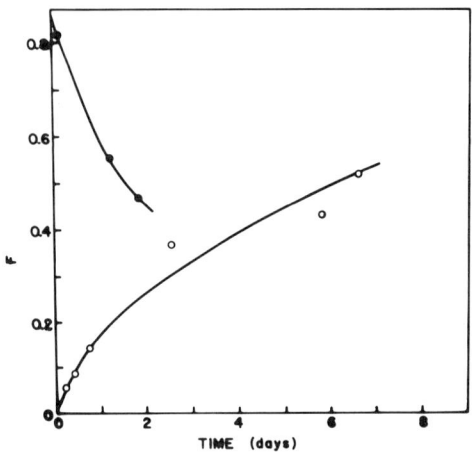

Fig. 1. The permeation of apoferritin channels by glucose. O, rate of ^{14}C-glucose entry. ●, rate of ^{14}C-glucose escape.

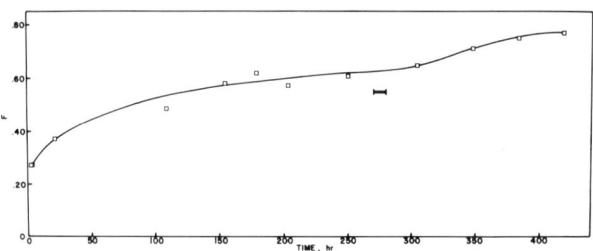

Fig. 2. The diffusion of methanol into apoferritin. The temperature was maintained at 10°C for the first 270 hr and then increased to 22°C.

temperature was increased to 22°C. The F values also rise at that time. If our predicted equilibrium values had been in error and the system had, in fact, reached diffusion equilibrium by ~250 hr, the increase in temperature would not have affected the total amount of diffusant associated with the protein, and the F values would have remained constant.

Table I summarizes what has been observed with respect to the relative rates of diffusion of the previously-mentioned compounds, as well as ferrous and ferric iron, into equine apoferritin.

Table I The Relative Rates of Diffusion of Small Molecules Through the Channels of Apoferritin[a]

Diffusant	Hydrodynamic Radius (Å)	Initial concentration (mM)	Time of Incubation (hr)	F[b]
Methanol	1.47	6.31	100	0.52
Glucose	3.6	0.99	100	0.27
Methylammonium ion	3.5	0.40	100	0.12
Acetate ion	4.5	1.47	100	0.01
Ferrous ion[c]	3.6-8.0[d]	25.0	1	6,200.[e]
Ferric ion-NTA[f] chelate	4.6-9.0[d]	30.0	1	0

a. Protein and diffusant were dissolved in 0.1 I sodium phosphate buffer, pH 7.4, which contained 0.02% (w/v)NaN$_3$. The temperature of incubation was 10°C.
b. F is the fraction of the amount of diffusant predicted to be associated with apoferritin at equilibrium.
c. Incubated under N$_2$ at 20°C. Reactants in 0.1 M sodium acetate buffer, pH 5.5.
d. As Fe(solvent ligand)$_6$(II) or as Fe(solvent ligand)(III). The lower values are based on an electroanalytical determination of the diffusion coefficient of FeCl$_3$ or FeCl$_2$ in 1 F HCl(18). The higher values are taken from Lange (16). The radius will depend upon the nature of the groups coordinating the iron atom.
e. A concomitant oxidation of the iron occurred with incorporation.
f. Incubated under atmospheric oxygen and 20°C. Reactants were in 0.1 M sodium acetate buffer pH 5.5.

These summarized results provide several interesting observations. First, even a molecule as small as methanol (which approaches the dimensions of water) experiences a great deal of steric hindrance by the apoferritin channels with respect to the small molecules' diffusive flux into the protein shell. Second, methylammonium cation diffuses slightly more slowly than uncharged glucose. This observation is complicated by the fact that glucose apparently interacts to some extent with the apoferritin protein fabric, especially at higher glucose concentration (17)[2]. Third, acetate anion is almost totally excluded from the protein. The size difference between methylammonium ion and acetate ion would not be predicted to produce as great a difference in the rate of diffusion as is experimentally observed, and a repulsion of the acetate ions by the overall negatively-charged protein is highly suspect.

Incubation of apoferritin with ferrous chelates resulted in a rapid binding and oxidation of iron beyond the equilibrium amount predicted for simple diffusion. On the other hand, ferric iron from the ferric nitrilotriacetate complex, which remains mononuclear at pH 7.0 (18), was not taken up by apoferritin.

Attempts were unsuccessful to produce and maintain totally anaerobic conditions of all solutions before and during incubation of ferrous ion with apoferritin to preclude the formation of Fe(III). Thus, we have not been able to establish if apoferritin can rapidly accumulate ferrous ion without a concomitant oxidation of the ion to the ferric state.

IV. DISCUSSION

The qualitative significance of the results of these permeation studies is that the rate of simple diffusive penetration of the apoferritin channels by small molecules is quite slow because of the steric restrictions imposed by the dimensions of the channels. This casts a great deal of uncertainty on the use of chemical modification reactions as a tool to probe the structure-function relationships of apoferritin. Most probably, amino acid side chains which are critical to the oxidation/reduction of iron or to the initial crystallite nucleation are located with the radial channels and/or the medial surface of apoferritin. It is very highly

---

[2]That there was a weak glucose-apoferritin binding was indicated by: (i) a rapid initial protein-associated radioactivity, the magnitude of which was inversely related to the specific activity of the glucose, and (ii) the rate of flux of glucose into apoferritin was concentration dependent. These are discussed in detail elsewhere(17).

probable that access to these residues by chemical modification reagents will be sterically limited.

The major biological ramification of the molecular discrimination imposed by apoferritin's channels is that mere passive diffusion into and out of the apoferritin shell by iron chelators, reductants, or oxidants cannot be invoked as the mechanism for iron uptake and release by ferritin. To explain the in vitro rates of iron accumulation and reductive release observed for ferritin in terms of a mechanism of entry into and exit from within the internal cavity of the apoferritin shell, it is necessary to require an interaction between the diffusant (or another molecule which acts as a modifier) and the protein which significantly modifies the channel dimesions. Though our results offer no new insights into the mechanism of the biological function of ferritin, the data support the hypothesis that iron uptake by ferritin is mediated by sites on the protein which posses two general characteristics: first, the sites strongly interact specifically with iron, and second, the sites link the translocation of iron into and out of the protein shell via oxidation-reduction reactions.

This work was supported by NIH grant HL19491 and an NSF Pre-doctoral Fellowship to M.E.M.

REFERENCES
1. Sirivech, S., Frieden, E., and Osaki, S., Biochem. J. 143, 311 (1974).

2. Harrison, P.M., Hoy, T.G., Macara, I.G., and Hoare, R.J., Biochem. J. 143, 445 (1974).

3. Dognin, J. and Crichton, R.R., FEBS Lett. 54, 234 (1975).

4. Pape, L., Multani, J.S., Stitt, C., and Saltman, P., Biochemistry 7, 606 (1968).

5. Niederer, W., Experientia 26, 218 (1970).

6. Bryce, C.F.A. and Chrichton, R.R., Biochem. J. 133, 301 (1973).

7. Hoare, R.J., Harrison, P.M., and Hoy, T.G., Nature 255, 653 (1975).

8. Harrison, P.M., Hoy, T.G., and Hoare, R.J., in "Proteins of Iron Storage and Transport in Biochemistry and Medicine" (Crichton, R.R. ed.), p. 271. North-Holland Publishing Co., Amsterdam, 1975.

9. Stuhrman, H.B., Haas, J., Ibel, K., Koch, M.H.J., and Crichton, R.R., J. Mol. Biol. 100, 399 (1976).

10. Levitt, D.G., Biophys. J. 13, 186 (1973).

11. Pappenheimer, J.R., Renkin, E.M., and Borrero, L.M., Am. J. Physiol. 167, 13 (1951).

12. May, M.E. and Fish, W.W., Arch. Biochem. Biophys. (in press) (1977).

13. Leach, B.S., May, M.E., and Fish, W.W., J. Biol. Chem. 251, 3856 (1976).

14. Babul, J. and Stellwagen, E., Annal. Biochem. 18, 216 (1969).

15. National Research Council, "International Critical Tables of Numerical Data, Physics, Chemistry, and Technology", Vol. V, p. 69. McGraw-Hill, New York, 1929.

16. Dean, J.A., Lange's Handbook of Chemistry, 11th Ed. p.5-5. McGraw-Hill, New York, 1973.

17. May, M.E., Ph.D. Thesis, Medical University of South Carolina, 1976.

18. Lingane, P.J., Anal. Chem. 36, 1723 (1964).

19. Spiro, T.G. and Saltman, P., Struct. Bonding (Berlin) 6, 116 (1969).

BINDING OF PROTONS AND OTHER MONOVALENT IONS BY FERRITIN AND APOFERRITIN

Esther Breslow and Susan T. Silk
Cornell University Medical College

I. INTRODUCTION

The unavailability of ferritin-sequestered iron is due principally to the marked insolubility of the ferric-hydroxyphosphate core and to physical protection by the protein; in the latter case, the width of the channels (1) between the interior and exterior of the protein imposes physical restrictions on the size of solutes which can diffuse into or out of the protein shell. A question which merits some consideration is whether the iron is shielded from the outside, not only physically by the protein shell, but also chemically via stabilizing interactions between the protein and core. Such interactions are of interest not only because of their potential consequences for the chemical reactivity of the core but also for their effects on the stability and possibly even the formation of protein structure. For example, the stimulation of apoferritin synthesis by iron has been attributed either to an effect on protein translation or to a post-translational event (2) and it has been suggested that this effect may be mediated by changes in subunit aggregation kinetics, or an increase in protein stability, in the presence of iron. The model of Pape et al. (3) for iron-incorporation into ferritin, although not generally accepted (4), also suggests the presence of stabilizing interactions between core and protein. In several in vitro studies of the relative susceptibilities of ferritin and apoferritin to either denaturation or enzymatic degradation, a greater stability of ferritin has been observed (5,6) and we and others have reported (6,7) that the CD-demonstrable change in α-helix content at low pH occurs at a slightly lower pH in ferritin than in apoferritin. Taken together, literature data are compatible with the presence of at least weak stabilizing interactions between core and protein, but it is also relevant that, in a study of denaturation by guanidine at neutral pH, a greater stability of apoferritin than of ferritin was noted (8).

Interactions between core and protein are potentially manifest not only by differences in the relative stability of ferritin and apoferritin, but also by differences in conformation between the two proteins. Recent circular dichroism studies (7,9) are not incompatible with small conformational differences between the two proteins, but it is also possible that the small CD differences between the two proteins arise from scattering or absorption artifacts produced by the ferritin core (7,9). In order to further probe the relative conformations of ferritin and apoferritin and to detect potential interactions between the protein and core, we have studied the relative reactivities of the sidechains of horse spleen ferritin and its derivative apoprotein to protons; aspects of side-chain reactivity towards group-specific modifying reagents were also probed.

II. RESULTS AND DISCUSSION

In order to interpret the titration behavior of a protein, the pH at which it has no net charge, exclusive of those contributed by bound ions, must be known; this pH is the isoionic pH (10) and can be determined by completely deionizing the protein. Both ferritin and apoferritin are largely insoluble in the absence of salt; for both proteins, the pH of a 1-2% suspension of deionized protein was found to be 4.58. On addition of low concentrations of salt, both proteins go into solution, but the change in pH of the system with change in salt concentration is markedly different for each (Fig. 1). For apoferritin, increasing concentrations of either NaCl or KCl lead to progressive increases in pH, signifying (11) the binding of chloride ions; small differences in response to NaCl and KCl are reproducible but (see also below) are of probably no major significance in the present context. The surprising feature of the data are the differences in response to salt of ferritin relative to apoferritin. For ferritin, addition of increasing concentrations of KCl to the deionized protein leads first to a large decrease in pH and then to a gradual increase in pH with increasing salt concentration. The initial decrease in pH suggests (11) that K^+ is being bound. This is verified by the marked difference between NaCl and KCl in their effects on ferritin; the smaller decrease in isoionic pH in the presence of NaCl than in the presence of KCl can only be explained (11) by a preference on the part of cation-binding sites for K^+ relative to Na^+. The gradual increase in the isoionic pH of ferritin as NaCl

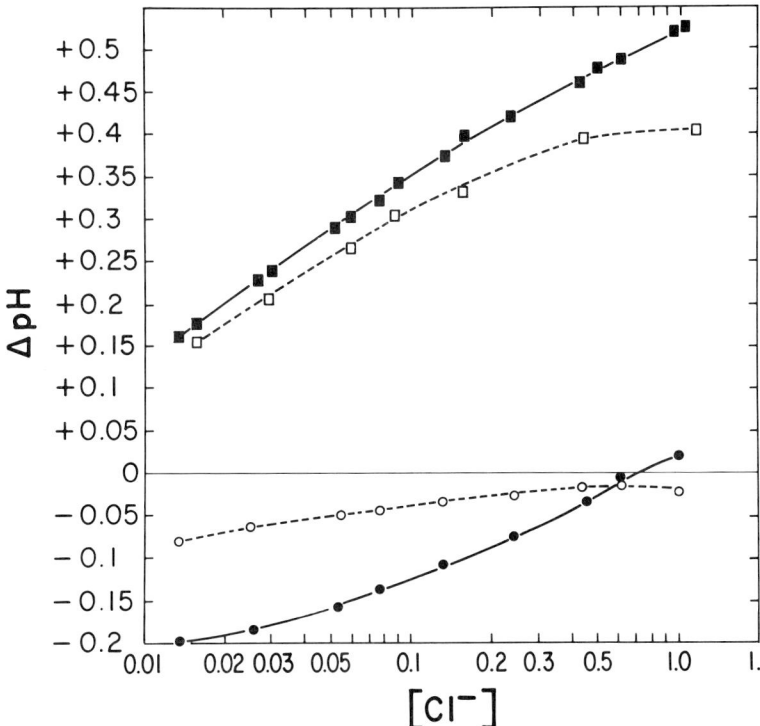

Fig. 1. Effect of KCl and NaCl on the pH at the isoionic point of 1.5 to 2.0% solutions of ferritin and apoferritin (semilogarithm plot). ■ , apoferritin + KCl; □ , apoferritin + NaCl; ●, ferritin + KCl; O, ferritin + NaCl. From Silk and Breslow (7).

and KCl concentrations are increased above 0.01 M indicates that anion-binding sites are also present on ferritin. The question then arises as to whether cation-binding sites are also present on apoferritin and whether the difference in response of ferritin and apoferritin to the lowest salt concentrations is due to the fact that there are a larger number of strong anion-binding sites on apoferritin than on ferritin and that the pH increases associated with binding to these extra sites mask the effects of cation-binding. The fact that NaCl dues not lead to a larger pH increase than KCl in its effects on apoferritin precludes the possibility that apoferritin has cation-binding sites similar to those on ferritin. Note, however, that the demonstrable presence of both cation-binding and anion-binding sites on ferritin does not permit, on the basis of these data alone, an evaluation of whether the same number of anion-binding sites are present on both proteins.

Potentiometric hydrogen ion titrations of ferritin and apoferritin were conducted by continuous addition of acid or base to the isoionic proteins in KCl; the extent of reversibility was determined by backtitration (7). In Fig. 2, titration curves of the two proteins are plotted at two ionic strengths; the ordinate, $\bar{h}$, represents the number of protons bound per subunit, setting $\bar{h} = o$ at the isoionic pH of apoferritin so that data for both proteins can be more readily compared. Only the forward titration data for apoferritin are shown (solid line). Backtitration of apoferritin from pH 8 was reversible, but a small degree of irreversibility was seen on backtitration from pH 11 or pH 3 and marked irreversibility accompanied backtitration from pH 2 (7); the irreversibility arising from exposure to low pH parallels CD and absorbance changes seen in this region (6,7). With the aid of spectrophotometric titration, side-chain modification studies, amide determination and studies of the effects of temperature, we have presented a preliminary self-consistent analysis of the apoferritin titration data (7). For the present purposes, the salient features of this analysis are that (per subunit), all of the tyrosine and cysteine residues, 2 of the 9 lysines, and at least 3 of the 6 histidines are unavailable for titration in the native protein; of the estimated 26 non-amidinated side-chain carboxyls per subunit, 3 are probably buried, and the remainder fall into at least two titration classes. Additionally the analysis suggests that the shape of the apoferritin titration curve below pH 3.5 is influenced by the known conformational changes in this region.

Comparison of the titration data of apoferritin and ferritin (Fig. 2) indicates that, between pH 3 and pH 5.5, at two ionic strengths, the titration curves of the two proteins are virtually identical. In addition, small differences between the two below pH 3 roughly parallel their differences in acid-stability; it is of interest, however, that the small degree of irreversibility in ferritin titration accompanying backtitration from pH 2 (a pH at which major CD changes have occurred) is much less than the degree of irreversibility of apoferritin backtitration from the same pH (7). The principal differences between native ferritin and apoferritin occur above pH 5.5. Between pH 5.5 and 7.5 two extra groups per subunit titrate in ferritin relative to apoferritin; the titra-

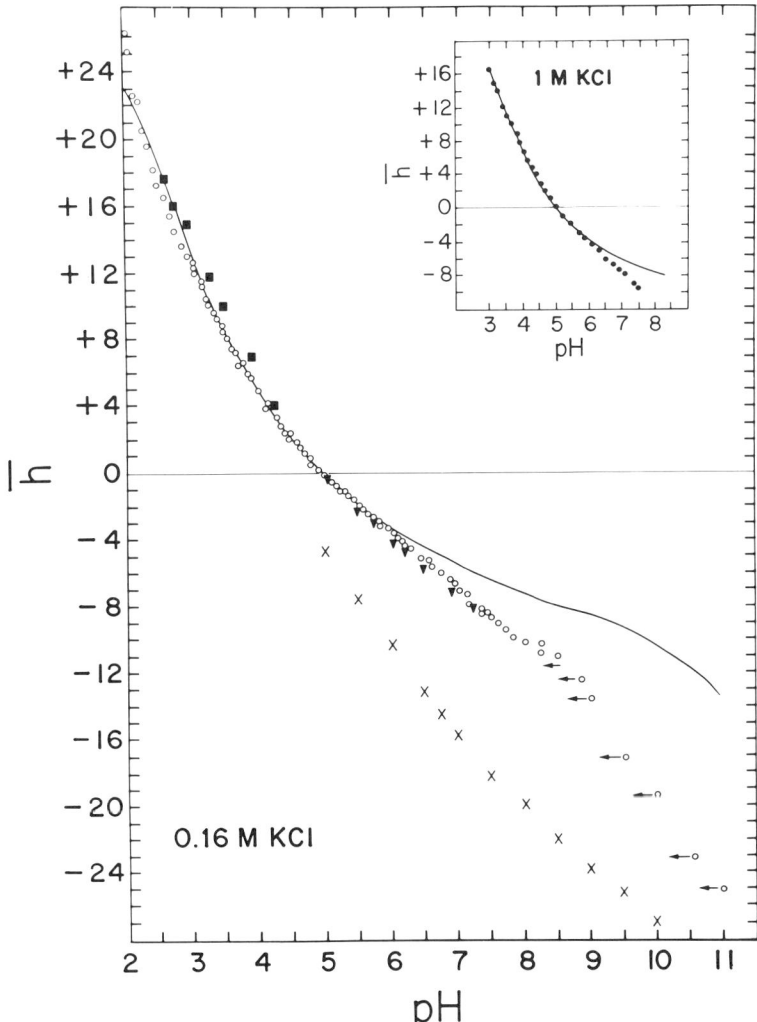

Fig. 2. Comparison of the titration behavior of apoferritin and ferritin at 25° C. ———, apoferritin forward-titration with acid or base from pH 5. In 0.16 M KCl, ferritin data are: ○, forward-titration with acid or base from pH 5 (arrows indicate presence and direction of drift); ■, rapid back-titration from pH 2; ▼, back-titration from pH 7.7; X, back-titration after 12 hours at pH 11. In 1 M KCl, ferritin data are: ●, forward-titration from pH 5. From Silk and Breslow (7).

tion of these two groups is time-independent and reversible (see back-titration from pH 7.7 in Fig. 2) and is invariant from one ferritin preparation to another provided (7) that the preparations have not been allowed to age. The probable identity of these residues will be discussed below. Above pH 8.5, titration of the ferritin core begins; this is a time-dependent process, manifest by considerable drift to lower pH on standing, and the rate of base uptake increases with increasing pH. After 12 hours at pH 11, an extra 43 groups per subunit were observed to titrate in ferritin relative to apoferritin. The exact nature of this process is uncertain. Clearly solvent is able to penetrate the core. In alkali, protein-free cores lose phosphate (12) and it has been suggested that this represents displacement of phosphate by hydroxyl ions. Granick and Hahn have also indicated (12) that in dilute alkali, both iron and phosphate can be leached from ferritin. It is therefore possible that the titration properties above pH 8.5 reflect actual displacement of phosphate both from the core and the protein, but we have no direct evidence to this point.

The extra two groups per subunit that titrate reversibly in ferritin relative to apoferritin above pH 5.5 offer an explanation of the difference in isoionic pH between the two proteins in KCl. The isoionic pH of 1% apoferritin in 0.16 M KCl is 4.96 and that of ferritin is 4.46. Inspection of the titration data indicates that 2.2 groups per subunit titrate between pH 4.96 and 4.46. This means that at pH 4.96, where apoferritin (exclusive of charge contributed by bound ions) has no net charge, ferritin has a net charge of -2.2 per subunit. Because the carboxyl titration regions of apoferritin and ferritin are indistinguishable (in the pH region where no conformational changes have occurred) and because we have also observed no significant differences between the two proteins in the reactivity of lysines, histidines or cysteines with formaldehyde and bromoacetic acid (7), we suggest that the extra negative charges on ferritin arise from the core. A self-consistent interpretation is that these charges represent inorganic phosphates at the core surface. The surface distribution of a large fraction of the core phosphate has long been argued (12). Inorganic phosphate can be expected to carry a negative charge at pH 5 and to lose a second proton with a pKa of approximately 7, thus explaining both the extra negative charges on ferritin at pH 5 and the extra titratable residues between pH 5.5 and 7.5. The number of such surface phosphates can be calculated as 2-3 per subunit, the uncertainty arising from the fact that horse spleen ferritin preparations contain a significant fraction of natural apoferritin.

The postulate that surface phosphates are the origin of the

additional negative charges and titratable residues on ferritin leaves unexplained the fact that both ferritin and apoferritin have identical isoionic points in the absence of salt. We have suggested (7) that this phenomenon might result from the unfavorable energetics of allowing the core to be polyanionic at the isoionic pH in the absence of counterions. In order to minimize inter-charge repulsion under these conditions, the surface phosphates are assumed to become protonated and therefore not to contribute to the net charge of the system; the decrease in pH when KCl is added to the isoionic ferritin solution is viewed as due to the displacement of protons by the binding of K^+ to phosphate. This explanation, as well as alternatives (7), are not without their limitations, but bring attention to the potential role of core surface phosphates in interacting either with solvent cations or with cationic residues on the inner surface of the protein shell. With respect to solvent cations, the apparent discrimination of ferritin between K^+ and Na^+, which we assume to reflect the preferences of the core surface phosphates, suggests that these phosphates have more specific bonding requirements than normally associated with inorganic phosphate (13) and that this property may be influential in the pattern, selectivity or rate of iron deposition. Potential interactions between the surface phosphates and cationic residues on the protein may be one of the factors contributing to the greater stability of ferritin than of apoferritin, as for example, to acid-denaturation.

In conclusion, titration studies reveal no differences between ferritin and apoferritin which cannot be accounted for by the assumed presence of a discrete number of phosphates on the core surface. If the presence of core materially affects the conformation of the ferritin shell, such perturbations are beyond the sensitivity of titration studies and probably represent local changes at the core-protein interface. Alternatively, the presence of core plays a demonstrable role in the ionic interactions and stability of ferritin; the functional significance of these effects merits investigation.

III. REFERENCES

1. Hoare, R.J., Harrison, P.M. and Hoy, T.G., Nature, 255, 653 (1975).

2. Drysdale, J.W. and Shafritz, D.A. in (Crichton, R.R., ed.), "Proteins of Iron Storage and Transport in Biochemistry and Medicine", p. 319, North-Holland Publishing Co., Amsterdam,

1975.

3. Pape, L., Multani, J.S., Stitt, C. and Saltman, P., Biochemistry, 7, 606 (1968).

4. Harrison, P.M. in (Crichton, R.R., ed.), "Proteins of Iron Storage and Medicine, p. 269, "North-Holland Publishing Co., Amsterdam, 1975.

5. Crichton, R.R., Biochim. Biophys. Acta, 229, 75 (1971).

6. Wood, G.C. and Crichton, R.R., Biochim. Biophys. Acta, 229, 83, (1971).

7. Silk, S.T. and Breslow, E., J. Biol. Chem., 251, 6963 (1976).

8. Listowsky, I., Blauer, G., Englard, S. and Betheil, J.J., Biochemistry, 11, 2176 (1972).

9. Leach, B.S., May, M.E. and Fish, W.W., J. Biol. Chem., 251, 3856 (1976).

10. Edsall, J.T. and Wyman, J., "Biophysical Chemistry", Vol. 1, pp. 599-605, Academic Press, Inc., N.Y., 1958.

11. Scatchard, G. And Black, E.S., J. Phys. Colloid Chem., 53, 88 (1949).

12. Granick, S. and Hahn, P., J. Biol. Chem., 155, 661 (1944).

13. Sillen, L.G. and Martell, A.E., "Stability Constants of Metal-Ion Complexes", p. 182, Special Publication No. 17, The Chemical Society, London, 1964.

(This study was supported by Grant GM-17528 from NIH and by NIH postdoctoral fellowship GM-55367 to S.T.S.)

Section II

ISOFERRITINS

SYNTHESIS OF FERRITIN SUBUNITS

BY FREE AND MEMBRANE

BOUND POLYSOMES

THOMAS G. ADELMAN
JAMES W. DRYSDALE

Department of Biochemistry and Pharmacology
Tufts University School of Medicine
Boston, Massachusetts

MINORU YOKOTA

Cancer Research Institute
Department of Medicine
Sapporo Medical College
Sapporo, Japan

I. INTRODUCTION

Ferritin, the eucaryotic iron storage protein, has a variety of structural and metabolic forms in mammalian tissues (reviewed in 1). These forms are composed of different combinations of two subunit types, MW 21,000 and 19,000, in a 24 subunit shell. Characteristic populations of the heteropolymeric isoferritins are found in different tissues or cell types (1).

Most ferritin is located intracellularly, but small amounts also occur in serum (2). This serum component is of biological and clinical interest since it varies in type and amount in various pathological conditions such as iron overload (3) and cancer (4). Its source is presently not known. While some serum ferritin might arise from tissue damage such as cirrhosis (5), findings of serum isoferritin populations distinct from those in major tissue stores suggest that some of these species could also arise from secretion.

For example, serum ferritin in iron overload has a very low iron content and consists mainly of a homopolymer of the L subunit (3). By contrast, most tissue ferritins in this disease have high iron contents and contain multiple isoferritins (6).

Such quantitative and qualitive alterations in serum ferritin point to specific mechanisms for its synthesis. Further support for this concept comes from two other observations. The first involves the site of ferritin synthesis. As might be expected for an intracellular protein, most of the cell's ferritin is made on free polysomes (7,8). However, it has been claimed that as much as 30% can be synthesized on membrane-bound polysomes (9) and that the two polysome classes synthesize different ferritin subunits, one of which is preferentially induced by iron (10). The second line of evidence comes from the recent reports that some ferritin is glycosylated (11, and M. A. Cynkin, this volume). Generally, glycosylation is initiated on nascent polypeptides on membrane bound polysomes and is usually associated with the production of secretory or membrane proteins (12).

It is difficult, however, to reconcile this apparent compartmentalization of ferritin subunit synthesis with the existence of heteropolymeric ferritins both intra- and extracellularly. We have attempted to resolve this apparent paradox by examining *in vitro* the synthesis of ferritin subunits by free and membrane bound polysomes.

II. METHODS

Ferritin synthesis was stimulated in male Sprague-Dawley rats (150-175 g) by intraperitoneal iron injection (400 µg iron as ferric ammonium citrate per 100 g body weight) 5 hr prior to preparation of free and membrane bound liver polysomes (13). Ferritin which co-sedimented with polysomes was quantitated by the Laurell "rocket" procedure (14) using antiserum to rat liver ferritin (15).

Nascent ferritin subunits on isolated polysomes were labelled in an *in vitro* run-off system using soluble translation factors prepared from post-microsomal liver supernatants (40 x 10^6 g-min supernatant) (13,16). Protein synthesis was carried out in a total reaction volume of 1 ml containing 50 mM Tris-HCl, pH 7.4, 75 mM KCl, 5 mM $MgCl_2$, 0.5 mM dithiothreitol, 10 µg/ml creatine phosphokinase, 2.5 mM ATP,

0.375 mM GTP, 10 mM creatine phosphate, 100 μCi ^{3}H-L-leucine (5 Ci/mmol), 20 μM of the remaining 19 amino acids, 0.2 to 1.8 mg polysomal protein, and 8-10 mg soluble translation factor protein. This mixture was incubated at 37° for 1 h followed by addition of 50 μg pancreatic ribonuclease and further incubation for 0.5 h. After standing at 4° for 16 h, denatured material was removed by centrifugation at 8,000 g for 4 min. ^{3}H-leucine incorporation into protein was determined on 25 μl of this supernatant after hot trichloroacetic acid precipitation (17). The remaining supernatant was adjusted to 0.05 M in sodium phosphate buffer pH 7.4, 0.15 M in NaCl, and 1% (w/w) in Triton X-100 (PBS-Triton); carrier rat liver ferritin was added when necessary to provide at least 20 μg ferritin per sample. A 1.5 fold excess of anti-rat liver ferritin antiserum was added and immunoprecipatation carried out for 1 h at 26° and 1 h at 4°. Precipitates were collected by centrifugation, washed twice with PBS-Triton, twice with PBS, and dissociated in 0.063 M Tris-HCl, pH 6.8, 1% SDS and 40 mM dithiothreitol at 100° 5 min. Electrophoresis of the dissociated immunoprecipitates was performed on SDS slab gels with an 8-22.5% acrylamide gradient (18). Gels were stained with Coomassie Blue to identify ferritin subunits. Sample tracks were sliced into 1 mm fractions and dissolved in 0.5 ml H_2O_2 by incubating for 16 h at 37°. Radioactivity was quantitated in a scintillation spectrometer after addition of 8 ml of cocktail (19) and storage in the dark for 24 h.

Ferritin synthesis in vivo was quantitated by labelling control or iron treated rats for 3 h with 100 μCi ^{3}H-L-leucine (5 Ci/mmol) (20). Labelled liver ferritin was purified by published procedures (15). Final purification was achieved by electrophoresis in a native gradient pore gel (21). Ferritin subunits were prepared by dissociation in SDS and analyzed as indicated above.

III. RESULTS AND DISCUSSION

A typical electrophoresis of immunoprecipitated ferritin subunits is shown in Figure 1; H and L are clearly separated from each other and from the immunoglobulin light chain. When ferritin synthesized by free polysomes is analyzed (Fig. 1), both H and L are labelled, indicating that the 21,000 and 19,000 MW subunits obtained from native holoferritin represent primary translation products.

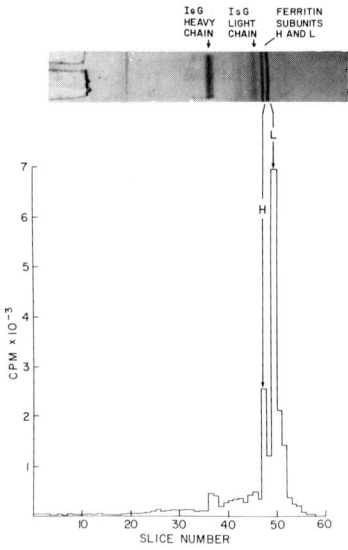

Fig. 1 Electrophoretic analysis of ferritin subunits synthesized by free liver polysomes from iron treated rats.

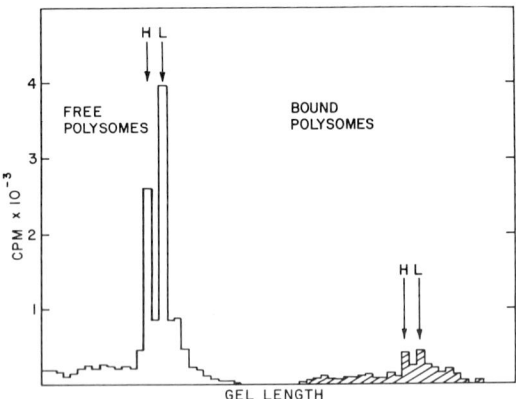

Fig. 2. Comparison of ferritin subunit synthesis by free and membrane bound liver polysomes from iron treated rats. Analyses as in Fig. 1; only the 25 fractions covering the 30,000 to 10,000 MW range are shown.

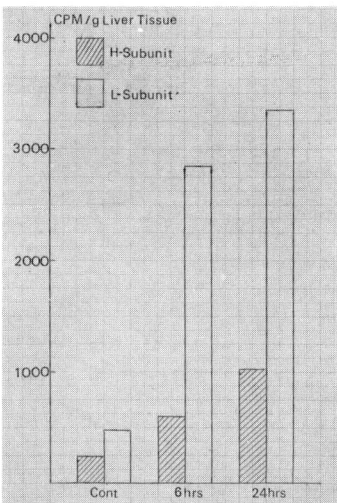 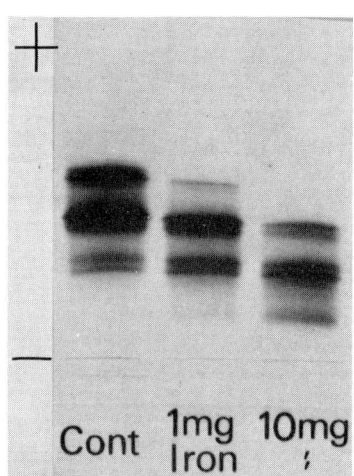

Fig. 3. (left panel) Ferritin subunit synthesis in control rats and in rats given 5 mg iron 6 h or 24 h prior to a 3 h ^{3}H-leucine pulse (see METHODS for details). Analysis performed as in Fig. 1.

Fig. 4. (right panel) Isoelectric focusing (15) of ferritin isolated from rats given injections of 0, 0.2, and 2.0 mg iron daily for 5 days prior to sacrifice.

The synthesis of ferritin subunits by free polysomes is compared to that of membrane polysomes in Figure 2. Although both classes of polysomes incorporate similar amounts of ^{3}H-leucine into polypeptides in the 15,000 to 90,000 MW range (18), the free polysomes synthesize approximately ten times as much total ferritin as do the membrane associated polysomes. In eight preparations of free polysomes, total ferritin synthesis averaged 0.25 $\pm$ 0.08% of total leucine incorporation into soluble protein; in four preparations of bound polysomes, the corresponding figure was 0.02% $\pm$ .01%. These estimates are similar to those obtained by Zahringer et al. (9). Much of the apparent synthesis of ferritin on bound polysomes could result from cross-contamination, since the preparations can contain up to 3% free polysomes (13).

Nevertheless, a portion of the detectable ferritin synthesis might represent subunits destined for secretion into the lumen of the endoplasmic reticulum prior to export. In either case, these experiments do not indicate preferential synthesis of either subunit on free and membrane bound polysomes.

Under the condition of iron stimulation used in these experiments, most of the ferritin radioactivity from free polysomes occurs in the L subunit. While some of the enhanced labelling of L might reflect a higher leucine content of that subunit (22), most of the increased incorporation probably results from preferential synthesis of the L subunit in response to iron. This conclusion is supported by findings that in vivo labelling of L in livers is enhanced up to five-fold over H after iron injection (Figure 3). Both these findings are consistent with our observation that iron causes a shift in cellular isoferritin populations, with increased accumulation of the less acidic heteropolymers (Figure 4).

Alhough 2 distinct products are obtained from the cell-free system, it is not clear that both are distinct gene products. Amino acid composition of the two subunits differs predominantly in only four residues and tryptic peptide maps (23) indicate that H and L contain common sequences. These features are suggestive of a possible precursor-product relationship. Conceivably, L might derive from proteolytic scission of some terminal peptides of H; subsequent selective modification by glycosylation could further enhance the differences between the two subunits. Glycosylation is unlikely to occur in our in vitro protein synthetic system, so that the differences in size between H and L probably reflects about 20 amino acid residues.

Regardless of the relationship between H and L, it is clear that iron enhances the production of L subunits and causes the consequent increase in L-rich isoferritins (6 and Fig. 4) in liver cells. It has been established that iron mediates the transfer of ferritin mRNA from mRNP to functional polysomes (9); but it is not known if there are discrete mRNAs for each subunit. If L is a post translational derivative of H, then iron would also be expected to enhance the conversion process. Further experiments with translation of ferritin mRNA in a heterologous translation system will be required to evaluate this latter possibility. Such studies might also resolve the paradoxical existence of specific extracellular isoferritins.

IV. REFERENCES

1. Drysdale, J.W., Adelman, T.G., Arosio, P., Casareale, D. Fitzpatrick, P., Hazard, J., and Yokota, M., Sem. in Hematol. 14, 71 (1977).
2. Lipschitz, D.A., Cook, J.D., and Finch, C.A., New Engl. J. Med. 290, 1213 (1974).
3. Arosio, P., Yokota, M., and Drysdale, J.W., Brit. J. Haematol. 36, 201 (1977).
4. Hazard, J. and Drysdale, J.W., Nature 265, 755 (1977).
5. Reissman, K..R. and Dietrich, M.R., J. Clin. Invest. 35, 588 (1956).
6. Powell, L.W., Alpert, E., Isselbacher, K.J. and Drysdale, J.W., Nature 250, 333 (1974).
7. Hicks, S., Drysdale, J.W., and Munro, H.N., Science 164, 584 (1969).
8. Redman, C.M., J. Biol. Chem. 244, 4308 (1969).
9. Zähringer, J., Baliga, B.S., Drake, R.L. and Munro, A.M., Biochim. Biophys. Acta 474, 234 (1977).
10. Konijn, A.M., Baliga, B.S. and Munro, H.N., FEBS Lett. 37, 249 (1973).
11. Shinjyo, S., Abe, A., Masuda, M., Biochim. Biophys. Acta 411, 165 (1975).
12. Campbell, P.N. and Biobel, G., FEBS Lett. 72, 215 (1976).
13. Ramsey, T.C. and Steele, W.J., Biochemistry 25, 1704 (1976).
14. Laurell, C.-B., Scand. J. Clin. Lab. Invest. 29 (Suppl. 124, 21 (1972).
15. Powell, P.W., Alpert, E., Isselbacher, K.J. and Drysdale, J.W., Brit. J. Haematol. 30, 47 (1975).
16. Baliga, B.S., Pronzcuk, A.W. and Munro, H.N., J. Mol. Biol. 34, 199 (1968).
17. Palmiter, R.D., J. Biol. Chem. 248, 2095 (1973).
18. Adelman, T.G. and Drysdale, J.W., Manuscript submitted for publication.
19. Anderson, L.E. and McClure, W.O., Anal. Biochem. 51, 173 (1973).
20. Drysdale, J.W. and Munro, H.N., J. Biol. Chem. 241, 3630 (1966).
21. Margolis, J. and Kenrick, K.G., Nature 221, 1056 (1969).
22. Linder, M.C., Moor, J.R., Munro, H.N., Morris, H.P., Biochim. Biophys. Acta 386, 409 (1975).
23. Arosio, P. and Drysdale, J.W., Manuscript submitted for publication.

ALTERATIONS IN SERUM AND TISSUE ISOFERRITINS
IN DISEASE STATES: I. DUODENAL FERRITIN CONTENT
AND ISOFERRITIN COMPOSITION

J.W.Halliday, U.Mack and L.W.Powell

*Department of Medicine, University of Queensland
Brisbane, Australia*

I. INTRODUCTION

The role that duodenal ferritin plays in the control of body iron balance remains uncertain. It is currently believed that ferritin in villous epithelial cells stores unwanted iron which is then lost from the body when the cells are sloughed into the lumen. In iron deficiency very little ferritin can be demonstrated by electron microscopy in the villous epithelial cell. The development of sensitive immunoassays[1,2] for ferritin has now made possible the direct measurement of ferritin protein in intestinal mucosa. Using such techniques we have measured simultaneously the ferritin concentration in plasma and in gut. We have also analyzed the isoferritin composition of ferritin from human and rat duodenum in order to correlate duodenal ferritin content and isoferritin composition with body iron status.

II. METHODS AND RESULTS

A. Human Studies

53 subjects were studied - 25 with normal iron stores, 15 with iron deficiency (I.D.), 11 with idiopathic hemochromatosis (IHC) and 2 with transfusional iron overload. Biopsies (10-30 mg wet tissue weight) were obtained from the duodenum via fiberoptic endoscopy after informed consent. The samples were incubated for 30 min. in a medium consisting of 92 mM NaCl, 4.3 mMKCl, 0.1 mM $CaCl_2$, 35 mMD(+)-mannose, 40 mM D(+)-glucose, 7 mM L-ascorbic acid, buffered to pH7.4 with Hepes/NaOH buffer and containing 10μCi of ^{59}Fe as ferric citrate with 2.5μg carrier iron per ml of medium. After incubation, the samples were washed rapidly in ice-cold bicarbonate-buffered-saline-KCl and homogenized. Ferritin was isolated and purified from the supernatants[3], its concentration

measured by radioimmunoassay[2], and the isoferritin composition determined[3]. There was a highly significant correlation (r=0.82) between serum ferritin and duodenal ferritin concentrations in normal subjects and patients with iron deficiency. Although a positive correlation was also observed in patients with iron overload, the relationship differed in that the duodenal ferritin concentration was lower at all levels of serum ferritin in comparison with normal and iron-deficient subjects (Fig.1). Patients with secondary iron-overload did not differ from those with IHC, indicating that any decrease in the concentration of duodenal ferritin was not due to a primary defect in IHC but rather was secondary to the excess body iron stores.

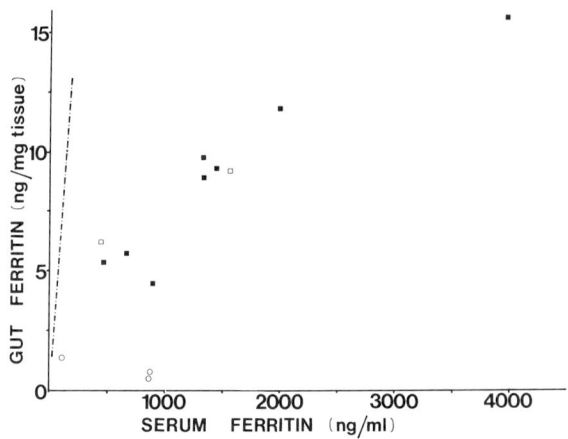

Fig. 1. Relationship of duodenal mucosal ferritin and serum ferritin concentrations in patients with (a) transfusional iron overload (□), (b) idiopathic hemochromatosis (■), (c) idiopathic hemochromatosis after venesection (o). Dotted line shows line of best fit for normal and iron deficient subjects.

Purified duodenal ferritin was also examined by isoelectric focusing on polyacrylamide gels[3]. The profile of normal duodenal ferritin showed two distinct isoferritins of pI 5.25 and 5.35 (Fig. 2). The profiles of duodenal ferritin from patients with secondary iron overload and IHC were identical with that of normal subjects. After exposure to oral iron, additional more basic isoferritins of pI 5.56 and 5.62 were detected, which resembled the major isoferritins of liver (Fig. 3).

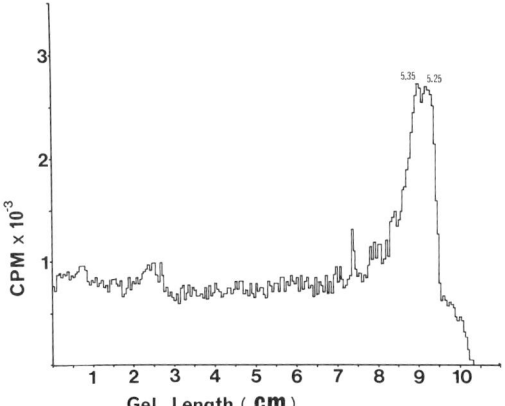

Fig. 2. Isoferritin profile of duodenal mucosal ferritin from a patient with normal iron stores.

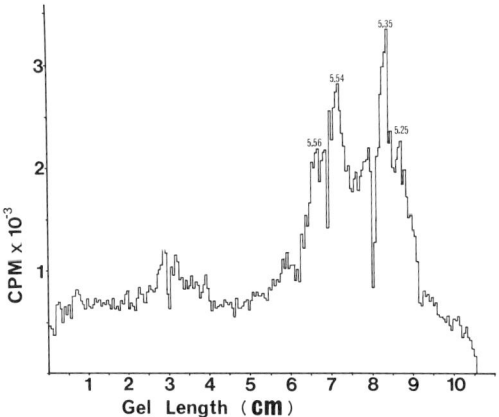

Fig. 3. Isoferritin profile of duodenal mucosal ferritin from a patient with iron deficiency who was receiving oral iron. In addition to the more acidic components demonstrated in normal mucosal ferritin there are more basic isoferritins of pI 5.54 and 5.56 which are similar to those demonstrated in human liver ferritin.

B. Rat Studies

40 normal, 40 iron-loaded and 20 dietary iron-deficient Sprague-Dawley rats were studied. Proximal duodenal

epithelial cells were obtained by scraping or mechanical vibration[4] and treated as above. Blood was obtained for serum ferritin concentration from the inferior vena cava. Acute iron loading (2 weeks) resulted in elevated intestinal ferritin levels (10-fold increase over normal) and increased serum ferritin concentrations (also 10-fold increase). Over a 3 month period the serum and intestinal ferritin levels gradually fell to 3 and 5 times normal respectively. In the iron-deficient rats, gut ferritin was reduced to one-fifth normal levels but the serum ferritin remained within the normal range. Duodenal isoferritin composition in the rat resembled that of man in that it differed from all other tissues studied and was unaffected by excess body iron stores.

III. CONCLUSIONS

(i) There is a linear relationship between duodenal ferritin concentration and body iron stores in normal and iron deficient human subjects. In rats there is a similar relationship in normal animals but the serum ferritin level is not reduced in iron deficiency. (ii) In long-standing primary and secondary iron overload states in man the intestinal ferritin content is disproportionately lower at all levels of serum ferritin concentration. This may reflect the rapid turnover of intestinal epithelial cells. (iii) Normal duodenal ferritin in both man and rat has a distinctive isoferritin profile with prominent acidic isoferritins. (iv) This profile is unchanged in iron overload states; however oral iron administration results in the appearance in the gut of isoferritins of pI 5.56 and 5.62 which may play a role in the control of iron balance.

IV. REFERENCES
1. Addison, G.M.,Beamish, M.R.,Hales, C.N.,Hodgkins, M., Jacobs, A. and Llewellin, P. J. Clin. Path., 29, 326-329 (1972).
2. Halliday, J.W.,Gera, K.L. and Powell, L.W. Clinica Chimica Acta, 58, 207-214 (1975).
3. McKeering, L.V.,Halliday, J.W.,Caffin, J.A.,Mack, U. and Powell, L.W. Clinica Chimica Acta, 67, 189-197 (1976)
4. Halliday, J.W. and Powell, L.W. Clinica Chimica Acta, 43, 267-276 (1973).

V. ACKNOWLEDGEMENTS

This work was supported by the National Health and Medical Research Council of Australia.

ALTERATIONS IN SERUM AND TISSUE ISOFERRITINS
IN DISEASE STATES: II.HEMOCHROMATOSIS AND MALIGNANT DISEASES

L.W.Powell, June.W.Halliday, L.V.McKeering and Rowan Tweedale

*Department of Medicine, University of Queensland
Brisbane, Australia*

I. INTRODUCTION

There is increasing evidence that changes occur in the tissue and serum isoferritin composition in disease states.

We have previously shown that in patients with primary or secondary iron overload the organ-specificity of tissue isoferritins is lost, all tissues exhibiting a preponderance of more basic isoferritins such as are found in liver[1]. This phenomenon is reversed by venesection therapy[2]. In the present study we have analyzed changes in human serum isoferritin composition during venesection to determine whether the changes which occurred in tissue isoferritins were also seen in serum.

Similarly, there is increasing evidence that ferritin produced by malignant neoplasms differs from that produced by corresponding normal tissues of the same individual[3-5]. Most investigators have inferred that the differences lie in the presence of isoferritins of more acidic isoelectric point (pI) in malignant tissues[3-5]. However, the precise nature of the abnormality and the possible tumor-specificity of these isoferritins are still debatable. We have studied the isoferritin composition of ferritin isolated and purified from sera and tissues of patients with renal, pancreatic and colonic cancers and compared these with the profiles of the corresponding normal tissues.

II. METHODS AND RESULTS

A. Hemochromatosis

Sera from 13 patients with uncomplicated idiopathic hemochromatosis (IHC) and 4 with secondary iron overload e.g. thalassaemia were studied. 7 of the subjects were studied serially during venesection therapy over periods of up to 18 months. Samples taken over this time were stored at $-20°$

until subjected to simultaneous isoelectric focusing (IEF) as previously described[6]. During the course of venesection therapy in patients with IHC there was a progressive increase in isoferritins of pI 5.02 to 5.06 relative to the more basic ones (Figs. 1 and 2). Serial studies in patients with secondary iron overload showed no change.

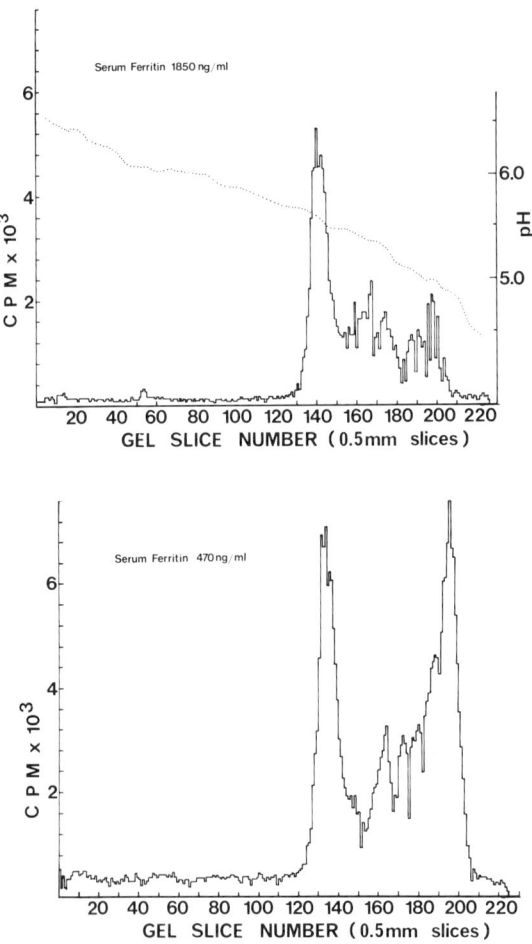

Figs. 1 and 2. Serum isoferritins before and during phlebotomy.

B. Malignant Diseases

A highly sensitive technique using ^{125}I-labelled monospecific anti-human-liver-ferritin-antibody for isoferritin detection after gel isoelectric focusing of 50 ng samples of ferritin has been described previously[6]. Samples of serum and tissue ferritin from the one subject were studied simultaneously. In all tumors studied (renal, pancreatic and colonic) the isoferritin composition differed from that of the corresponding normal tissue in that major isoferritins with pI more <u>basic</u> than those of the normal tissues were consistently detected (Fig. 3). Abnormal acidic isoferritins were not apparent in these tumors despite the fact that the anti-ferritin antibody used recognized isoferritins in the pH range 5.04 to 5.62. The isoferritin composition of purified ferritin from metastases closely resembled that of the primary tumor. Analysis of the <u>serum</u> isoferritin profiles of patients with cancer did not reveal any changes which could be interpreted as tumor-specific. Rather, the isoferritins demonstrated in human serum encompass the whole range of isoferritins present in normal and malignant tissues.

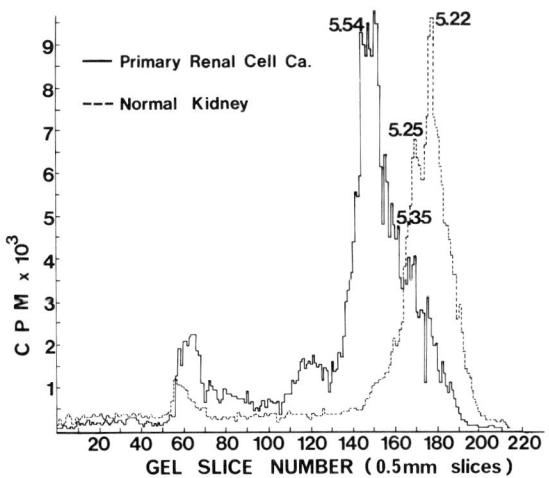

Fig. 3. Isoferritin profiles of ferritin purified from a primary renal carcinoma.

III. DISCUSSION AND CONCLUSIONS

These studies provide further evidence for a possible biological role of the individual isoferritins. The results in patients with IHC are compatible with the hypothesis that the more basic isoferritins correspond to a 'storage' ferritin and the more acidic to a 'secretory' ferritin, and that phlebotomy results in the mobilization of more acidic isoferritins from within tissues. The findings in patients with malignant disease, together with our earlier observations[2] which suggested that the more basic isoferritins correspond to those synthesized on free polysomes, are consistent with the hypothesis that tumor tissues contain a large quantity of free ribosomes which synthesize isoferritins of pI 5.54 to 5.56. Thus, it is possible that the appearance of these isoferritins merely reflects a change in the site of synthesis of the major component isoferritins by tumor tissue rather than the synthesis of a tumor-specific isoferritin.

IV. REFERENCES

1. Powell, L.W., Alpert, E., Isselbacher, K.J. and Drysdale, J.W. Nature, 250, 333-335 (1974).
2. Powell, L.W., McKeering, L.V. and Halliday, J.W. Gut, 16, 14-23 (1975).
3. Alpert, E., Coston, R.L. and Drysdale, J.W. Nature, 242, 194-196 (1973).
4. Linder, M.C., Moor, J.R., Munro, H.N. and Morris, H.P. Biochim. Biophys. Acta 386, 409-421 (1975).
5. Marcus, D.M. and Zinberg, N. Arch. Biochem. Biophys., 162, 493-501 (1974).
6. McKeering, L.V., Halliday, J.W., Caffin, J.A., Mack, U. and Powell, L.W. Clinica Chimica Acta, 67, 189-197 (1976)

V. ACKNOWLEDGEMENTS

This work was supported by the National Health and Medical Research Council of Australia and by the Queensland Cancer Fund.

ASSEMBLY OF FERRITIN INTERSPECIES AND INTERTISSUE HYBRIDS

Yoshiro Niitsu, Shinobu Ohtsuka, Naoki Watanabe
Junichi Koseki, Utaka Kohgo and Ichiro Urushizaki

Department of Medicine
Cancer Research Institute
Sapporo Medical College
Sapporo, Japan

I. INTRODUCTION

Ferritin from many tissues exhibits a heterogeneous profile when examined by isoelectric focusing (1-3). This heterogeneity of ferritin was recently ascribed to differences in the relative proportion of two different subunit types (4). Furthermore, an increase in the more acidic components in tumor isoferritins was demonstrated, and this acidic shift in isoelectric point was considered to reflect a change in the proportion of these different subunits (5, 6). In the present paper, we report the assembly in vitro, of hybrids of subunits from different species, and hybrids from different tissues of a single species.

II. EXPERIMENTAL PROCEDURE

Horse spleen ferritin, (6x crystalized), was purchased from Pentex Co. and further purified by gel filtration on Sepharose 6B. Ferritins from human liver and horse heart were isolated by standard procedures, followed by repeated gel filtrations on Sephadex G-200 and Sepharose 6B. Before gel filtration of the heart supernatant the supernatant was passed through a millipore membrane. Reduced apoferritin was prepared by the method of Granick (7). All apoferritin preparations were extensively dialyzed against distilled water and lyophilized. These lyophilized apoferritins were then degraded into subunits by treatment with sodium dodecyl sulfate (10x of the protein weight) to ensure complete degradation (Fig. 1) in 0.01 M tris acetate buffer, pH 8.3, containing 6M urea and 1% mercaptoethanol.

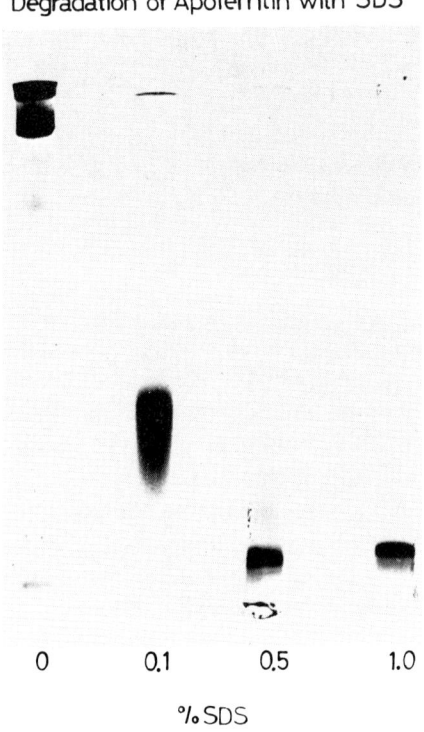

Fig. 1. Degradation of horse spleen apoferritin with SDS examined by polyacrylamide gel electrophoresis. The 0.1% protein solutions were incubated with 0.1%, 0.5% and 1.0% SDS as indicated. Complete degradation of apoferritin into subunits was attained at concentrations of SDS above 0.5%.

For hybrid formation, subunits of different ferritin species (horse spleen ferritin and human liver ferritin or horse spleen ferritin and horse heart ferritin) were used in the proper proportions indicated in the text. The solutions were first applied to Sephacryl S-200 columns (1.5 x 30 cm) which was previously equilibrated to 6M urea and 0.1% mercaptoethanol in 0.01 M tris acetate buffer, pH 8.3. The second peak just after the void volume was collected and dialyzed for 3 days against the same buffer in the absence of urea. The solution was concentrated using amicon membranes (UM 10) and subjected to preparative gel electrophoresis (in 5% polyacrylamide gels). The main band with a mobility expected for ferritin was cut from the gel and isolated electrophoretically. The total recovery of protein was approximately 50 percent. The isolated protein (hybrid molecule) was examined for its electrophoretic and immunological properties.

SDS gradient gel electrophoresis was performed according to O'Farrel's Method (8). Gel electrofocussing was carried out using 2% phisolyte, in 4% polyacrylamide gels containing 1% Triton X 100 with a constant voltage of 300 V for 9 hours. After electrophoresis, the gels were stained with Coomassie brilliant blue R-250.

III. RESULTS

The gel focusing profile of dissociated and reconstituted horse spleen ferritin treated according to the conditions used for the hybridization study was compared with the original horse spleen ferritin (Fig. 2). There was essentially no difference in mobility and in the heterogeneous pattern between these two samples. The hybrids prepared by combining subunits of human liver and horse spleen ferritins were first examined by slab gel electrophoresis. Horse spleen ferritin migrated faster than human liver ferritin and the hybrid ferritin migrated to an intermediate position between the two ferritins.

The mobility of mixtures of horse spleen and human liver ferritin was also intermediate between that of the two pure proteins. However two dimer bands were observed in the mixture of ferritins whereas the hybrid ferritin showed a single dimer band. The latter result indicates that the assembled hybrids are not mixtures but are indeed single molecules.

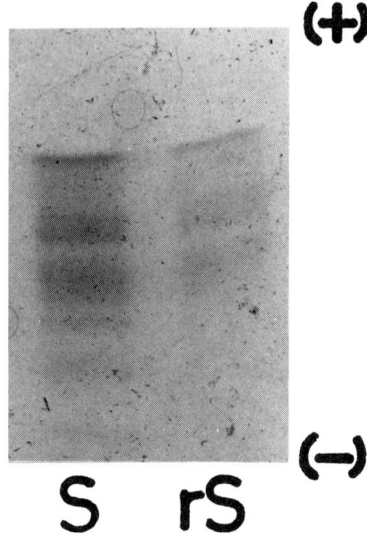

Fig. 2. Isoelectric focusing profile of horse spleen ferritin (S) and horse spleen ferritin (rS) reconstituted from its subunits. There are no differences in mobility and in the heterogeneous pattern.

In order to confirm the above observations, the hybrid horse spleen-human liver ferritin was then examined by isoelectric focusing methods and gel focusing pattern of ferritins from horse spleen, human liver and hybrid ferritins are shown in Fig. 3. Multiple bands were observed in all ferritin preparations, in accordance with previous reports (2). Differences in mobility between the hybrid and pure ferritins were more pronounced as compared to ordinary slab gel electrophoretic methods described above. The hybrid ferritins (containing 2 parts human liver and 1 part horse spleen ferritin) showed a distribution of multiple bands over wide pH range. These bands included those with mobilities identical to that of human liver ferritin, and a distribution of isoproteins ranging from the acidic portion of human liver ferritin to basic side of horse spleen ferritin.

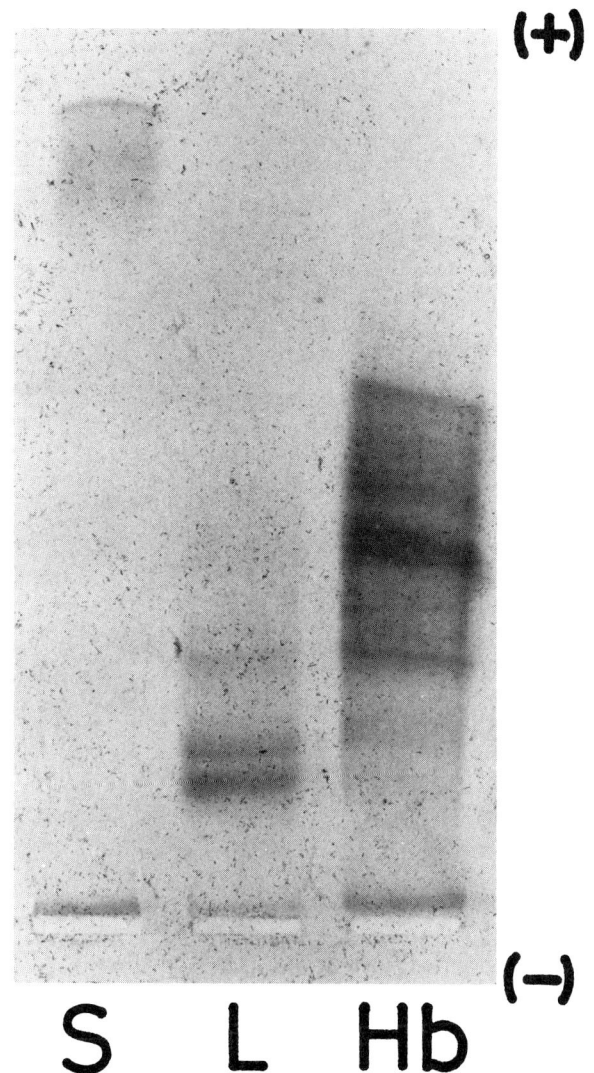

Fig. 3. Isoelectric focusing profile of horse spleen ferritin (S), human liver ferritin (L) and hybrid horse spleen-human liver ferritin (Hb). Distribution of hybrid isoferritins ranged from the acidic portion of human liver to basic side of horse spleen ferritin.

Subunits of the hybrid molecule were also analyzed by SDS gradient gel electrophoresis. Horse spleen ferritin and human liver ferritin showed two types of subunits with molecular weights of 19,000 and 21,000, and 20,000 and 22,000 respectively. In both ferritins, subunits with the lower molecular weight predominated. Hybrid ferritins also showed two distinct bands with molecular weights of 19,000 and 20,000. These corresponded to the predominant subunits of each original ferritin and the proportion of these two bands reflected the corresponding amount of the original ferritin preparations hybridized (2 parts human liver and 1 part horse spleen in this case). Similar hybrid like isoferritin components were observed by reconstitution of intraspecies subunits from horse spleen and horse heart ferritins.

IV. CONCLUSIONS

The present study appears to confirm the hypothesis that heterogeneity in the isoelectric focusing profiles of ferritin originates from molecules containing different combinations of different subunit types. Some of the synthetic hybrids assembled here from combinations of horse spleen and human liver subunits, focused at intermediate pI's between the two individual components from which they were prepared. These data also imply that there is appreciable conservation in subunit-subunit interaction sites, since interspecies hybrids are readily formed.

V. REFERENCES

(1) Drysdale, J. W., Biochim. Biophys. Acta, 207, 256 (1970).
(2) Urushizaki, I., Niitsu, Y., Ishitani, K., Matsuda, M. and Fukuda, M., Biochim. Biophys. Acta, 243, 187 (1971).
(3) Drysdale, J. W., Biochem. J., 141, 627 (1974).
(4) Adelman, T. G., Arosio, D. and Drysdale, J. W., Biochem. Biophys. Res. Commun., 63, 1056 (1975).
(5) Arosio, P., Yokota, M. and Drysdale, J. W., Cancer Res., 36, 1735 (1976).
(6) Niitsu, Y., Kohgo, Y., Ohtsuka, S., Watanabe, N., Koseki, J., Shibata, K., Ishitani, K., Nagai, T., Gocho, Y. and Urushizaki, I., in "Oncodevelopmental Gene Expression", Academic Press, p. 757, (1977).
(7) Granick, S. and Michaelis, L., J. Biol. Chem., 147, 91 (1943).
(8) O'Farrel, P. H., J. Biol. Chem., 250, 4007 (1975).

FERRITIN AND APOFERRITIN FROM HUMAN LIVER: ASPECTS OF HETEROGENEITY

Daniel J. Lavoie, Donald M. Marcus,
Kunitsugu Ishikawa and Irving Listowsky

*Departments of Biochemistry and Medicine
Albert Einstein College of Medicine
Bronx, New York, 10461, USA*

I. INTRODUCTION

The microheterogeneity of ferritin, which has attracted considerable attention recently, is not fully understood. Evidence has been presented that assembled ferritin molecules may consist of different proportions of two or more (1-4) different subunit types, which give rise to a family of isoferritins. The underlying structural basis for the subunit heterogeneity has not been defined. Additional complexity has been observed in subunit analyses, and 19,000 molecular weight subunits and some smaller peptides (11,000 and 8,000 molecular weight) are usually present in ferritin preparations (5-7). In this report, the appearance of the two smaller polypeptides has been attributed to the action of a serine protease, and the fragmentation process may be prevented by substances that inhibit proteolytic enzymes. The electrofocusing pattern of natural apoferritin from human liver is defined, and the possibility that ferritin may be a glycoprotein is presented.

II. EXPERIMENTAL PROCEDURE

Human liver ferritin was prepared according to methods described earlier (8). Modifications using immuno-adsorbents (9), and additions of phenyl methylsulfonyl fluoride were employed as indicated. For carbohydrate analysis, reduced, iron free, apoferritins were hydrolyzed in 0.5 N HCl in dry methanol under reflux for 24 hrs. After evaporation of the methanol, the residue was dried in a vacuum dessicator over phosphorus pentoxide and KOH pellets for an additional 6 hrs. The syrup obtained (methyl glycosides) was triturated for 5-10 min with 0.1 ml pyridine-trimethylsilyl chloride-hexamethyl disilazane (5:1:1 v/v). Samples (0.5-1.0 μl) of the TMS reaction mixture were injected into a Hewlett Packard Model 7610A high efficiency gas chromatograph equipped with a 3370B integrator. Nitrogen was used as the carrier gas, and the flow rate was adjusted for optimum column efficiency. The column was maintained at $160°$. Alternatively, alditol acetates were prepared, primarily to detect mannose and to verify the results obtained with the TMS derivatives. For these experiments, the protein was hydrolyzed in 1 N HCl at $100°$ for 4 hrs. The hydrolysate was then passed through Dowex 50 x 4 (H^+) and Dowex 1 x 8 (formate) columns sequentially. The sugars were reduced in 0.25 M $NaBH_4$ solutions at $4°$. After 12 hrs the pH of the solution was adjusted to pH 4-5 with 4 N acetic acid, and it was passed through a Dowex 50 resin. The samples were lyophilized, 150 ml of methanol added, and the borates evaporated under vacuum at $40°$. The remaining alditols were dried and then acetylated in 0.25 ml pyridine by adding 0.25 ml of acetic anhydride and heating to $100°$ for 1 hr in a sealed tube. A Perkin Elmer model 910 Gas Chromatograph was used to quantitate the alditol acetates. For amino sugar analysis, the proteins were hydrolyzed in 2 N HCl for 4-8 hrs, in a sealed evacuated tube. The hydrolysate was then applied to a Joelco Model JLC-6AH amino acid analyzer with LCR-2 acidic ion exchange resin. The column (21 cm length) was eluted with 0.4 M citrate buffer pH 5.28. Pure glucosamine and galactosamine standards were used as markers.

III. RESULTS AND DISCUSSION

A. Subunit Fragmentation

Analysis of the subunits of ferritins obtained from various sources showed consistently that the protein dissociated into a 19,000 molecular weight component and varying amounts of two smaller peptides with molecular weights of about 11,000 and 8,000 (5-7). Each of these polypeptides was purified. Since the sum of amino acid compositions of the two smaller peptides were identical to that of the larger peptide, it was suggested that the 11,000 and 8,000 molecular weight components originated from the 19,000 molecular weight subunit (7). To determine if a cleavage occurs as a result of manipulations commonly employed during the isolation and purification procedures, we have introduced some modifications in the standard procedure for purification of ferritin.

Omission of the 75-80° heating step, by substituting immuno-adsorbent columns to selectively bind ferritin, had no pronounced effect on the subunit patterns and the smaller peptides remained. Incorporation of a serine protease inhibitor into the procedure, however, did have a dramatic effect. If 0.1 mM phenyl methylsulfonyl fluoride (PMSF) was incubated with the tissue homogenates and was present at all stages of the purification process, the selective fragmentation of the 19,000 molecular weight component was effectively prevented. A recent study found protease contaminants in commercial ferritin preparations (10). SDS polyacrylamide gels comparing peptide patterns of ferritins prepared in the presence or absence of PMSF (Fig. 1), clearly implicate proteolytic cleavages in the fragmentation of the 19,000 molecular weight subunit.

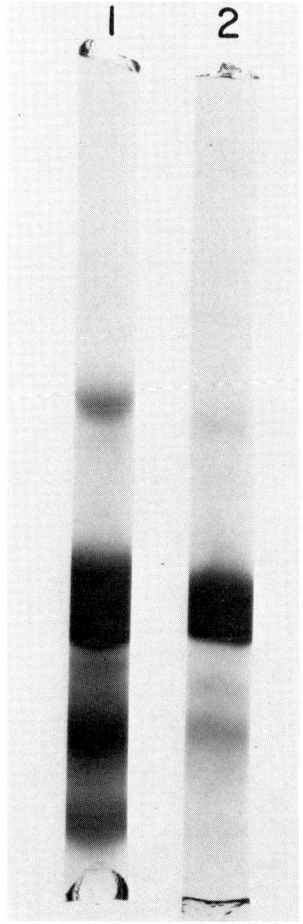

Fig. 1. The effect of phenyl-methylsulfonyl fluoride (PMSF) on subunit patterns of ferritin. Ferritin in gel 1 was prepared in the absence of PMSF and ferritin in gel 2 in the presence of PMSF. Electrophoretic conditions were in SDS-tris-acetate system as described previously (7).

Evidently, the proteolytic cleavage is not extensive or a random process, since discrete fragments are always produced. There appears to be a particularly susceptible peptide bond(s) in the ferritin subunit, and the cleavage process is highly specific. Thus, comparisons of the fragment size induced in ferritins from various different tissues and species (7) show a consistent pattern.

It has been shown that the 11,000 and 8,000 molecular weight fragments may be assembled to form a multimer that retains many of the properties and morphological appearance of authentic apoferritin samples (7). At present it is not clear if this selective fragmentation of ferritin has any physiological significance for the metabolism or function of the protein or is a fortuitous event without functional meaning. The process must however, be considered in chemical or structural analysis of the protein, and preventive measures such as those discussed here, may simplify analyses of heterogeneity and comparisons of results from different laboratories.

B. Natural Apoferritin

To determine the relation between the isoelectric point and iron content of ferritins, natural apoferritin from human liver was examined. In earlier studies we showed that natural apoferritin from horse spleen focused primarily as the most acidic components of the isoferritin spectrum (3). The natural apoferritin component from human liver ferritin was isolated by a density gradient contrifugation procedure described previously (11). A gel scan of the electrofocusing pattern of this apoferritin is compared to that of iron containing human liver ferritin in Fig. 2. The apoferritin component contains all 7 of the major isoferritins found in the iron containing molecules, but, in contrast to the horse spleen apoferritin, is greatly enriched in the more basic isoferritins (Components 5, 6 and 7 in Fig. 2 predominate). The iron rich ferritins isolated from the gradient had correspondingly less of these basic components. Indeed, the iron saturated ferritin components obtained from the bottom of the gradient, were devoid of the most basic isoferritins.

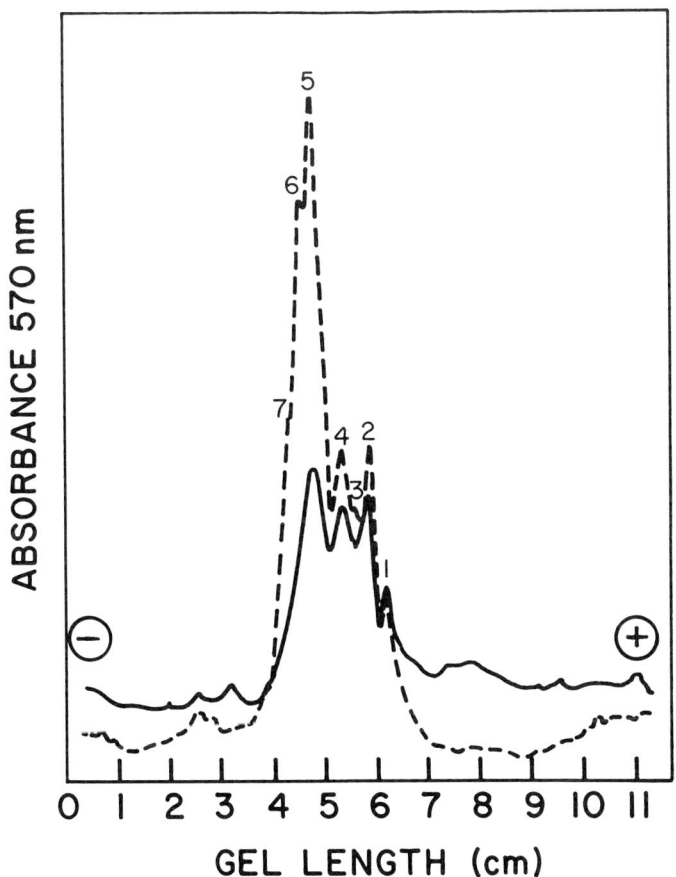

Fig. 2. Isoelectric focusing profile of human liver ferritin (solid line) and apoferritin (dotted line). Electrofocusing was carried out using pH 4-6 ampholytes, and the gels were scanned using a Gilford Spectrophotometer equipped with a model 2410 Linear Transport.

The structural differences between natural apoferritin and ferritin that result in differences in their isoelectric focusing profiles are presently obscure. Iron content alone cannot account for these differences since apoferritin prepared by reduction and removal of the iron from ferritin exhibits an isoelectric focusing profile that is almost identical to that of the iron loaded ferritin from which it is derived. Natural apoferritin should therefore be considered as a unique component in the context of the isoferritins. It is conceivable that the electrofocusing differences are related to the differences in metabolic state of natural apoferritin as compared to ferritin. It is also interesting to note that serum ferritin, which is relatively iron poor, is enriched in basic isoferritin components (12). Further understanding of the function, and definition of the role assumed by apoferritin, must await elucidation of the underlying structural basis for the overall microheterogeneity of ferritin.

C. Sugar Components

In view of preliminary reports from this and other laboratories that carbohydrates are associated with ferritin (13,14), hydrolysis or methanolysis products of horse spleen, rat liver, and human liver ferritins were examined in an effort to detect sugars. Peaks with retention times corresponding to glucose, galactose, and mannose were obtained by gas-liquid chromatography analysis. Glucosamine and galactosamine were detected by ion exchange chromatography, and the amino sugars were also found for purified subunits of horse spleen ferritin. The latter result rules out the possibility that these sugars originate from a glycoprotein contamination or from residual iron micelles. Sialic acid was not detected by the thiobarbituric acid method (15), and in contrast to the results of Shinjyo et al., no fucose was found. A summary of the carbohydrate composition of horse spleen ferritin is shown in Table I.

TABLE I

CARBOHYDRATES IN HORSE SPLEEN FERRITIN

Sugar	Residues per 19,000 Dalton Subunit
Glucosamine	2.3
Galactosamine	1.1
Glucose	1.6
Galactose	1.3
Mannose	0.9
Fucose	Not Detected
Sialic Acid	Not Detected

These findings have broad implications for many areas of ferritin research. For example glycosylation may influence microheterogeneity patterns of individual ferritins. In addition the sugars could affect processes such as iron loading, and metabolism of ferritins, and determine immunological properties of the protein. Further studies in this context are necessary to define the role of the sugars.

IV. ACKNOWLEDGEMENTS

We wish to thank Drs. K. Ishitani and Y. Niitsu who carried out the initial experiments to detect sugars in ferritin. This work was supported by grant HL 11511 from the National Institutes of Health. Dr. Lavoie is a trainee on an Immunooncology training program (5 T32 CA 09173).

V. REFERENCES

(1) Adelman, T. G., Arosio, P. and Drysdale, J. W., Biochem. & Biophys. Res. Commun. 63, 1056 (1975).
(2) Arosio, P., Yokota, M. and Drysdale, J. W., Cancer Res., 36, 1735 (1976).
(3) Ishitani, K., Listowsky, I., Hazard, J. and Drysdale, J. W. J. Biol. Chem., 250, 5446 (1975).

(4) Ishitani, K., Niitsu, Y. and Listowsky, I., in "Proteins of Iron Storage and Transport in Biochemistry and Medicine" (R. R. Crichton, Ed.), North-Holland/American Elsevier, p. 245, (1975).
(5) Niitsu, Y., Ishitani, K. and Listowsky, I., <u>Biochem. Biophys. Res. Commun.</u>, <u>55</u>, 1134 (1973).
(6) Bjork, I. and Fish, W. W., <u>Biochemistry</u>, <u>10</u>, 2844 (1971).
(7) Ishitani, K., Niitsu, Y. and Listowsky, I., <u>J. Biol. Chem.</u>, <u>250</u>, 3142 (1975).
(8) Niitsu, Y. and Listowsky, I., <u>Biochemistry</u>, <u>12</u>, 4690 (1973).
(9) Marcus, D. M. and Zinberg, N., <u>Arch. Biochem. Biophys.</u>, <u>162</u>, 493 (1974).
(10) Freedman, M.L., Cohen, H. S., Rosman, J. and Forte, F. J., <u>British J. Haemotology</u>, <u>32</u>, 579 (1976).
(11) Niitsu, Y. and Listowsky, I., <u>Arch. Biochim. Biophys.</u>, <u>158</u>, 276 (1973).
(12) Worwood, M., Aherne, W., Dawkins, S. and Jacobs, A., <u>Clin. Sci. & Molec. Med.</u>, <u>48</u>, 441 (1975).
(13) Ishitani, K., Niitsu, Y. and Listowsky, I., <u>Federation Proc.</u>, <u>33</u>, 1478 (1974).
(14) Shinjyo, S., Abe, H. and Masuda, M., <u>Biochim. Biophys. Acta</u>, <u>411</u>, 165 (1975).
(15) Warren, L., <u>J. Biol. Chem.</u>, <u>234</u>, 1971 (1959).

BIOCHEMICAL AND IMMUNOLOGICAL PROPERTIES OF HUMAN ISOFERRITINS

Mark Worwood, Michael Wagstaff, Brian M Jones, Sarah Dawkins and Allan Jacobs

Department of Haematology, Welsh National School of Medicine

Isoelectric focussing separates ferritin into a number of isoferritins which appear to be composed of different proportions of two types of subunit (1). Heart muscle has a predominance of the most acidic isoferritins found in normal tissues and the least acidic predominate in liver and spleen. In pathological states changes in the isoferritin composition may occur - for example the shift to less acidic ferritins in iron loaded tissues (2) and the appearance of more acidic isoferritins in tumours taken from tissues such as liver (3). These "carcino-foetal" ferritins (4) appear to be immunologically and structurally similar to acidic isoferritins found in the heart (3). Immunoassays for ferritin have been widely applied to the measurement of serum ferritin concentrations (5) and high concentrations are found not only in cases of iron overload and tissue damage but also in serum from patients with malignancies. Immunoassays for ferritin appear to be relatively specific for particular isoferritins. Thus, assays employing antibodies to liver or spleen ferritin react only poorly with the more acidic heart ferritin (6,7) or with the ferritin from HeLa cells (8). This paper describes the application of immunoradiometric assays for spleen and heart ferritin to various tissues, cells and serum with particular reference to the biochemical and immunological properties of acidic isoferritins.

1. IMMUNORADIOMETRIC ASSAYS

Iron-rich spleen ferritin (9) and acidic isoferritins of normal heart (7) were prepared as described previously. Antibodies to these proteins were extracted from rabbit antisera with horse spleen ferritin coupled to amino cellulose, in the case of spleen ferritin (10), or acidic heart ferritin coupled to CNBr activated Sepharose 4B (Pharmacia Fine Chemicals) in the case of heart ferritin. The antibodies were iodinated with ^{125}I and eluted from the immunoadsorbent (10). Two-site immunoradiometric assays (11) using either anti-spleen or anti-heart ferritin were carried out in 0.4ml polyethelene tubes (Sarstedt) coated with appropriately diluted antiserum (the anti-heart ferritin was previously absorbed with spleen ferritin). In some cases spleen ferritin assays were carried out by the method of Jones and Worwood (12) with very similar results to the two-site assay.

II. OTHER METHODS

Isoelectric focussing was carried out in slabs of polyacrylamide gel with apparatus supplied by MRA Corp., Boston according to techniques of Righetti and Drysdale (13). Focussing on a preparative scale was carried out as previously described (9) after which the gels were cut into 5mm sections and ampholytes were eluted in 1.5ml boiled distilled water. The pH of the eluates was measured at the running temperature of the gels (0-4°C) and elution was then continued with 25ml of 0.05M phosphate buffer pH 7.5 for a further 48 h. Anion exchange chromatography was carried out as described by Worwood et al,(9) Con A-Sepharose 4B was purchased from Pharmacia Fine Chemicals and columns of length 50 mm and diameter 11 mm were prepared. Partially purified serum ferritin (maximum, 40mg protein) was passed through the column and eluted with 0.05 M acetate buffer pH 6.0 containing 0.5 NaCl and 1 mM, Mn^{++}, Mg^{++} and Ca^{++}. Bound proteins were then eluted from the column with 0.05M α-D-methylglucoside in the same buffer. The eluates were concentrated to small volumes by ultrafiltration before isoelectric focussing.

III. APPLICATION OF IMMUNORADIOMETRIC ASSAYS TO TISSUES, CULTURED CELLS AND SERA

The specificities of the assays were demonstrated with spleen ferritin and the acidic isoferritins of heart (Fig.1). In addition, isoferritins from liver were examined (Table I). The anion exchange affinity varies with isoelectric point as does the immunoreactivity of the isoferritins. In order to study ferritin in other tissue extracts isoferritins were separated by anion exchange chromatography. That separation occurred was confirmed by isoelectric focussing of various fractions eluted from the columns. The heart ferritin assay reacts with acidic isoferritins in kidney and reticulocyte extracts as well as with heart ferritin (Fig.2) whereas the spleen ferritin assay reacts preferentially with the less acidic isoferritins. Cultured human cells of the Chang and HeLa lines also contain a predominance of acidic ferritin (Fig.3). The apparently higher concentrations of ferritin measured with the heart assay as well as the high affinity of the ferritin for the anion exchange column should be noted. However, application of the heart-type assay to serum did not demonstrate the presence of high concentrations of circulating acidic isoferritins in normal or in pathological sera (Table II). Many of the pathological sera contained high concentrations of ferritin measured with the spleen-type assay and only in the case of myocardial infarction was there any evidence of an increase in the heart/spleen ratio.

TABLE I Properties of liver and heart isoferritins

Ferritin	pI	Anion exchange affinity *	Immunoreactivity+ spleen assay	heart assay
Heart	5.05	0.290	0.036	1.0
Liver	5.29	0.248	0.43	0.029
	5.40	0.240	0.84	0.019
	5.58	0.214	1.18	0.014
	5.64	0.190	1.40	0.007
	5.75	0.155	1.53	0.005

* [Cl⁻] mol/l for the fraction containing the highest ferritin concentration, + μg assayed ferritin/μg protein. Iron-rich ferritin was purified from normal liver and heart (9). Preparative isoelectric focussing, measurement of pH and elution of ferritin are described in the text. Ampholytes were removed from the concentrated isoferritin solutions by chromatography on Sepharose 6B before anion exchange chromatography. Protein concentrations were measured by the method of Lowry et al (23) with bovine serum albumin as standard.

IV. PROPERTIES OF ACIDIC ISOFERRITINS IN SERUM

The results obtained with the heart ferritin assay for serum were surprising as high concentrations of acidic isoferritins are often present in serum from untreated patients with acute myeloid leukaemia and also in patients with haemochromatosis (Fig.4). These acidic isoferritins were therefore further studied in serum obtained from two patients undergoing venesection therapy for haemochromatosis and in one patient with transfusional iron overload.

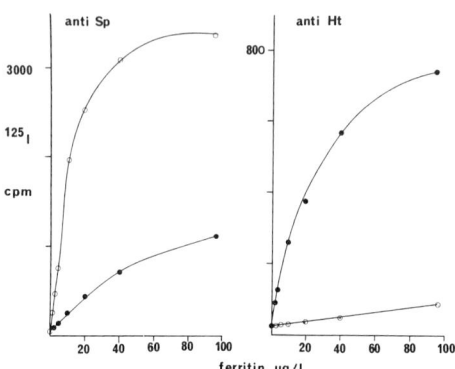

Fig.1 Immunoradiometric assays for heart and spleen ferritin. Tubes coated with anti-spleen (sp) or anti-heart (ht) ferritin were washed and incubated with 200µl 0.05M veronal buffer (12) containing spleen (o——o) or heart (•——•) ferritin for 3 h at 37°C. After washing the tubes were incubated overnight at 4°C with ^{125}I-labelled antibodies to spleen or heart ferritin. The tubes were then washed and bound radioactivity determined.

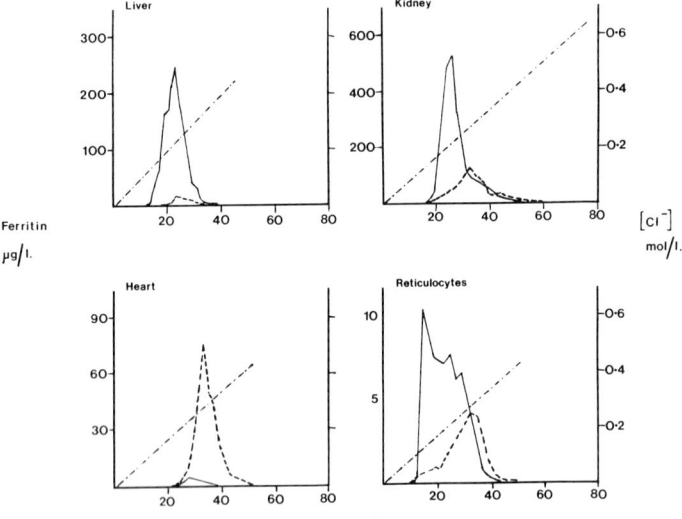

Fig.2 Anion exchange chromatography of ferritin. Spleen (———) and heart (-----) ferritin concentrations are indicated along with the chloride ion gradient (-·-·-·-). Ferritin from normal tissues was purified (9) without ultracentrifugation. Reticulocyte-rich red cells were obtained from a patient with pernicious anaemia. Ferritin was extracted (22) by affinity chromatography (9).

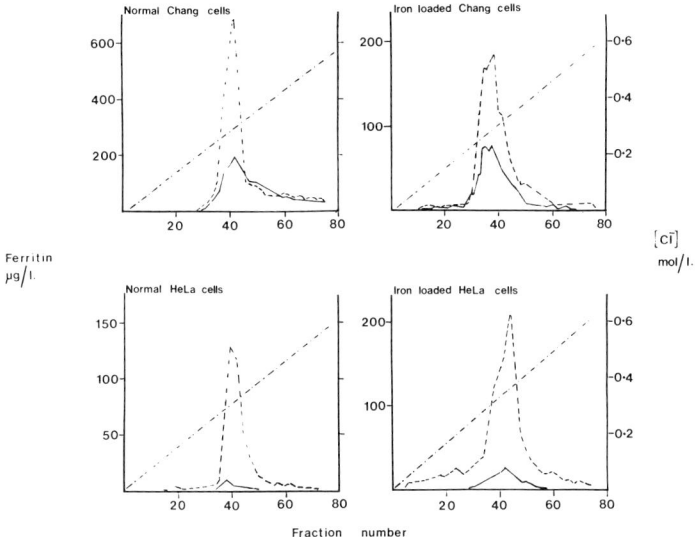

Fig.3. Anion exchange chromatography of cultured cell extracts. Spleen (———) and heart (- - - -) ferritin concentrations are indicated along with chloride ion gradient (-·-·-·-). Cells were loaded with iron by incubation with medium containing ferric-NTA (10μg Fe/ml) for one week. Cells were harvested, washed 3 times in saline and sonicated. After heating to 70°C for 10 min denatured protein was removed by centrifguation at 25,000g for 30 min.

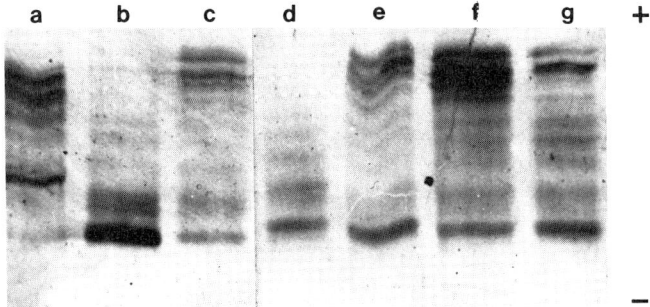

Fig.4. Isoelectric focussing of partially purified serum ferritin. Tissue ferritins were prepared without ultracentrifugation. Serum ferritin was partially purified by the heat treatment method (9) and concentrated to a small volume after anion exchange chromatography. Ferritin was detected in the gels by immunoprecipitation (9) and staining with Coomassie blue. a) ferritin from iron loaded heart; b) and c) serum ferritin, acute myeloblastic leukaemia, d) normal liver ferritin, e,f and g) serum ferritin haemochromatosis.

TABLE II   Serum ferritin concentrations

Diagnosis (no)	Ferritin concn.µg/l (Mean ± S.D.)		Heart/spleen ratio for each serum (Mean ± S.D.)
	heart assay	spleen assay	
Normal (10)	< 10	74+60	-
Breast cancer (10)	33 + 16	480 + 369	0.11+ 0.09
Acute leukaemia (10)	79 + 53	2459 + 3051	0.055+ 0.053
Myocardial infarction (10)	43 + 24	259 + 226	0.31 + 0.33
Transfusion, iron overload (4)	36 + 15	3520 + 3180	0.017 + 0.013
Haemochromatosis during treatment (6)	23 + 15	1690 + 1000	0.015 + 0.007

Spleen ferritin assays were carried out by the method of Jones and Worwood (12). For the heart assay standard ferritin was diluted in 1:20 normal horse serum in buffer. Sera were diluted 1:20 in buffer. The minimum detectable ferritin concentration in serum is approximately 10µg heart ferritin /l. The mean recovery of heart ferritin added to 8 sera was 113 + 11%. Patients with breast cancer were untreated. Of the patients with leukaemia there were 5 with acute ALL and 5 with acute AML. Most of the sera were obtained before treatment began. Serum samples were taken from patients with myocardial infarction within 24h of admission to hospital. Serum creatine kinase activities were at least twice the upper limit of normal.

Ferritin was partially purified and subjected to preparative isoelectric focussing. Ferritin concentrations were measured in the eluates from the sliced gels with both spleen and heart-assays. For each of the serum preparations (Fig.5) ferritin was present in relatively high concentration when measured with the spleen assay but was barely detectable with the heart assay. Furthermore, serum isoferritins all had a low affinity for the anion exchange column when compared with tissue isoferritins of similar pI (Fig.6). Earlier studies on α_2-H globulin, which appears to be a "carcinofoetal" ferritin, described this protein as containing carbohydrate residues (14)   It seemed possible that the properties of serum ferritin might be influenced by the presence of carbohydrate or by binding to glyco-proteins and for these reasons the interaction of serum ferritin with concanavalin-A was examined. Serum extracts, prepared by heat treatment from 2 patients with haemochromatosis and another patient with secondary iron overload, were loaded on columns of Con A-Sepharose. In each case it was found that the most basic isoferritin did not bind while the acidic isoferritins bound to the column and could then be released with α-D-methylglucoside (Fig.7). Thus the circulating acidic isoferritins found in serum from patients with iron overload appear to contain carbohydrate and are not recognised by the heart ferritin assay. The presence of carbohydrate may be responsible for the anomalous ion exchange affinity of the more acidic isoferritins of serum when compared with tissue isoferritins.

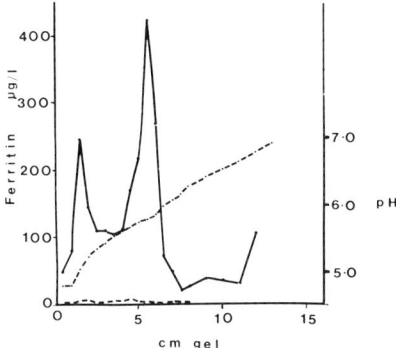

Fig.5. Preparative isoelectric focussing of serum ferritin from a patient with haemochromatosis. Experimental details are given in the text and in the legend to Fig.4. Spleen (———) and heart (- - - -) ferritin concentrations are indicated along with the pH gradient (-·-·-).

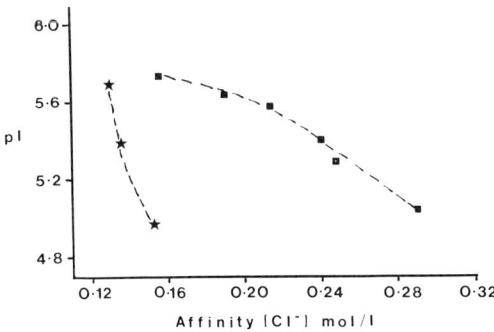

Fig.6. Anion exchange affinity of serum ferritin. For experimental details see legend to Table I. Serum ferritin from a patient with haemochromatosis was prepared as described in the legend to Fig.4. Serum ferritin ★, tissue ferritin ■ (See Table I).

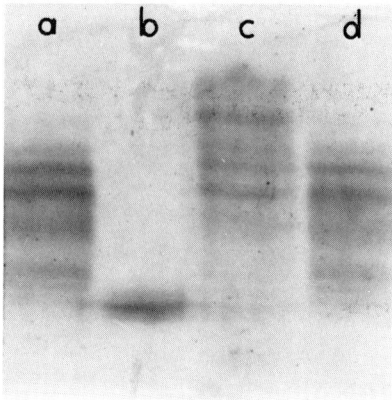

Fig.7. Isoelectric focussing of partially purified serum ferritin from a patient with haemochromatosis undergoing treatment (See Fig.4e).
a and d) normal human liver ferritin. b) serum ferritin which did not bind to Con A-Sepharose 4B c) serum ferritin binding to the Con A-Sepharose 4B and eluted with α-D-methylglucoside

V. DISCUSSION

The results presented here show that it is possible to develop an immunoradiometric assay which is specific for the acidic isoferritins of heart and which, in addition, recognises acidic isoferritins in a number of other tissues. HeLa cell ferritin (15) also reacts in the heart-type assay but has low reactivity in the spleen assay. These findings provide further support the multiple subunit hypothesis of Drysdale and colleagues (1). Cultured Chang liver cells appear to contain ferritin of predominantly acidic type but in this case there is a higher reactivity in the spleen assay.

Application of the heart-type assay to serum has not demonstrated the presence of acidic isoferritins in high concentration in cases of myocardial infarction, breast cancer, or in acute leukaemia. Although raised concentrations were found in the pathological sera when compared with normal sera a similar rise can be measured with the spleen ferritin assay. The heart contains very little ferritin compared to the liver so that it is perhaps not surprising that high concentrations of circulating heart ferritin were not found in cases of myocardial infarction. In serum from patients with various malignancies, including breast cancer, Hazard and Drysdale (16) have recently reported that an immunoassay based on HeLa cell ferritin detected higher ferritin concentrations than one based on crystalline liver ferritin. The difference between our results and those of Hazard and Drysdale may be due to the specificities of the assays. Hazard and Drysdale may have developed assays which are specific for either liver or heart subunit types but which can detect them in heteropolymers. The heart ferritin assay described here employs an anti-heart ferritin which has been absorbed with spleen ferritin containing both liver and heart-type subunits and may therefore recognise a complex antigenic site involving two or more subunits rather than the heart type subunit. However, both heart and HeLa cell ferritin assays fail to recognise acidic isoferritins in serum from patients with acute leukaemia or haemochromatosis.

In sera from untreated patients with haemochromatosis and in some cases of transfusional iron overload the only circulating isoferritins present correspond in pI to the more basic isoferritins of liver (9,17). However, during treatment of haemochromatosis by phlebotomy circulating acidic isoferritins appear (18). Acidic isoferritins are also found in sera from normal subjects (18). These isoferritins are not artefacts caused by the process of purification as Halliday et al (18), using a sensitive technique for detecting ferritin after isoelectric focussing, have shown that acidic isoferritins are present in untreated sera. We have found that the serum isoferritin distribution was the same when the ferritin was purified by heating serum to $70°C$ at pH 4.8 or extracted with an immunoadsorbent (9). These acidic isoferritins are recognised by the spleen ferritin assay but not by the heart assay and have a low affinity for the anion exchange column compared with tissue ferritins of similar pI. Serum ferritin from patients with leukaemia also has a very low anion-exchange affinity (19) although ferritin from leukaemic cells shows the relationship between pI and affinity seen in other tissue ferritins (20). Our preliminary findings suggest that these properties of serum ferritin are related to the presence of carbohydrate residues on the more acidic isoferritins. The curious isoferritin distribution in patients with iron overload or acute myeloblastic leukaemia should also be noticed (Fig.4). The low concentrations of intermediate isoferritins may indicate that two types of ferritin molecule have a prolonged survival in the plasma. The first type is the most basic and which corresponds to the 'natural apoferritin' of liver (17). Serum ferritin is of low iron content (9) and different rates of clearance for iron-poor and iron-rich ferritin (21) may lead to a predominance of iron-poor ferritin in the plasma after release of ferritin from cells. However, ferritin molecules containing, or associated with, carbohydrate may also have a prolonged survival time.

REFERENCES

1. Drysdale, J.W. Adelman., T.G., Arosio, P. et al., Sem.Hematol. 14, 71 (1977)
2. Powell, L.W., McKeering, L.V., Halliday, J.W., Gut 16, 909 (1975)
3. Arosio, P., Yokota, M., Drysdale, J.W., Cancer Res. 36, 1735 (1976)
4. Alpert,E., Coston, R.L., Drysdale, J.W.,Nature 242, 194 (1973)
5. Jacobs,A., Worwood, M., Prog. Hematol. 9, 1 (1975)
6. Marcus, D.M., Zinberg, N., J. Natl. Cancer Inst. 55,791 (1975)
7. Worwood, M., Jones, B.M., and Jacobs, A.Immunochem. 13, 477 (1976)
8. Hazard, J.T., Yokota, M., Arosio, P. and Drysdale, J.W., Blood 49, 139 (1977)
9. Worwood, M., Dawkins, S., Wagstaff, M. et al., Biochem. J. 157, 97 (1976)
10. Addison, G.M., Beamish, M.R., Hales, C.N. et al., J. Clin. Pathol. 25, 326 (1972)
11. Miles, L.E.M., Lipschitz, D.A., Bieber, C.P., et al., Anal. Biochem. 61, 209 (1974)
12. Jones, B.M., Worwood, M., J. Clin. Pathol. 28, 540 (1975)
13. Righetti, P.G., and Drysdale, J.W., Ann. N.Y. Acad.Sci. 209, 163 (1973)
14. Buffe, D., Rimbaut, C., Fuccaro, C., et al., Ann. Inst. Pasteur 123, 29 (1972)
15. Drysdale, J.W., and Singer, R.M., Cancer Res. 34, 3352 (1974)
16. Hazard, J.T., and Drysdale, J.W., Nature 265, 753 (1977)
17. Arosio, P., Yokota, M., and Drysdale, J.W., Brit. J. Haematol. (in press)
18. Halliday,J.W., McKeering, L.V., Tweedale, R., et al., Brit. J. Haematol. (in press)
19. Worwood, M., Aherne,W., Dawkins, S., et al. Clin. Sci.Molec. Med. 48, 441 (1975)
20. Cragg, S.J., Jacobs, A., Parry, D.H., et al.,Brit.J.Cancer (in press)
21. Lipschitz, O.A., Pollack, A., Savin, M.A., et al. Clin.Res. 24, 571A (1976)
22. Worwood, M., Summers, M., Miller, F., et al. Brit. J. Haematol. 28, 27 (1974)
23. Lowry, O.H., Rosebrough, N.J., Farr, A.L., et al. J. Biol. Chem. 193, 265 (1951)

Section III
FERRITIN METABOLISM

ASSEMBLY OF THE FERRITIN MOLECULE IN RAT HEPATOMA CELLS IN VITRO

S. S. C. Lee
G. W. Richter

Department of Pathology
University of Rochester, Rochester, N.Y.

To learn more about (a) the biosynthesis of protein subunits of ferritin by free and bound polyribosomes, (b) the sites of assembly of subunits in cells, and (c) the binding of iron by subunits and by ferritin, we have made time sequence studies with rat hepatoma cells. In this work, cell fractionation was combined with radioimmunoassay. The evidence bears not only on the three points raised above, but also indicates vectorial transport of subunits and ferritin through cytoplasmic compartments. It extends earlier investigations (1) in which subunits were localized in situ in hepatoma cells by immunofluorescence.

METHODS

Since technical details of the work are being published elsewhere (2,3), only a summary of the methods is given. The general procedure was to prepare cell fractions from a rat hepatoma cell clone, M-5123-Cl (1), at progressive time intervals after biosynthetic labeling of the protein subunits and of ferritin with [^{14}C] leucine or with [^{14}C]leucine and ^{59}Fe, and then to determine amounts of labeled subunits and labeled ferritin plus apoferritin in the various cell fractions by radioimmunoassay. Subunit-specific antibodies did not react with either ferritin or apoferritin, but ferritin-specific antibodies reacted with apoferritin as well as ferritin (4). In pulse experiments, the hepatoma cells were incubated with the radioactive precursors during chosen time intervals. In pulse-chase experiments, the cells were incubated with the labeled precursors for 3 minutes, washed, then reincubated with non-radioactive precursors for set intervals before subcellular fractionation.

By procedures described elsewhere (2,3) four cell fractions were prepared: free polyribosomes, membrane-bound polyribosomes, smooth membranes, and final supernatant. Subunits, apoferritin and ferritin were extracted from each fraction and their quantities determined by radioimmunoassay with purified subunit-specific or ferritin-specific rabbit IgG antibodies (2). In the graphs, specific activities are expressed either as dpm per mg RNA (for the two polyribosomal fractions), or as dpm per mg protein (for the smooth membrane and supernatant fractions). Changes of specific activities are plotted as a function of time after pulsing. The purity of cell fractions was checked by transmission electron microscopy of pelleted, fixed, embedded and thin-sectioned aliquots.

RESULTS

At least 80% of protein subunits of ferritin were synthesized on free polyribosomes, the remainder on membrane-bound polyribosomes. This is in agreement with the earlier findings of Redman (5) and of Puro and Richter (6).

After *pulsing without chase*, ^{14}C *radioactivity* associated with *subunits* increased in both *free* and *membrane-bound polyribosomal fractions* for 20 minutes and leveled off after 30 minutes, while ^{14}C *radioactivity* associated with *ferritin* increased gradually and continuously. In *pulse-chase* experiments, *maximal* ^{14}C-*subunit radioactivity* in the two polyribosomal fractions was reached 10 minutes after pulsing, *maximal* ^{14}C-*ferritin radioactivity* $\sim$ 30 minutes after the pulse (Figure 1). After the hepatoma cells had been simultaneously exposed to ^{59}Fe and ^{14}C-leucine, the $^{59}Fe/^{14}C$ ratios of ferritin in the two polyribosomal fractions were considerably greater than those of subunits at the same time points (Table I). This suggests that relatively little iron becomes attached to a newly synthesized subunit.

On the other hand, *in pulse-experiments* ^{59}Fe *radioactivity* rose relatively early in *ferritin* in the two *polyribosomal fractions* (in 5 to 10 minutes) by comparison to ^{14}C *radioactivity*, and reached a maximum within 30 minutes (Figure 2). This indicates early uptake of ^{59}Fe by preëxisting apoferritin or ferritin.

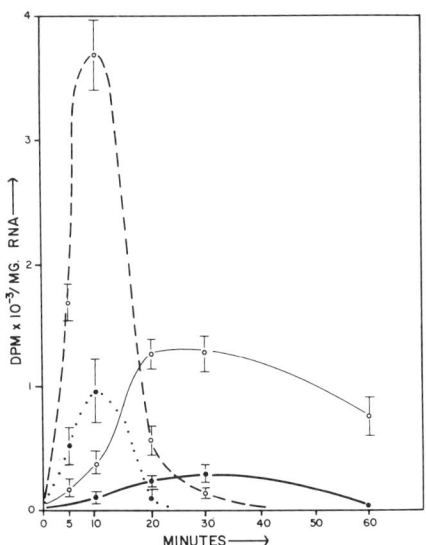

Fig. 1. Counts of DL-[1-^{14}C]leucine-labeled protein subunits and ferritin/apoferritin in the free polyribosomal and the membrane-bound polyribosomal fractions of M-5123-Cl hepatoma cells at various times after addition of radioactive precursors to culture medium. Chase with "cold" precursors started 3 minutes after pulse. Each point represents the mean of two independent experiments. Bars, standard deviations. o---o, protein subunits in free polyribosomal fraction; o—o, ferritin and apoferritin in free polyribosomal fraction; •⋯•, protein subunits in membrane-bound polyribosomal fraction; •—•, ferritin and apoferritin in membrane-bound polyribosomal fraction.

Table I

^{59}Fe/^{14}C ratios of ferritin and subunits in different subcellular fractions of M-5123-Cl hepatoma cells (pulse and chase experiments)
Mean and standard deviation of two independent determinations

Time Minutes	Free Polyribosomes		Membrane-bound Polyribosomes	
	Subunits	Ferritin	Subunits	Ferritin
5	0.41 ± 0.09	37.0 ± 6.4	2.3 ± 0.37	(2500 ± 198)/0
10	0.59 ± 0.01	3.2 ± 0	1.1 ± 0.16	22.5 ± 11.2
20	0.63 ± 0.01	1.5 ± 0.1	0.99 ± 0.87	2.1 ± 0.12
30	0.47 ± 0.3	0.9 ± 0.045	0	2.7 ± 0.14
60	0	0.87 ± 0.05	0	(66 ± 32)/0

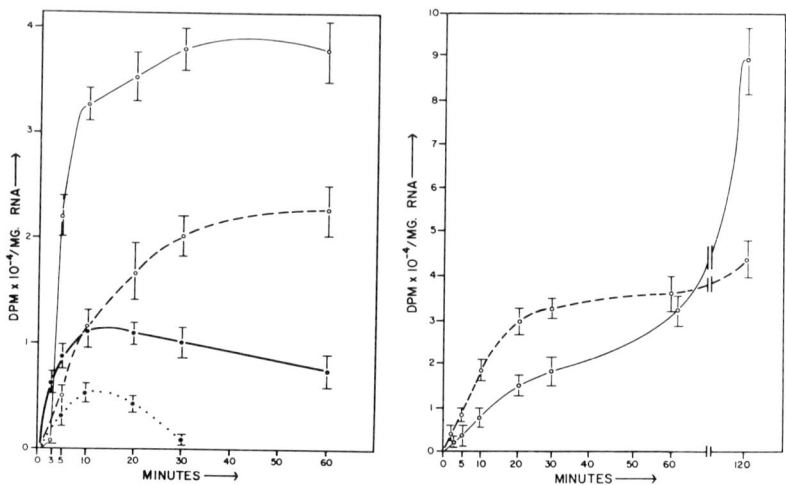

Fig. 2. Time course of appearance of ^{59}Fe-labeled and [^{14}C]leucine-labeled protein subunits and ferritin in free polyribosomal and membrane-bound polyribosomal fractions of M-5123-Cl hepatoma cells after addition of ^{59}Fe and [^{14}C]leucine to culture medium. Left, ^{59}Fe radioactivity; right, [^{14}C]leucine radioactivity. Each point represents the mean of three independent experiments. Bars, standard deviations. o---o, protein subunits in free polyribosomal fraction; o——o, ferritin and apoferritin in free polyribosomal fraction; ●···●, protein subunits in membrane-bound polyribosomal fraction; ●——●, ferritin and apoferritin in membrane-bound polyribosomal fraction.

Pulse-chase experiments gave similar results and also showed that 5 to 10 minutes after pulsing the ^{59}Fe *radioactivity* attributable to *ferritin* in the *polyribosomal fractions* dropped rapidly (Figure 3), presumably because of movement of ferritin from the polyribosomal compartments. A later, but smaller rise of ^{59}Fe *ferritin radioactivity* in the *free polyribosomal fraction* (Figure 3) coincided with the drop of ^{59}Fe *subunit radioactivity* in this fraction and may have been due to assembly of subunits into ferritin.

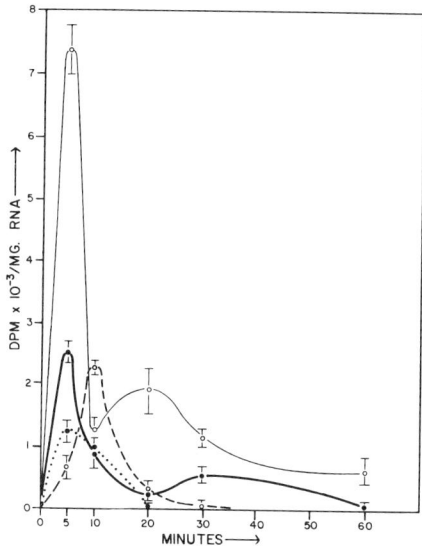

Fig. 3. Time course of appearance of ^{59}Fe-labeled protein subunits and ferritin in the free and membrane-bound polyribosomal fractions of M-5123-Cl hepatoma cells after addition of ^{59}Fe to culture medium in pulse-chase experiments. Chase with "cold" precursor started 3 minutes after pulse. Each point represents the mean of two independent experiments. <u>Bars</u>, standard deviations. o---o, protein subunits in free polyribosomal fraction; o——o, ferritin and apoferritin in free polyribosomal fraction; •···•, protein subunits in membrane-bound polyribosomal fraction; •——•, ferritin and apoferritin in membrane-bound polyribosomal fraction.

As can be seen by comparing Figures 1 and 4, *^{14}C-labeled ferritin* rose earlier in the *smooth membrane* and *supernatant fractions* than in the two polyribosomal ones, then decreased until 30 minutes after pulsing; thereafter it increased gradually. This result suggests the existence of a basal concentration of unassembled subunits in the smooth membrane cell compartment as well as in the cytosol. Influx of newly synthesized subunits from polyribosomes may push the assembly reaction. The late increase of ^{14}C ferritin radioactivity could have been due to transfer from the membrane-bound polyribosomal fraction. Preferential binding of *iron* to *ferritin* (or *apoferritin*) in the *smooth membrane* and *supernatant fractions* was also evident (Figure 5). In the smooth-membrane fraction ^{59}Fe was presumably taken up by preëxisting ferritin or apoferritin soon after addition, a peak being reached in 20 minutes.

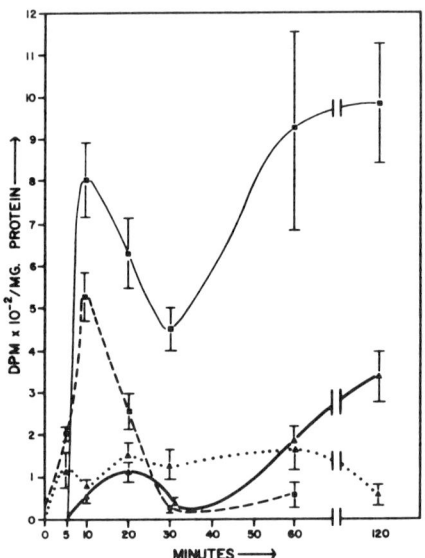

Fig. 4. Counts of DL-[1-^{14}C]leucine-labeled protein subunits and ferritin/ apoferritin in the smooth membrane and the supernatant fraction of M-5123-Cl hepatoma cells at various times after addition of radioactive precursors to culture medium. Chase with "cold" precursors started 3 minutes after pulse. Each point represents the mean of two independent experiments. *Bars*, standard deviations. ■---■, protein subunits in smooth membrane fraction; ■——■, ferritin and apoferritin in smooth membrane fraction; ▲···▲, protein subunits in supernatant fraction; ▲——▲, ferritin and apoferritin in supernatant fraction.

Moreover, it appears likely that ^{59}Fe which had become part of ferritin in the two polyribosomal compartments was later transported to the smooth membrane and supernatant compartments, thus accounting for the late increase of ^{59}Fe ferritin radioactivity in these two fractions and for the concomitant drop of ^{59}Fe ferritin radioactivity in the polyribosomal fractions (Figures 3 and 5). Thus, our data indicate that both subunits and ferritin are transported from the two polyribosomal compartments to the smooth membrane compartment and the cytosol and that nearly all of the iron in ferritin is incorporated after assembly of subunits.

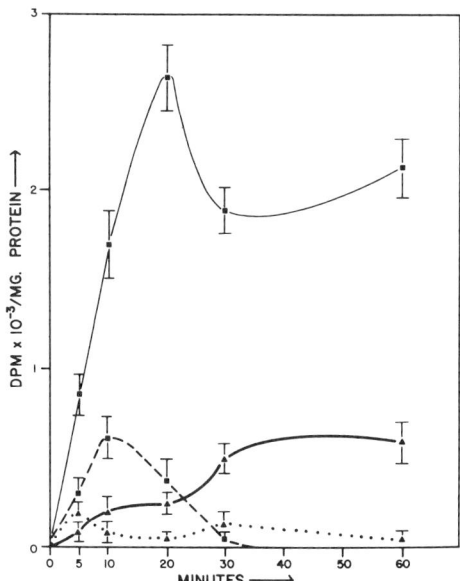

Fig. 5. Time course of appearance of ^{59}Fe-labeled ferritin and protein subunits in smooth membrane and supernatant fractions of M-5123-Cl hepatoma cells after addition of ^{59}Fe to cultured medium in pulse-chase experiments. Chase with "cold" precursor started 3 minutes after pulse. Each point represents mean of two independent determinations. Bars, standard deviations. ■---■, protein subunits in smooth membrane fraction; ■——■, ferritin and apoferritin in smooth membrane fraction; ▲···▲, protein subunits in supernatant fraction; ▲——▲, ferritin and apoferritin in supernatant fraction.

DISCUSSION

These findings, together with results of earlier work, in which the location of subunits and ferritin was studied in situ in cultured M-5123-Cl cells by immunofluorescence (1) and, in the case of ferritin, also by electron microscopy, lead us to the following conclusions:

(a) There is a basal concentration of free, unassembled subunits in the cytosol. When the concentration rises in consequence of synthesis of new subunits by free polyribosomes, subunits will assemble.

(b) Assembly into apoferritin or iron-poor ferritin molecules is immediately followed by incorporation of iron.

(c) The much smaller fraction of subunits synthesized by membrane-bound polyribosomes is released into the cisternal channels of the rough endoplasmic reticulum and moved into channels of smooth endoplasmic reticulum where assembly and insertion of iron take place.

The later fate of ferritin can only be inferred from a large body of electron microscopic evidence, published over the past twenty years and obtained from diverse sorts of cells, principally hepatocytes. It is clear that ferritin is eventually sequestered in vacuoles delimited by single membranes, but the mechanism by which this occurs is unclear. After sequestration, lysosomal enzymes are added, a process that converts the vesicles to secondary lysosomes. Here covalently linked oligomers of ferritin may be formed (7) and here ferritin seems to be degraded and hemosiderin formed.

REFERENCES

(1) Lee, J.C.K., Lee, S.S.C., Schlesinger, K.J., Richter, G.W. Am. J. Path. 80, 235 (1975).

(2) Lee, S.S.C., Richter, G.W. J. Biol. Chem. 252, 2046 (1977).

(3) Lee, S.S.C., Richter, G.W. J. Biol. Chem. 252, 2054 (1977).

(4) Lee, J.C.K., Lee, S.S., Schlesinger, K.J., Richter, G.W. Am. J. Pathol. 75, 473 (1974).

(5) Redman, C.M. J. Biol. Chem. 244, 4308 (1969).

(6) Puro, D.G., Richter, G.W. Proc. Soc. Exp. Biol. Med. 138, 399 (1971).

(7) Lee, S.S.C., Richter, G.W. Biochemistry 15, 65 (1976).

ACKNOWLEDGMENT

This research was supported by NIH Grants 5 R01 AM 12381 and 5 T01 GM 00133.

RED CELL FERRITIN AND IRON STORAGE
DURING ANIMAL DEVELOPMENT

Elizabeth C. Theil and J. Edward Brown
Department of Biochemistry
North Carolina State University

I. INTRODUCTION

Iron storage is a specialized function of cells found mainly in the liver, spleen and marrow of adult animals (1). Iron is acquired by the storage cells from transferrin via specific surface receptors (2) or from the catabolism of red cells, and is deposited in ferritin and hemosiderin (3). In adults, general iron storage occurs mostly in the reticuloendothelial cells and the liver parenchymal cells. Erythroid cell iron stores appear to be used intracellularly, since ferritin accumulation is normally restricted to immature cells synthesizing hemoglobin and the mature red cell contains very little storage iron (1). In contrast, the mature red cell of a developing animal, the bullfrog tadpole, has a high concentration of ferritin which contains 30% of the cell iron (4,5). This observation suggests that early in life red cells have the dual functions of oxygen transport and iron storage.

The results of additional investigations of the role of red cells in iron storage during animal development described here indicated that the mature red cells derived from the yolk-sac of the embryonic mouse also contained large amounts of ferritin, that mature circulating red cells of developing animals rapidly incorporated exogenous iron into their iron stores *in vivo* and that red cell ferritin was depleted during animal development while liver iron concentrations remained relatively constant. In addition, the ferritin-rich red cells of early development were used as a source of ferritin from a pure cell type to show that ferritin heterogeneity was independent of isolation procedures, multiple cell types and cell age.

II. EXPERIMENTAL METHODS

Bullfrog tadpoles were obtained from Howe Brothers Minnow Farm, Atlanta, Texas, and mice (ICR strain) from the Mouse Colony at North Carolina State University. The ferritin content of circulating red cells of the embryonic mouse was determined with a rabbit antiserum to purified mouse liver ferritin (6). Yolk-sac-derived red cells remain nucleated and contain unique hemoglobins; they were assayed as the percentage of nucleated red cells and the percentage of embryonic hemoglobin (7). The distribution of exogenous iron in developing animals containing ferritin-rich circulating red cells was determined after injection of ^{59}Fe-nitriloacetic acid (NTA) (10 ng/gm) (8) into Stage 13 tadpoles followed in an hour by the injection of Fe-NTA (1.0 μg/gm) to dilute any unbound ^{59}Fe-NTA; the combined doses of iron injected were less than 2% of the total body iron. Analogous experiments were performed with young frogs six months or more after metamorphosis in the laboratory. Blood was collected, plasma separated from red cells, the washed red cells fractionated into heme and non-heme fractions and the samples assayed for radioactivity. Purified red cell ferritin or red cell lysates were fractionated by isoelectric focusing in 4% polyacrylamide gels, pH 4-6, at 4° for 24 hours (9). Ferritin was visualized by the formation of Prussian Blue with ferrocyanide.

III. RESULTS

The ferritin content of the circulating red cells of embryonic mice declined as the yolk-sac-derived cells were replaced by hepatic-derived red cells (Figure 1). Concentrations of ferritin in the yolk-sac-derived cells were as high as 0.65% (6) of the soluble protein and approached that of the red cells of bullfrog tadpoles (10). The increase in the ferritin content of red cells of the embryonic mouse between day 11 and 12 of gestation (Figure 1) may represent an accumulation of ferritin by the yolk-sac-derived red cells, since fractions of circulating red cells enriched in non-nucleated (hepatic-derived) red cells contained less ferritin than unfractionated red cells (6).

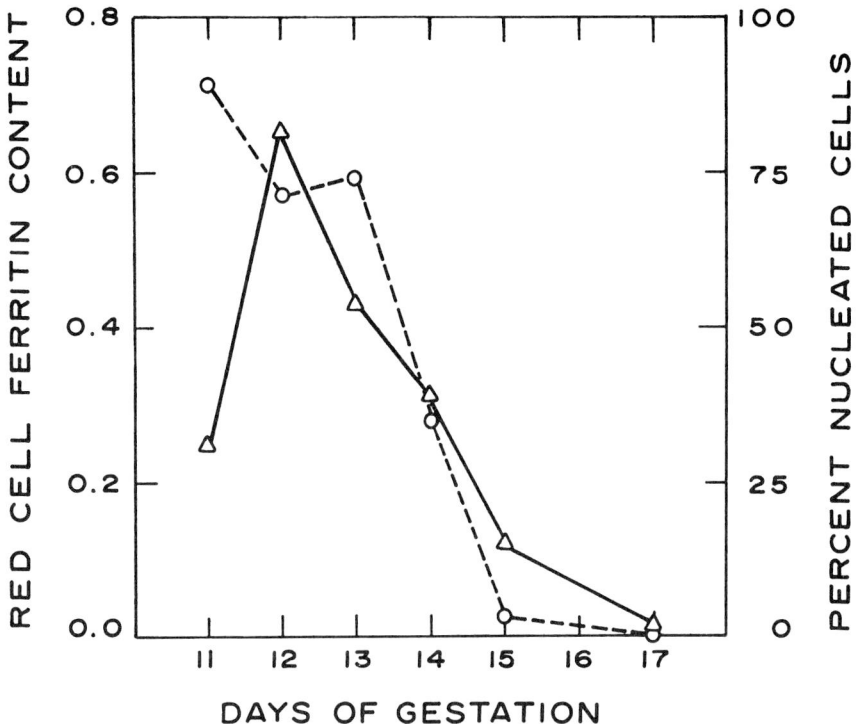

Fig. 1. The change in the ferritin content of circulating red cells of the embryonic mouse during the replacement of yolk-sac-derived cells (nucleated) by hepatic-derived cells (non-nucleated), from Theil (1976), ref. 6. △, Red Cell Ferritin Content, in µgm/100 µgm protein; O, Percent Nucleated Cells.

Radioactive iron was cleared from the plasma of adult and developing bullfrogs (Figure 2) at a rate that was comparable to that of adult mammals. The appearance of ^{59}Fe in the red cells of adult frogs was also similar to that of adult mammals (11) reaching 47% of the injected iron in 48 hours (Figure 2). In contrast, the ^{59}Fe content of tadpole red cells rose very rapidly; four hours after injection, the red cells contained 23% of the ^{59}Fe. Fractionation of the red cells revealed that the radioactivity in heme and non-heme fractions of cells from adult frogs increased at similar rates. On the other hand, most of the radioactivity incorporated into tadpole red cells was in the non-heme fraction (Figure 3). The incorporation of ^{59}Fe into heme occurred

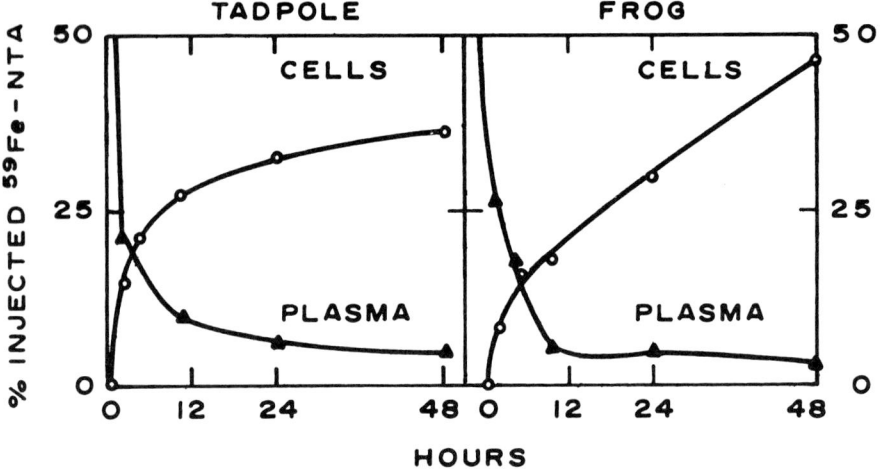

Fig. 2. The radioactivity of the plasma and circulating red cells of bullfrog tadpoles and adults after the injection of ^{59}Fe-NTA. Each curve represents the average of data collected from 2 or 4 experiments with 15 animals each.
▲—▲—▲, *Plasma;* O—O—O, *Red Blood Cells.*

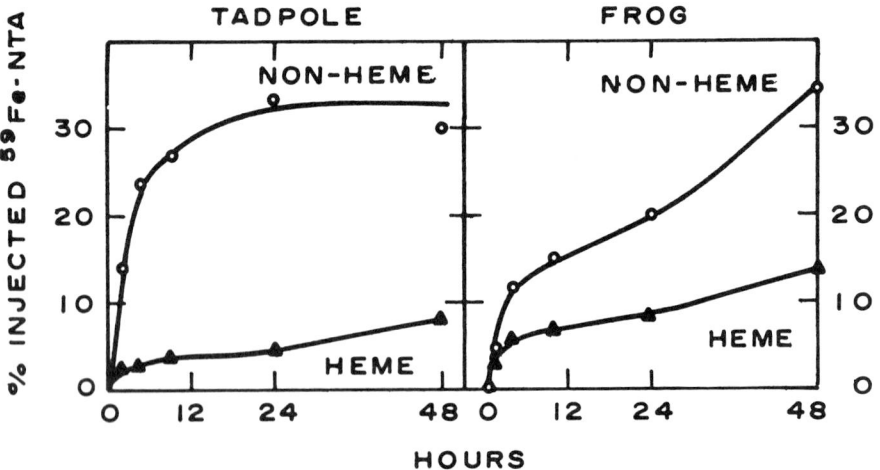

Fig. 3. The distribution of radioactivity between heme (extracted with ethylacetate/acetic acid and HCl) and non-heme (storage) fractions of the circulating red cells of bullfrog tadpoles (high ferritin) and adults (low ferritin) after injection of ^{59}Fe-NTA. Each curve represents the average of data collected from 2 or 4 experiments with 15 animals each.
▲—▲—▲, *Heme;* O—O—O, *Non-heme.*

more slowly, and the rate was similar to that for adult cells (Figure 3). Much of the radioactivity of the non-heme fraction was in ferritin since previous studies showed that $\geq 95\%$ of the ^{59}Fe in tadpole red cell lysates was in ferritin (4), and the lysate contained 80-90% of the total cellular ^{59}Fe.

Replacement of larval or embryonic red cells by new adult red cells coincided with the decline in the iron storage capacity of circulating red cells. Increased iron demands for new red cell synthesis could deplete iron stores both in the liver and red cell. However, the results in Table 1 show that the concentration of iron in the liver was relatively constant when the ferritin content of the circulating red cells declined during metamorphosis or gestation.

TABLE I The Concentration of Red Cell Ferritin and Liver Iron During Early Development

	Bullfrog Metamorphosis				
Developmental Stage	18	20	22	23	24
Red Cell Ferritin[a] mg/100 mg protein	1.1	0.66	0.74	0.44	0.15
Liver Iron mg/liver		.426 ±.138	.461 ±.117	.495 ±.071	.627 ±.212	.501 ±.050
	Mouse Gestation				
Day after Conception	12	13	14	15	17
Red Cell Ferritin[b] mg/100 mg protein	.65	.43	.31	.12	.007
Liver Iron mg/gm	.134 ±.030	.116 ±.038	.134 ±.046	.120 ±.042	.68 ±.13

[a]The data are taken from Theil (1973); the red cell ferritin protein concentration was constant from Stages 5-18.

[b]The data are taken from Theil (1976).

The liver iron concentration was determined after digestion of the tissue in HNO_3-$HClO_4$ (3:1). Data were obtained from the pooled livers of single litters of embryonic mice, analyzing 3-8 litters for each day of gestation or using groups of 4-6 tadpole livers and analyzing 6 groups for each developmental stage. The error is presented as the standard deviation.

A storage iron content of 228 μgm/ml of cells can be computed for tadpole (Stage 13) red cells using the total iron content of 713 μgm/ml (5), the ferritin content of 1% of the protein with iron/protein of 0.13 (4,10) and a hemoglobin content of 90% of the protein. Using the storage iron content of tadpole red cells, the blood volume of 10.4% (determined by the dilution of injected Evans' Blue), the hematocrit of 22% and the average body weight of 8.5 gms, the total amount of storage iron in the red cell mass of a premetamorphic tadpole (State 13) can be calculated to be 52 µgm. Assuming that most of the storage iron is contained in liver (126 µgm) and red cells (the spleen iron content is 3-4 µgm), red cells contain 29% of the storage iron in tadpoles. Using analogous calculations for froglets (Stage 24), red cells contain 0.6% of the iron stores. (Note that at both developmental stages, the liver is the main erythropoietic tissue.)

Embryonic or larval red cells of early development provided a source of ferritin from a pure cell type which could be analyzed by isoelectric focusing directly in lysates without extensive purification. The ferritin in lysates of tadpole or mouse yolk-sac red cells was heterogeneous and contained at least 7 or 4 isoferritins with apparent pI ranges of 4.8-5.4 or 4.8-5.2, respectively (9). Comparison of purified tadpole red cell ferritin with lysates of fresh cells or frozen cells stored 9 months, or with lysates of cells pooled from many animals or from single animals showed no qualitative differences. Ferritin, analyzed in concentrated lysates of frog red cells, had a heterogeneity similar to that of tadpole red cell ferritin. Five isoferritins were observed, and each corresponded to an isoferritin in the tadpole red cells. Red cell ferritins were less acidic than horse spleen ferritins.

IV. DISCUSSION

The presence of large amounts of ferritin in mature, circulating, hemoglobinized red cells early in animal development occurs not only in free-swimming, larval amphibia but also in mammalian embryos. In amphibia, the circulating cells with a large iron storage capacity rapidly accumulate exogenous iron (Figures 2 and 3). If the appearance of exogenous iron into heme is equated with new cell formation, then the red cells which incorporated iron into iron stores during the first 2-10 hours after the injection of the iron tracer were old cells. Iron incorporation in the adult red cells appeared to be restricted to young cells, since the incorporation of iron into heme and storage fractions increased at similar rates (Figure 3). Apparently, the circulating red cells formed early in animal development store iron in excess of the need for hemoglobin synthesis; in the case of the tadpole, approximately 29% of the storage iron is in circulating red cells.

The loss of ferritin from the circulating red cells of bullfrog larvae or mouse embryos accompanies a period of increased erythropoiesis (7,12) and consequently increased iron demands. In the case of the bullfrog, the maturation of adult red cells occurs during metamorphosis, a period of starvation when the iron for new red cells must come from iron stores. Since the iron content of the liver, the site of erythropoiesis, remains relatively constant or increases slightly during this period (Table 1), the loss of ferritin from the red cells suggests that red cell storage iron was used either for new red cell synthesis or to replenish liver stores. The observation that liver ferrireductase activity is enhanced during metamorphosis (13) supports the latter hypothesis. In contrast to the amphibian larva during metamorphosis, the developing mammalian embryo has an exogenous supply of iron through the placenta. Nevertheless, the red cells formed early in gestation store iron. Since a replacement of embryonic transferrin by adult transferrin (14) is coincident with the loss of iron storage by the circulating red cells, it is possible that the combination of embryonic transferrin and yolk-sac red cells provides a source of iron more readily available to the embryo than that obtained through the placenta.

Ferritin preparations are composed of multiple components which differ in apparent isoelectric point. Different tissues have characteristic groups of isoferritins (1,3). The heterogeneity observed in lysates of embryonic or larval red cells indicates that isoferritins can occur even after minimal preparative manipulation and in a pure cell type (9). The heterogeneity of ferritin in yolk-sac red cells, a cohort of cells of similar age (15), demonstrates that isoferritins do not necessarily result from variations in cell age.

During amphibian metamorphosis, a fifteen-fold decrease in red cell ferritin content occurred; but there was no qualitative change in isoferritins. Nor did the immunologic reactivity of ferritin from tadpole and frog red cells differ (10). Thus, the change in red cell ferritin content that occurs during amphibian development is a quantitative change in the expression of ferritin genes. The developmental events which alter the expression of ferritin genes during red cell maturation have not yet been discovered.

REFERENCES

1. Jacobs, A., Worwood, M., Prog. in Hematol. 9, 1 (1975).
2. Aisen, P., Prog. in Hematol. 9, 25 (1976).
3. Harrison, P. M., Hoare, R. J., Hoy, T. G., Macara, I. G., in "Iron in Biochemistry and Medicine" (Jacobs, A., Worwood, M. eds.), p. 73, Academic Press, New York, 1974.
4. Theil, E. C., J. Biol. Chem. 248, 622 (1973).
5. Theil, E. C., in "Proteins of Iron Storage and Transport in Biochemistry and Medicine" (Crichton, R. R. ed.), p. 371, North-Holland, Amsterdam, 1975.
6. Theil, E. C., Brit. J. Haem. 33, 437 (1976).
7. Craig, M. L., Russell, E. S., Dev. Biol. 10, 191 (1964).
8. Bates, G. W., Wernicke, J., J. Biol. Chem. 246, 3679 (1971).
9. Brown, J. E., Theil, E. C., Brit. J. Haem. 34, 663 (1976).
10. Theil, E. C., Dev. Biol. 34, 282 (1973).
11. Cheney, B. A., Lothe, K., Morgan, E. H., Sood, S. K., Finch, C. A., Amer. J. Physiol. 212, 376 (1967).
12. Theil, E. C., Comp. Biochem. Physiol. 33, 717 (1970).
13. Osaki, S., James, G. T., Frieden, E., Dev. Biol. 39, 158 (1974).
14. Renfree, M. B., Hensleigh, H. C., McLaren, A., J. Embryol. exp. Morph. 33, 435 (1975).
15. Fantoni, A., De La Chapelle, A., Marks, P. A., J. Biol. Chem. 244, 675 (1969).

ACKNOWLEDGEMENT

This investigation has been supported in part by the National Science Foundation and the North Carolina Agricultural Experiment Station.

INDUCTION OF FERRITIN FORMATION IN HEPATOCYTES

H. KIEF, R. R. CRICHTON, H. BÄHR,
K. ENGELBART and R. LATTRELL

Experimentelle Pathologie und Pharma Synthese Hoechst AG, Frankfurt, Germany and Unité de Biochimie, Université de Louvain, Louvain-la-Neuve, Belgium

An enhanced iron uptake into hepatocyte ferritin can be achieved either by increasing the level of transferrin saturation, or by specific release of iron within hepatocytes from iron-containing compounds which are metabolised in the liver. We will only deal here with the second mechanism and in particular with iron release in the liver from ferrocene derivatives. Since the discovery of ferrocene[1,2] (dicyclopentadienyliron) many derivatives have been synthesised. The great majority of them are mono- or dicarboxylic-substituted ferrocenes. These derivatives are very different with respect to their absorption and metabolism. Most of them are well absorbed from the gastrointestinal tract. Some are eliminated in the urine, others partly or totally retained in the body and metabolised, leaving their iron in a form which can be incorporated into liver ferritin.

To study the metabolic capacity of the liver and the excretion of non-metabolised derivatives we conducted several short-term experiments. High doses of different ferrocene derivatives with an equivalent iron content were administered to normal rats and mice for three consecutive days. The excretion of unchanged compound was easily observed by the progressive discoloration of the wood shavings in the cages. On the other hand for compounds that were absorbed and metabolised a gradual increase in faint blue coloration of the cytoplasm, and finally the formation of hemosiderin granules in the liver cells were regarded as good criteria for making a rough estimation of the rate of metabolism of the compound. Such preliminary experiments are consistent with the hypothesis that the iron contained in the derivative is incorporated directly into liver cell ferritin.

In order to establish this pathway more clearly we have carried out electron microscopic investigations and biochemical analysis of liver ferritin protein and ferritin iron. In these experiments a special ferrocene derivative, HOE 117, was administered once or twice to normal rats at different dose levels. The animals were killed 8 and 24 hours after one treatment. In a parallel study the animals received two administrations of the compound and were sacrificed 24 hours after the second treatment. Because of the well known sex difference in iron metabolism in rats for these and the subsequent experiments, only male rats were used.

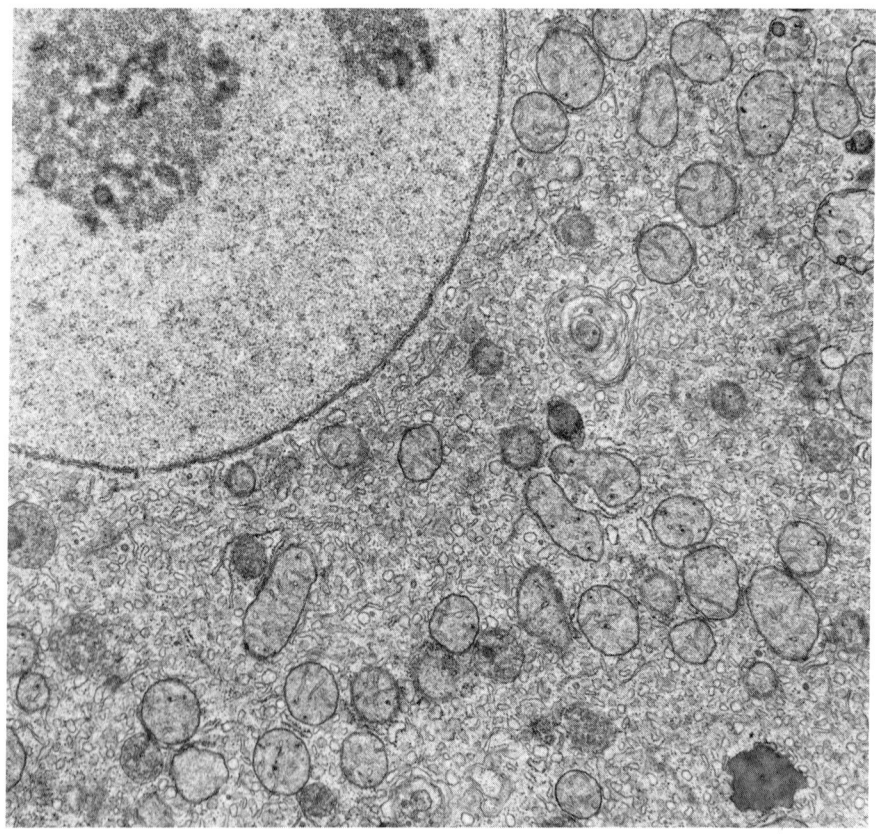

Figure 1: Rat liver cell 8h after one administration of HOE 117 (15o mg/kg). Magnification x 15,ooo. Uranyl acetate/lead hydroxide stain.

The ultrastructural changes, which can be observed 8 hours after one administration of 15o mg/kg body weight reflect predominantly the metabolic activation of the liver cells (Fig.1). The rough endoplasmic reticulum is markedly reduced, numerous vesicles of the smooth endoplasmic reticulum are present instead, and the nucleoli are enlarged. 24 hours after one administration the activation of cell metabolism has nearly returned to normal. The lamellae of the rough endoplasmic reticulum are completely reconstructed. Only the nucleoli are still enlarged. At this time the amount of ferritin particles in liver cells is increased. The same effects are seen, more markedly, after two administrations (Fig. 2). The ferritin particles are irregularly

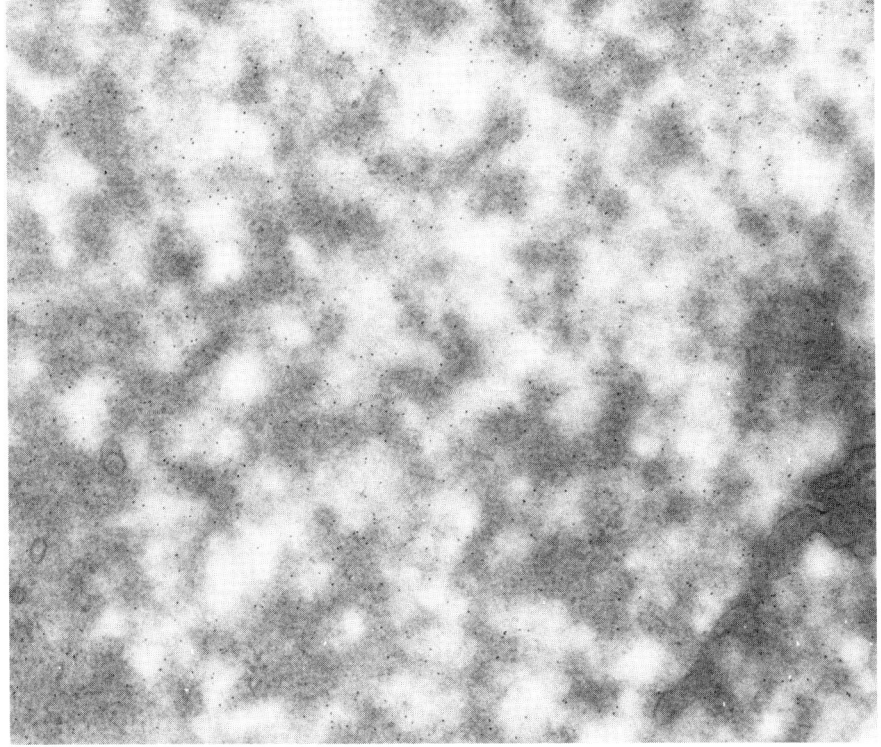

Figure 2: Rat liver cell 24h after two administrations of HOE 117 (15o mg/kg). Magnification x 6o,ooo. Unstained.

distributed in the cell sap; only few are attached to membranes[3-5]. The biochemical results are in accordance with the morphological findings. Ferritin iron and the iron/protein ratio are increased in a time- and dosage-dependent manner. With one and two administrations the ferritin iron is increased after 24 hours three and four fold respectively compared to controls. However, the ferritin protein values after 24 hours with one or two administrations are not significantly different from the control values. Yet we see many more ferritin particles by electron microscopy than in control animals. This seeming discrepancy between morphological and biochemical results can be explained by the steady increase in iron-rich ferritin, which is more easily visible by electron microscopy.

One or two administrations of 3oo mg/kg body weight increase ferritin protein and ferritin iron significantly compared with controls (Table 1). This result is already demonstrable as early as 8 hours after one administration. There is an increase in ferritin protein of about 80% and of ferritin iron of two and a half fold. With two administrations the iron content and the iron/protein ratio are unusually high. Light and electron microscopy show both ferritin and hemosiderin in liver cells (Fig. 3). In addition, it appears that in about half of the ferritin particles the electron dense cores are enlarged. We can therefore conclude from these experiments that the iron released from the compound stimulates ferritin synthesis in liver cells. Ferritin protein increases rapidly, and reaches a maximum of around two fold compared with control values. Ferritin iron increases more slowly until a point is reached where the ferritin particles with the highest iron content enter lysosomes, and are converted to hemosiderin[6,7].

Having established that HOE 117 increases liver ferritin iron, we wanted to examine whether this increased pool of liver ferritin iron could be used for hemoglobin synthesis in anaemic animals. Rats were rendered anaemic by repeated, standardised bleeding over a period of 12 days. On the 12th day the animals were divided into three groups and were

Table 1: Liver ferritin levels in normal rats after administration of HOE 117 and FeSO4.

Rat male	n=6		Control	HOE 117 300 mg/kg 1x 8 h	HOE 117 300 mg/kg 2x 24 h	Fe SO4 164 mg/kg 2x 24 h
Body weight in g	$\bar{x}$		142.66	136.29	236.62	148.84
	s		4.80	3.66	2.42	7.87
	t		–	2.56+	2.88+	0.28
Liver weight in mg	$\bar{x}$		7470	6195	9298	7678
	s		754	638	521	704
	t		–	3.16++	4.88+++	0.49
Ferritin protein ug/g liver	$\bar{x}$		365.34	666.34	761.33	378.33
	s		68.15	74.15	80.28	43.08
	t		–	7.32+++	9.20+++	0.37
Ferritin iron ug/g liver	$\bar{x}$		59.16	161.50	652.01	59.10
	s		6.24	16.08	102.42	7.49
	t		–	14.52+++	14.14+++	0.14
Ratio			0.1619	0.2424	0.8563	0.1559

$P = 0.05+$ $P = 0.01++$ $P = 0.001+++$

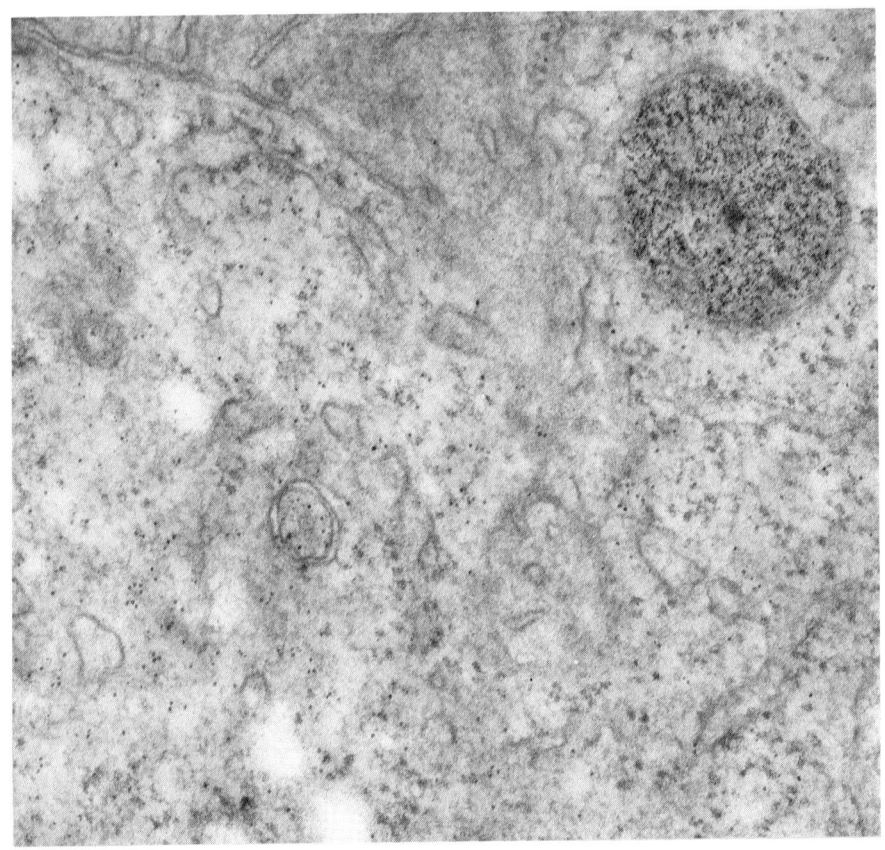

Figure 3: Rat liver cell 24 h after two administrations of HOE 117 (3oo mg/kg). Magnification x 94,5oo. Unstained.

subsequently (a) fed an iron free diet for 14 days (b) as group (a) but supplemented by three administrations of HOE 117 on day 13, 15 and 17 (c) as group (a) but supplemented by three administrations of $FeSO_4$ on day 13, 15 and 17. On day 26 the animals were sacrificed. Table 2 presents the results of this experiment, together with data for a control group maintained throughout the period on a normal diet. Hemoglobin levels were measured on day 12 and day 26. Administration of HOE 117 to anaemic rats completely restores the hemoglobin level to normal within two weeks. Ferritin protein

Table 2: Hemoglobin and liver ferritin levels in anaemic and control rats

Rat male n=6		Control	Anaemic iron free diet 150 mg/kg HOE 117 (3x)	Anaemic iron free diet 82 mg/kg FeSO4 (3x)	Anaemic iron free diet no treated
Body weight in g	x̄ s t	347.50 16.85 -	333.33 21.83 1.25	335.83 34.12 0.75	310.83 20.35 3.39++
Liver weight in mg	x̄ s t	11,293 575 -	11,712 1,267 0.74	10,802 964 1.07	9,789 846 3.64++
Hemoglobin in g%	12th d 26th d	15.29 15.19	9.42 14.95	9.30 11.13+++	9.13 7.80+++
Ferritin protein μg/g liver	x̄ s t	472.83 59.06 -	682.00 125.52 3.69++	438.66 33.35 1.23	458.34 66.04 0.40
Ferritin iron μg/g liver	x̄ s t	148.82 35.82 -	349.66 90.74 5.04+++	14.16 8.06 8.98+++	5.50 1.04 9.79+++
Ratio		0.3147	0.5128	0.032	0.012

$P = 0.05+$ $P = 0.01++$ $P = 0.001+++$

and ferritin iron in liver cells are markedly increased at the end of the experiment and it appears that 65% of the iron administered as HOE 117 passes through the liver ferritin pool and ends up either in hemoglobin or in liver ferritin. In contrast to these findings, an equivalent dose of iron sulphate is not nearly so effective as the ferrocene derivative in restoring the hemoglobin level. Furthermore, iron sulphate does not bring about any effect on ferritin protein and ferritin iron in liver cells. The iron/protein ratio remains extremely low and is even lower in anaemic rats fed with an iron-free diet without any treatment.

Our investigations lead to the conclusion that appropriate ferrocene derivatives induce ferritin synthesis in liver cells, and that the iron is processed through the liver ferritin pool. In normal animals (mice, rats, guinea pigs and dogs) this results in iron overload and in anaemic rats the iron is utilised for hemoglobin synthesis.

Acknowlegements: We thank Francine Brouwers and Dorothea Kardos for technical assistance.

References

1.) Kealy, T. J. and Pauson, P. L., Nature 168, 1o39 (1951).
2.) Miller, S. A., Tebboth, J. A. and Tremaine, J. F., J. Chem. Soc. 632 (1952).
3.) Hick, S. J., Drysdale, J. W. and Munro, H. N., Science 164, 584 (1969).
4.) Redman, C. M., J. Biol. Chem. 244, 43o8 (1969).
5.) Puro, D. G. and Richter, G. W., Proc. Soc. Exp. Biol. Med. 138, 399 (1971).
6.) Sturgeon, P. and Shoden, A. in "Pigments in Pathology". (M. Wolman, ed.) Academic Press, New York, p. 93 (1969).
7.) Trump, B. F., Valigorsky, J. M., Arstila, A. U., Mergner, W. J. and Kinney, T. D., Amer. J. Path. 72, 295 (1973).

STUDIES ON THE CARBOHYDRATE COMPONENTS

OF FERRITIN

MORRIS A. CYNKIN
MARGARET KNOWLTON

Department of Biochemistry and Pharmacology
Tufts University School of Medicine
Boston, Massachusetts

I. INTRODUCTION

Ferritin, an iron-containing protein, consists of a multimeric protein shell containing approximately 24 subunits, which can sequester variable amounts of iron in its interior. By the use of electrophoretic techniques it has been shown that ferritin is a heterogeneous population of molecules, isoferritins, whose proportional distribution varies from tissue to tissue. Recent evidence indicates that much of this observed heterogeneity is due to different sub-unit populations and that many isoferritins are hybrid molecules consisting of different populations of dissimilar subunits. However, superimposed upon the heterogeneity due to differences in the primary structure of the subunits may be other elements, such as the post-translational modification of polypeptides. The reports of Buffe et al. (1,2) that $\alpha_2 H$ globulin, a form of ferritin found in serum, contains carbohydrate, suggested to us that the presence or absence of carbohydrate components might provide another basis for the differentiation of isoferritins. With this possibility in mind, we have examined highly purified ferritin from human liver and from horse spleen, heart and liver. All samples were found to contain galactose, mannose, fucose, and glucose. Horse spleen and human liver ferritins also contain hexosamines. The other ferritins have not yet been tested for hexosamine content. While this study was in progress, Shinjyo et al. (3) reported fucose, mannose, galactose, and hexosamine (10:16:8:3) in crystalline horse spleen ferritin, although they did not find glucose.

II. MATERIALS AND METHODS

Crystalline horse spleen ferritin (Cd salt) was obtained from Nutritional Biochemicals Co. In some studies, this ferritin was converted to apoferritin (depleted of iron) by the method of Granick and Michaelis (4).

Other preparations of ferritin, from horse spleen, liver and heart, were prepared essentially according to the procedure of Drysdale and Munro (5).

As a final purification step, ferritin samples were subjected to gradient pore polyacrylamide gel electrophoresis essentially by the method of Margulis and Kenrick (6). In this procedure, an equilibrium state is reached, and fine resolution can be attained, assuring high purity. Samples were eluted with water and lyophilized.

Total neutral sugars were determined by the anthrone reaction (7). Fucose was determined by the cysteine-sulfuric acid reaction (8). The procedure of Aminoff (9) was used for determination of sialic acid.

In preparation for gas chromatographic analysis of the carbohydrate components, samples of ferritin were hydrolyzed in 2N constant boiling HCl for 4 hours, neutralized by passage through Dowex-1-(CO_3^-) and Dowex 50-(H^+). This procedure also removed amino sugars, amino acids and other ions. Neutralized eluates were lyophilized and converted to trimethylsilyl (TMS) derivatives by treatment with Tri-Sil (Pierce Chemicals).

Gas chromatography was performed using a Hewlett-Packard 7620A Gas Chromatograph with a 3% SE30 column and a programmed temperature gradient.

SDS-gel electrophoresis under conditions which dissociate ferritin into subunits was performed by the method of Niitsu et al. (10). Protein bands were located by the use of Coommassie Blue stain. The periodic acid-Schiff (PAS) reaction, as described by Segrest and Jackson (11) or by Glossman and Neville (12), was used to detect the presence of carbohydrate in gels.

III. RESULTS

In our initial study of the carbohydrate components of ferritin, we determined that crystalline horse spleen apoferritin contained at least 0.5% neutral sugar, including at least 0.1% fucose. No sialic acid was detected (less than .01%). Samples of crystalline horse spleen ferritin and ferritin obtained from normal and tumorous human liver were hydrolyzed in 2N HCl at 105° under N_2 for 4 hours, neutralized by flash evaporation, treated with Tri-Sil, and analyzed by gas chromatography. In all samples, fucose, galactose, mannose, glucose, and hexosamine were detected. (M. A. Cynkin and R. Siegel, unpublished). The presence of glucose was surprising, since this sugar is not usually found as a component of soluble glycoproteins in association with fucose, mannose and hexosamine.

Since the possibility existed that the carbohydrate components were present in a non-ferritin contaminant, samples of apoferritin prepared from horse heart and spleen ferritin and from human liver ferritin were subjected to polyacrylamide gel electrophoresis (PAGE) by several different procedures.

Ferritin samples were subjected to PAGE either in cylindrical gels or in gradient pore slab gels for improved resolution. In addition, ferritin was subjected to PAGE in the presence of DTT and SDS under conditions which dissociate ferritin into subunits. Known glycoproteins - ovalbumin and bovine fibrinogen - and bovine serum albumin, a protein known to lack carbohydrate, were similarly treated as controls. Carbohydrate was found to be associated with the protein components of both native ferritin and its subunits under all conditions of electrophoresis. The presence of carbohydrate in ferritin bands after SDS-gel electrophoresis eliminates any possibility that the carbohydrate was present in a non-covalently-bound form.

Because of the unusual presence of glucose extreme pains were taken to avoid the use of carbohydrate-containing materials, such as Sephadex. As the final purification step, all samples were subjected to gradient pore gel electrophoresis. Ferritin bands were eluted and analyzed as before. As controls, bovine serum albumin and ovalbumin were also hydrolyzed and analyzed by gas chromatography. The neutral sugar compositions of highly purified ferritin obtained from horse spleen liver and heart are shown in

Table I. Under these conditions, BSA was devoid of carbohydrate as expected and ovalbumin yielded only mannose, also as expected (13). Although hexosamine was detected in early analyses, the current procedure, involving passage of the hydrolyzates through ion exchange resins, prevents detection of hexosamines. However, it is assumed that hexosamine is present.

TABLE I

Neutral Sugar Composition of Ferritin

Residues per Mole Ferritin

Tissue Source	Fucose	Mannose	Galactose	Glucose	Total Neutral Sugar % Weight
Spleen	7	6	10	38	2.4
Liver	9	9	7	37	2.4
Heart	19	21	14	72	4.9

IV. DISCUSSION

The results of our analyses, demonstrating the presence of fucose, mannose, and galactose (and hexosamine, shown earlier) are in qualitative agreement with the report of Shinjyo et al. (3). These monosaccharides are typical of most soluble glycoproteins in animals. With respect to the presence of glucose, whereas in collagen and basement membrane proteins, a disaccharide, glucosyl-α-(1$\rightarrow$2)-galactose, is found O-glycosidically linked to the hydroxyl group of hydroxylysine (13), the large quantity of glucose relative to galactose present in ferritin, and the absence of hydroxylysine in ferritin, eliminate the collagen-type carbohydrate from consideration as a possibility. The simplest hypothesis is that in ferritin there are at least two types of carbohydrate components, the first containing fucose, galactose, mannose, and hexosamine, and the second being a glucose homopolymer, or glucan. Such a structure has not been found in glycoproteins. Although the subunits contain carbohydrate, as shown by the PAS-positive reaction of ferritin on SDS gels, it is possible that at least part of the carbohydrate component, such as the glucan, might be sequestered in the interior of the native ferritin. If this were true, the isolated sub-units would be expected to lack glucose.

The amount of carbohydrate present in ferritin is probably not sufficient to account for a carbohydrate chain for each subunit. We are currently examining the carbohydrate composition of isolated subunits to determine whether the carbohydrate components are associated with a particular subunit type. It will also be of interest to determine if the glucose-containing component is on the same sub-unit, or even on the same peptide chain, as the heteropolysaccharide component. Also unanswered at this time is the possible role of the carbohydrate components in altering the electrophoretic mobility of ferritin and generating, at least in part, the isoferritins.

The fact that ferritin is a glycoprotein at all is surprising, since most glycoproteins are either extracellular, such as the majority of the plasma proteins and collagen, or are membrane proteins (13). Since ferritin is for the most part an intracellular cytoplasmic protein, the presence of carbohydrate is anomalous. It has been suggested that the function of carbohydrate in most glycoproteins is to serve as a recognition signal in interactions involved in transport, secretion or other membrane-associated activities. It remains to be seen if the carbohydrate components of ferritin are related to either the secretion of ferritin into the plasma, or to the storage or mobilization of iron. In addition, the carbohydrate components may be partly responsible for the immunological properties of the ferritins.

The higher carbohydrate content of heart ferritin may arise from the different ratio of sub-units in heart ferritin compared to liver, for example. If the H sub-unit were richer in carbohydrate than the HL sub-unit, than not only would heart ferritin be expected to contain more carbohydrate because of the higher H:HL ratio, but tumor ferritins would also be expected to be higher in carbohydrate content, since it has been shown that ferritin from tumor cells is composed predominantly of H sub-units (14). This prediction is consistent with the observations of Buffe and Rimbaut that serum ferritins from tumor patients have higher carbohydrate content than ferritin from normal sera (2).

Our current studies on the structure and distribution of the carbohydrate components of the isolated sub-units of ferritin may serve to resolve these issues.

V. REFERENCES

1. Buffe, D., Rimbaut, C., Fuccaro, C. and Burtin, P., Ann. Inst. Pasteur, 123, 29 (1972).
2. Buffe, D. and Rimbaut, C., Ann. N. Y. Acad. Sci. 259, 417 (1975).
3. Shinjyo, S., Abe, H. and Masuda, M., Biochem. Biophys. Acta 411, 165 (1975).
4. Granick, S. and Michaelis, L. J. Biol. Chem. 147, 91 (1943).
5. Drysdale, J. W. and Munro, H. N., J. Biol. Chem. 241, 3630 (1966).
6. Margulis, J. and Kenrick, K. G. Nature, 221, 1056 (1969).
7. Roe, J. H., J. Biol. Chem., 212, 335 (1955).
8. Dische, Z. and Shettles, L. B., J. Biol. Chem. 175, 595 (1948).
9. Aminoff, D., Biochem. J., 81, 384 (1961).
10. Niitsu, Y., Ishitani, K. and Listowsky, I., Biochem. Biophys. Res. Commun. 55, 1174 (1973).
11. Segrest, J. and Jackson, R. L. in Meth. Enzymol. XXVIII (V. Ginsburg, Ed.) p. 54, Academic Press, New York (1972).
12. Glossmann, H. and Neville, D.M., Jr., J. Biol. Chem. 246, 6339 (1971).
13. Sharon, N., Complex Carbohydrates. Their Chemistry Biochemistry and Functions, Addison-Wesley Publishing Co., Reading, Mass. (1975).
14. Arosio, P., Yokota, M. and Drysdale, J. W. Cancer Research 36, 1735 (1976).

VI. ACKNOWLEDGEMENTS

This research was supported by funds from the Charlton Fund of Tufts University School of Medicine and from the Dean of the School of Medicine.

SITES OF FERRITIN SYNTHESIS AND NATURE OF SUBUNIT PRODUCTS

M.C.Linder, J.Zahringer, B.S.Baliga, R.L.Drake,
B.Barres and H.N.Munro
*Massachusetts Institute of Technology, Cambridge, Mass.
and California State University, Fullerton, Cal.*

I. INTRODUCTION

Based on migration in polyacrylamide gels (1), peptide maps (2), reaction with ferritin antibody (3) and isoelectric focussing (4), it is evident that ferritin occurs in multiple forms, both within the same tissue and in different tissues. Interpretation of this molecular heterogeneity depends on identifying how many types of ferritin subunit are present. Listowsky et al. (5) obtained subunits of 19,000, 15,000, 11,000 and 8,000 daltons from horse spleen, human spleen and liver and rat liver ferritins, and we (2) observed two major components of 19,000 and 11,000 to 14,000 daltons following dissociation of a variety of normal and malignant rat tissue ferritins with SDS followed by electrophoresis in SDS. In our study, the proportions of these subunits varied from tissue to tissue and also in the liver following iron administration (6). In addition, turnover of these two subunit groups after labeled amino acid injection was found to differ (6). Nevertheless, Crichton et al. (7) have concluded that the smaller subunits are products of breakdown of the larger subunit due to activation of proteolytic enzymes during heating of the tissue homogenate in the first step of the standard ferritin isolation procedure.

In order to resolve the status of the subunits in ferritins of different origins, we have studied first the size of the products formed by in vitro translation of ferritin mRNA, and second the effect of inhibitors of proteolytic enzymes on recovery of subunits of different sizes from ferritin isolated with or without heat treatment. In addition, we address ourselves to the question raised by our earlier studies (8) that polyribosomes in different intracellular locations may make ferritin subunits of different sizes and respond differently to iron administration (9).

II. IN VITRO TRANSLATION OF FERRITIN mRNA FROM FREE AND MEMBRANE-BOUND RIBOSOMES

To determine the intracellular sites of ferritin biosynthesis and their responsiveness to iron administration, we compared the in vitro synthesis of ferritin by rat liver free and membrane-bound polyribosomes and also by the mRNA extracted from these (Table 1). Free polysomes made proportionately more ferritin than membrane-bound polysomes, but the mRNA extracted from each ribosome population showed a similar capacity for ferritin synthesis in a wheat germ translation system. In agreement with Shafritz (10), this implies some non-translated ferritin mRNA in the bound population. After iron injection into the rats, the proportion of total in vitro protein synthesis represented by ferritin increased markedly in the free ribosome fraction and its mRNA, but not in the bound fraction (Table 1). Other studies (11) indicate that the additional ferritin mRNA in the form of free polysomes comes from mRNP for ferritin latent in the cytosol.

In these studies, the sizes of the products of translation of ferritin mRNA in the wheat germ system were determined by SDS gel electrophoresis. Following incubation, nascent

TABLE I Effect of Iron Injection on the Amount
of Ferritin Synthesized in Vitro by Rat Liver
Free and Membrane-Bound Polysomes and by
mRNA Extracted from Them.[a]

Type of Polysomes or mRNA	Ferritin mRNA x 100/total protein mRNA	
	Normal rats	Iron-treated rats
Polysomes		
Free	0.087±0.012	0.325±0.069
Membrane-bound	0.031±0.004	0.039±0.007
mRNA		
Free	0.110±0.018	0.349±0.065
Membrane-bound	0.100±0.016	0.120±0.030

[a] Experimental details from Zahringer et al. (12).

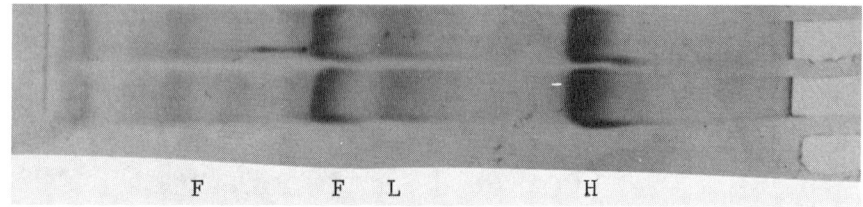

Fig. 1 Separation of subunits made by precipitating liver ferritin with a ferritin antiserum, and followed by dissociation in SDS and resolution on a SDS polyacrylamide gradient gel (12). From right to left, the stained protein bands are (1) the heavy immunoglobulin chain (H), (2) the light immunoglobulin chain (L), (3) ferritin subunit of about 19,000 daltons (F), (4) several fainter bands of ferritin subunits ranging from 14,000 to 11,000 daltons (F).

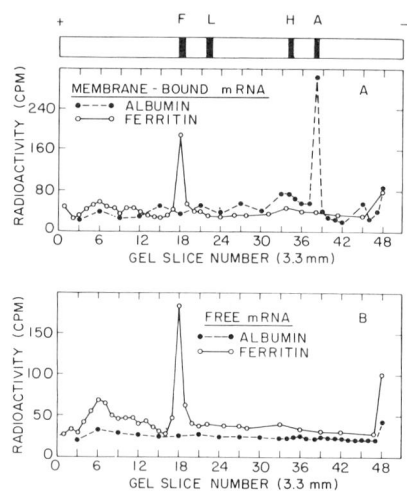

Fig. 2 Synthesis of ferritin and albumin by the mRNAs extracted from free and membrane-bound polysomes in the wheat germ S_{30} cell-free system. Isolation and translation of the mRNAs, immunoprecipitation of the synthesized ferritin and albumin chains, and purification of the immunoprecipitate and its analysis on dodecyl sulfate-polyacrylamide slab gels (10-15%) were performed as described (6). F, ferritin; A, albumin; L, L-chain; H, H-chain of antibody. (Reproduced from Zahringer et al., 12)

chains attached to ribosomes were removed and released chains were harvested with ferritin antiserum. The immunoprecipitates were then dissociated into the heavy and light immunoglobulin chains and ferritin subunits on gradient SDS gels. Fig. 1 shows the distribution of ferritin subunits when chemically isolated pure rat liver ferritin is subjected to this dissociation procedure. In addition to the heavily stained heavy and light immunoglobulin chains, there is a 19,000 dalton subunit of ferritin and a number of smaller subunits of lower molecular weight. Fig. 2 shows the separation of labeled products obtained after in vitro translation of ferritin mRNA from free and bound ribosomes. In each case, the major radioactive product coincides with the 19,000 dalton subunit but the products of free ribosomal mRNA translation also include small amounts of radioactivity coinciding with the locations of the smaller subunits. Thus while the primary product of translation is about 19,000 daltons in each case, we cannot exclude small amounts of lesser subunits.

III. PURIFICATION OF FERRITIN BY PROCEDURES TO LIMIT BREAKDOWN

Approaching the problem of ferritin subunits from a different angle, we purified ferritin from livers of iron-injected rats (a) by a series of procedures avoiding heat treatment, and (b) by standard procedures involving heat treatment but in the presence of α_2-macroglobulin, a trap for proteases (13), or the two inhibitors of seryl proteases, phenylmethylsulfonylfluoride (14) and ε-amino caproic acid (15). In the case of purification without heating, liver homogenates and extracts were kept at 4° throughout and

TABLE 2. Purification of Rat Liver Ferritin Without Heat or in the Presence of Proteolytic Enzyme Inhibitors[a]

			Ferritin Characteristics		
	Yield[b] (%)	Ratio[c] Fe:prot.	Electro- phoretic Migration (R_f)[d]	Subunits[e] 19,000	11- 14,000
Standard (+heating)	59	0.25	1.00	++++	++
No heating (4°)	3	0.31	0.95[f]	++++	++
+ α_2M (62 µg/ml)	65	0.28	1.03	+++++	+
+ PMSH (0.1 mM)	88	0.31	1.03	+++++	+
+ ε-AC (20 mM)	59	0.31	1.04	+++++	+

[a] Fresh, ice-cold livers from adult female Fischer rats injected 5 days previously with 25 mg Fe (as iron dextran) were minced, mixed and subjected to purification (a) by a series of parallel procedures with and without a heating step (see text), and (b) in the presence or absence of proteolytic enzyme inhibitors (α_2-macroglobulin; phenylmethylsulfonyl-fluoride; ε-amino caproic acid), using the standard method for purifying ferritin (16).

[b] antibody-precipitable iron content of preparation relative to starting supernatant.

[c] µg/µg, as determined by assays of iron and protein (16), using bovine serum albumin as protein standard.

[d] migration of undissociated ferritin in disc polyacrylamide gel electrophoresis (relative to tracking dye); Rf values shown were recalculated relative to those of standard ferritin (Rf = 1.00) determined in the same gel run.

[e] holo- and apoferritins dissociated with 1% SDS-1% dithio-threitol in phosphate at pH 7, 60 min at 60° and then subjected to standard SDS electrophoresis (6).

[f] $p < 0.01$ for difference from standard ferritin.

purified by centrifugation at 18,000 g for 60 min, followed by Sepharose 6B filtration, ammonium sulfate precipitation, filtration on Sepharose 4B and carboxymethylcellulose chromatography, following the procedure recommended by Crichton (7). In the case of the proteolytic enzyme inhibitors, these were added directly to the liver during homogenization at 4° followed by the standard purification with heating to 70°, then treatment at pH 4.8, ammonium sulfate fractionation and Sephadex G200 treatment (16), inhibitor concentrations being kept constant.

As shown in Table 2, purification without heating produced a ferritin of slightly higher iron content but in much lower yield than by the standard procedure. In agreement with previous results (16), most of the losses occurred at the CM cellulose step. This ferritin had a slightly but significantly lower rate of electrophoretic migration than standard ferritin. Nevertheless, following dissociation with SDS, this ferritin still contained smaller subunits in the same proportion as ferritin purified in the standard way, suggesting that heating was not the primary cause of the appearance of the small-molecular-weight subunits. However, when ferritin was purified in presence of proteolytic enzyme inhibitors, much less of the smaller subunits were observed. thus implying that proteolysis was responsible for the appearance of smaller subunits. These series of experiments also demonstrate that ferritin isolated from different groups of rats of the same strain, sex and age, and under identical standard conditions could produce ferritin with different proportions of the larger and smaller subunits. This suggests that there are variations among animals in the amounts of inherent protease activity responsible for production of fragments of the larger subunits of ferritin found upon SDS dissociation. Nevertheless, even in the presence of proteolytic enzyme inhibitors, some residual small molecular weight material in the ferritin was apparent.

IV. CONCLUSION

Evidence from in vitro ferritin synthesis and from purification by various procedures agrees with the premise that ferritin is composed primarily or even exclusively of subunits in the 19,000 dalton range, and that the occurrence of smaller subunits is essentially attributable to the occurrence of limited proteolysis, during purification of the protein. Nevertheless, the question remains whether a small fraction of the ferritin shell consists of smaller units, synthesized on the free polyribosomes. Final resolution of this question should be forthcoming as elucidation of the structure of isoferritins continues.

REFERENCES

1. Linder-Horowitz, M., Ruettinger, R.T., Munro, H.N., Biochim. Biophys. Acta 200, 442 (1970).
2. Linder, M.C., Moor, J.R., Munro, H.N., Morris, H.P. Biochim. Biophys. Acta 386, 409 (1975).
3. Linder, M.C., Munro, H.N. Am. J. Pathol. 72, 263 (1973).
4. Arosio, P., Yokota, M., Drysdale, J.W. Cancer Res. 36, 1735 (1976).
5. Ishitani, K., Niitsu, Y., Listowsky, I. J. Biol. Chem. 250, 3142 (1975).
6. Linder, M.C., Moor, J.R., Munro, H.N. J. Biol. Chem. 249, 7707 (1974).
7. Collet-Cassart, D., Crichton, R.R., in "Protein or Iron Storage and Transport in Biochemistry and Medicine", (R.R. Crichton, eds) Amsterdam: North-Holland, p. 185.

8. Konijn, A.M., Baliga, B.S., Munro, H.N. FEBS Lett. 37, 249 (1973).
9. Zahringer, J., Konijn, A.M., Baliga, B.S., Munro, H.N. Biochem. Biophys. Res. Commun. 65, 583 (1975).
10. Shafritz, D.A. J. Biol. Chem. 249, 81 and 89 (1974).
11. Zahringer, J., Baliga, B.S., Munro, H.N. Proc. Natl. Acad. Sci. U.S. 73, 857 (1976).
12. Zahringer, J., Baliga, B.S., Drake, R.L., Munro, H.N. Biochim. Biophys. Acta 474, 234 (1977).
13. Barrett, A.J., Starkey, P.M. Bioch. J. 133, 709 (1973).
14. Farney, D.E., Gold, A.M., J. Am. Chem. Soc. 85, 997 (1963)
15. Alkjaersig, N., Fletcher, A.P., Sherry, S. J.Biol. Chem. 234, 832 (1959).
16. Linder, M.C., Munro, H.N. Analyt. Biochem. 48, 266 (1972).

Section IV

TRANSFERRIN STRUCTURE

STRUCTURE AND EVOLUTION OF SERUM TRANSFERRIN

ROSS T.A. MACGILLIVRAY
ENRIQUE MENDEZ
KEITH BREW

Department of Biochemistry
University of Miami School of Medicine
Miami, Florida, U.S.A.

I. INTRODUCTION

A number of fundamental structural features of the transferrin molecule have been clarified by recent physical and chemical studies. There is now overwhelming evidence that the two metal-binding sites are contained within a single polypeptide chain (1, 2, 3) yet no indication of internal homology was found in the sequences around the two glycosylation sites within the same protein. Sequences around the glycosylation sites in transferrins from different species also show little sequence similarity (4). In contrast, the amino-terminal sequences of transferrins from a number of species (4, 5) and human lactoferrin (6) are strikingly homologous, confirming the common ancestry of all members of the transferrin group. The relative conformational independence of the two metal-binding sites has been demonstrated by the isolation of active fragments corresponding to the amino- and carboxyl-terminal halves of the molecule following limited proteolysis of chicken ovotransferrin (7, 8). The existence of two homologous domains, each presumably associated with a single metal-binding site was also confirmed by our previous sequence studies with human transferrin (9).

Nevertheless, the elucidation of the complete covalent structure of transferrin is essential to aid in determining the molecular basis of iron-binding and release, as well as the details of the evolutionary development of the transferrins. We report here the current status of our structural studies of human transferrin. The evolutionary and functional implications of a preliminary partial complete structure for the protein will be discussed.

II. PROCEDURES AND RESULTS

The strategy used in determining the complete sequence of transferrin is summarized in Figs. 1 and 2. Cleavage of human transferrin with CNBr gives rise to five peptide fragments. (10, 11): CN-A, which was formerly supposed to contain four disulfide-bonded peptides (CN-1, CN-2, CN-3, CN-4) but which has recently been found to contain a fifth peptide (CN-10); CN-B, which contains the amino-terminal peptide (CN-6) disulfide bonded with CN-5; and three cystine-free peptides (CN-7, CN-8, CN-9). CN-1 is glycosylated and contains the C-terminal 179 residues of transferrin, while the remaining oligosaccharide prosthetic group is carried by CN-3. Contrary to previous reports (10, 12), sequence studies have shown that CN-2 is devoid of carbohydrate.

The structures of these ten fragments are being assembled by the analysis of peptides from digests with various proteases (trypsin with and without citraconylation, chymotrypsin, thermolysin, pepsin and the dicarboxylic acid-specific protease from S. aureus.). For the most part, the aminoethyl derivatives of cystine-containing fragments have been utilized, following previous problems of cysteine assignments encountered, using the carboxamidomethyl derivative of cysteine (see 13 and 6). Sequences of peptides have been determined by the manual dansyl-Edman procedure and using a Beckman 890C automatic sequencer.

In this way, complete sequences of five fragments (CN-3, CN-6, CN-7, CN-8, CN-9) have been elucidated (11, 12, 13, 14). The structures of CN-4 and CN-5 are complete, apart from some amide assignments and one confirmatory overlap (14). CN-1, the largest fragment requires two confirmatory overlaps and several amide assignments (14). The sequences of the remaining two fragments (CN-2 and CN-10) are incomplete at present.

To assemble the complete sequences, peptides overlapping methionyl residues have been sought in two ways:

1. Transferrin was oxidized with perfomic acid to convert methionyl residues to the sulfone, as well as cystine to cysteic acid, and the protein digested with thermolysin. Cleavage at methionyl residues by this enzyme should be prevented by conversion to the charged sulfone derivative. Unfortunately, an extremely complex mixture of peptides was generated in this way, from which only two pure methionine sulfone-containing peptides were isolated, denoting overlaps from CN-7 to CN-9, and from CN-7 to CN-9 to CN-4, as reported previously (9).

Human serum transferrin
CNBr
CN-A
CN-B
CN-C
CN-D
Reduction, alkylation
Reduction, alkylation
CN-1
CN-2
CN-3
CN-4
CN-10
CN-5
CN-6
CN-7
CN-8
CN-9
No. residues
179
150
73
69
31
85
26
53
7
4
Sequence status
PC
P
C
PC
P
PC
C
C
C
C
Comments
CHO
CHO
N-term.
C-term.

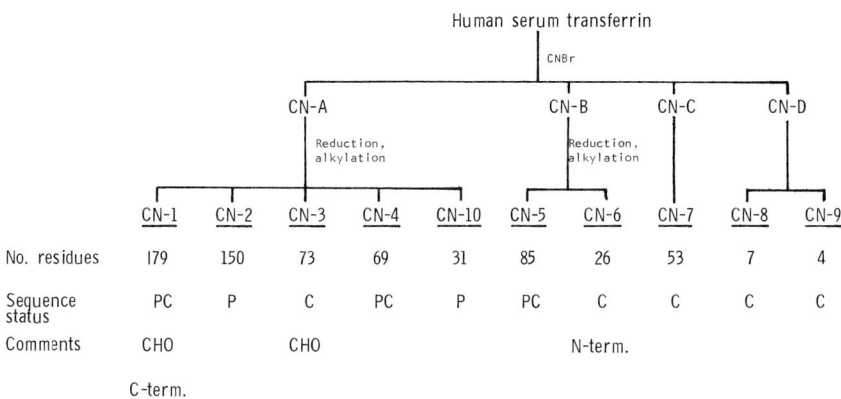

Fig. 1. Isolation and characterization of the ten CNBr fragments of human transferrin. P, partial; C, complete.

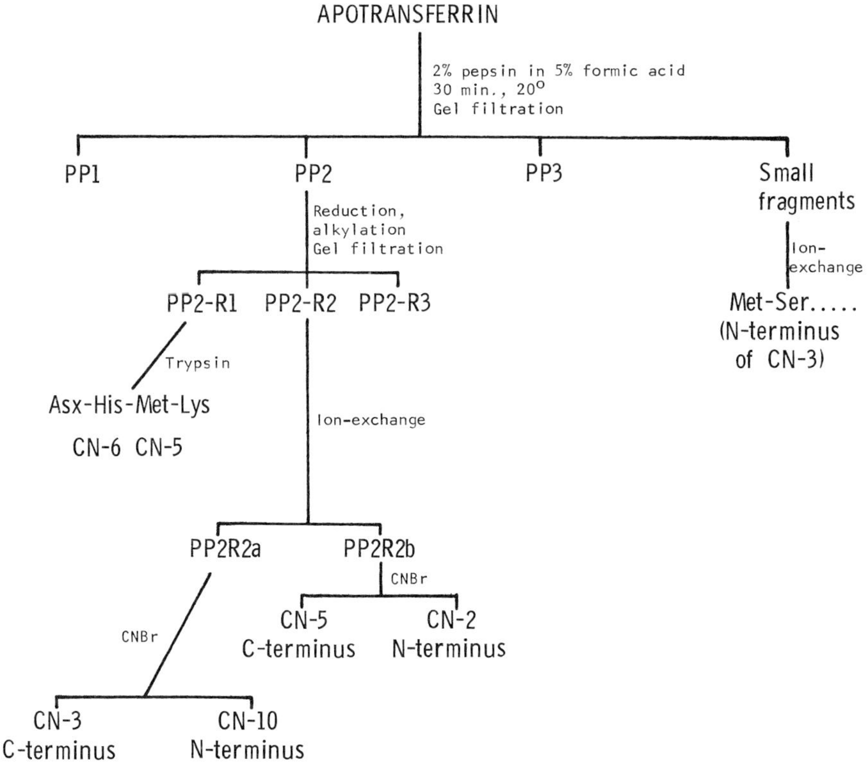

APOTRANSFERRIN
2% pepsin in 5% formic acid
30 min., 20°
Gel filtration
PP1
PP2
PP3
Small fragments
Reduction, alkylation
Gel filtration
Ion-exchange
PP2-R1
PP2-R2
PP2-R3
Met-Ser.....
(N-terminus of CN-3)
Trypsin
Asx-His-Met-Lys
CN-6 CN-5
Ion-exchange
PP2R2a
PP2R2b
CNBr
CN-5
C-terminus
CN-2
N-terminus
CNBr
CN-3
C-terminus
CN-10
N-terminus

Fig. 2. Preparation of the partial peptic fragments.

2. Apotransferrin was subjected to partial peptic proteolysis under conditions that were found to generate large fragments. A number of disulfide-bonded fragments were generated (Fig. 2) which, following reduction, carboxymethylation and reseparation, gave rise to a number of pure methionine containing peptides. Digestion of one of these with trypsin gave rise to the peptide Asp-His-Met-Lys, indicating that CN-5 (the only CNBr fragment with N-terminal Lys) follows CN-6, the N-terminus, in the sequence. Two other pure methionine-containing fragments (PP2R2a and PP2R2b) have been isolated to date. Cleavage of PP2R2a with CNBr gave the C-terminal 34 residues of CN-3 and N-terminal di- and tripeptides from CN-10 (Gly-Leu and Gly-Leu-Leu). Similar treatment of PP2R2b generated the C-terminal 10 residues of CN-5 and the N-terminal 25 residues of CN-2. These regions have been identified by direct sequence analysis as well as by subcleavage with trypsin.

III. DISCUSSION

A. Assembly of Sequence.

While the structural studies reported here are incomplete, they can be used, along with other data, to assemble a provisional partial structure for the polypeptide chain of human transferrin (Fig. 3). Besides lacking the details of regions of sequence in fragments CN-2 and CN-10, we have not isolated peptides overlapping the amino-terminal regions of CN-7, CN-8, CN-3 and CN-1 or the carboxy-terminal regions of CN-2, CN-8 and CN-10. However, the alignment of CN-10 and CN-1 in the structure is suggested by a region of sequence from chicken ovotransferrin (15), which is strikingly homologous with the C-terminus of CN-3, the presently known sequence of CN-10 and the N-terminus of CN-1 (Fig. 4). We have used this to place CN-3 and CN-10 before CN-1 in the sequence and have positioned CN-8 between CN-4 and CN-3 to maximize the internal homology revealed by the rest of the sequence. Confirmatory overlaps across these methionyl residues are being sought in the partial peptic fragments.

B. Internal Homology

From the available sequence information, the transferrin polypeptide chain contains 676 residues, giving a molecular weight of about 81,000. A comparison of residues 1-339 with residues 340-676 (Fig. 3) reveals strong internal homology,

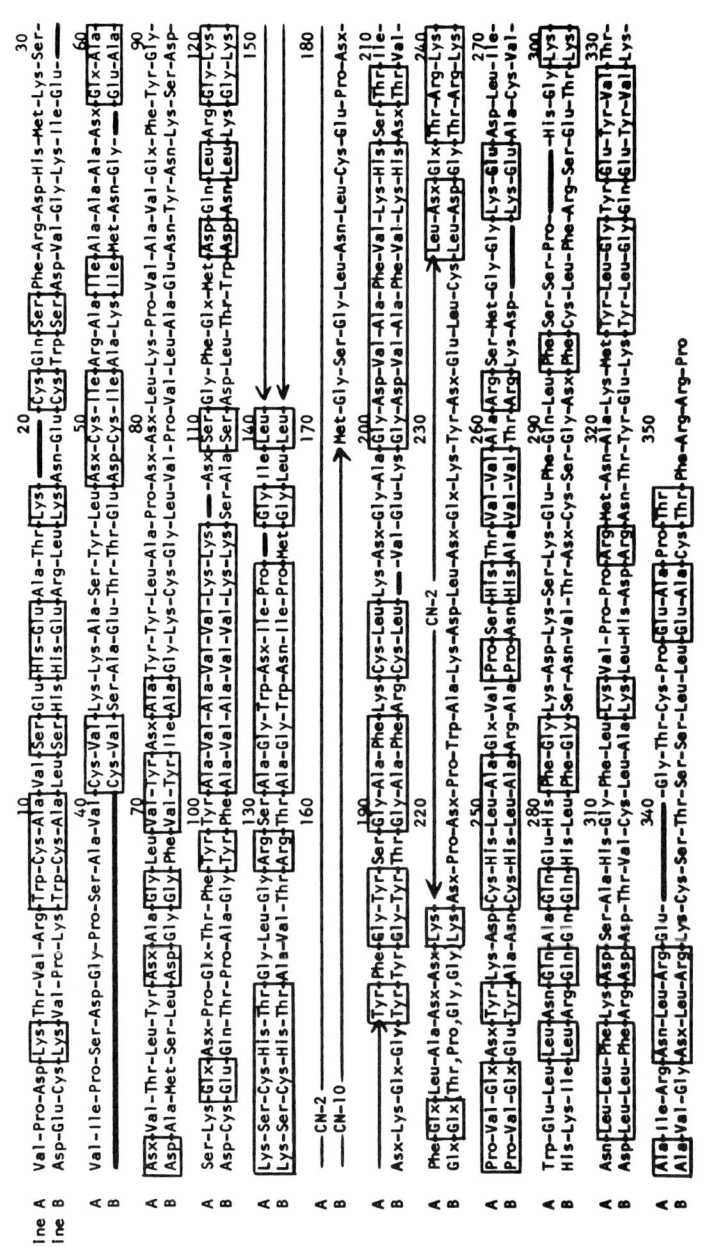

Fig. 3. The sequence of human transferrin arranged to demonstrate internal homology. Line A contains residues 1-339. Line B contains residues 340-676. Identical residues in corresponding positions are boxed. Thick bars indicate gaps inserted to maximize the homology. See text for details.

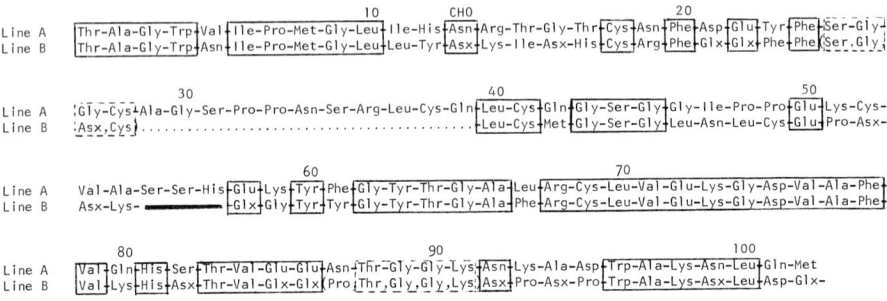

```
                              10            CHO              20
Line A    Thr-Ala-Gly-Trp Val Ile-Pro-Met-Gly-Leu Ile-His Asn Arg-Thr-Gly-Thr Cys Asn Phe Asp Glu Tyr Phe Ser Gly
Line B    Thr-Ala-Gly-Trp Asn Ile-Pro-Met-Gly-Leu Leu-Tyr Asx Lys-Ile-Asx-His Cys Arg Phe Glx Glx Phe Phe Ser Gly

                  30                         40                       50
Line A    Gly-Cys Ala-Gly-Ser-Pro-Pro-Asn-Ser-Arg-Leu-Cys-Gln Leu-Cys Gln Gly-Ser-Gly Gly-Ile-Pro-Pro Glu Lys-Cys
Line B    Asx,Cys ........................................ Leu-Cys Met Gly-Ser-Gly Leu-Asn-Leu-Cys Glu Pro-Asx-

                         60                        70
Line A    Val-Ala-Ser-Ser-His Glu Lys Tyr Phe Gly-Tyr-Thr-Gly-Ala Leu Arg-Cys-Leu-Val-Glu-Lys-Gly-Asp-Val-Ala-Phe
Line B    Asx-Lys-━━━━━━━━━━━ Glx Gly Tyr Tyr Gly-Tyr-Thr-Gly-Ala Phe Arg-Cys-Leu-Val-Glu-Lys-Gly-Asp-Val-Ala-Phe

              80                    90                   100
Line A    Val Gln His Ser Thr-Val-Glu-Glu Asn Thr-Gly-Gly-Lys Asn Lys-Ala-Asp Trp-Ala-Lys-Asn-Leu Gln-Met
Line B    Val Lys His Asx Thr-Val-Glx-Glx Pro Thr,Gly,Gly,Lys Asx Pro-Asx-Pro Trp-Ala-Lys-Asx-Leu Asp-Glx-
```

Line A : Ovotransferrin fragment BCd plus octapeptide at N-terminus (from Kingston & Williams (15))

Line B : Human serum transferrin fragments CN-3 (residues 1-8), CN-10 (residues 9-42) and CN-1 (residues 43-102). The C-terminal region of CN-3 and the N-terminal region of CN-1 only are shown. The complete sequence of CN-10 is unknown at present.

Fig. 4. A comparison of the sequence of a region from ovotransferrin (15) with regions of sequence of CN-3, CN-10 and CN-1 of human transferrin.

with approximately 40% of the amino acids in corresponding positions being identical. A number of gaps have been inserted in the sequence to maximize the homology. This high level of internal homology reflects a doubling of an ancestral structural gene during the phylogenetic development of the protein, resulting in the production of a transferrin with two metal-binding sites from a precursor with a single metal-binding site.

C. Functional Correlation with Structure

A comparison of the two homologous domains of transferrin reveals features that are of functional and structural interest. Thus, a number of half-cystinyl residues in the two halves are identical, including four in the amino-terminal region of each domain. Studies of the disulfide bond arrangement in CN-B (i.e. the N-terminal 111 residues of transferrin) have revealed a loop structure for this region with Cys^9 joined to Cys^{48} and Cys^{19} bonded with Cys^{39} (14). As the conformations of the two domains would be expected to be closely similar, the same disulfide bond arrangement might be expected to be present in the C-terminal domain, although the loop is shortened by a major deletion of 11 residues.

There are some major differences between the two domains that may be of functional significance. Thus, both glycosylation sites in human transferrin are in the carboxy-terminal domain (residues 415 and 608), which would be expected to produce a marked difference in the surface properties of the two regions. That this may be a universal feature of transferrins is suggested by the report by Williams (8) that the carboxy terminal half of ovotransferrin contains all the carbohydrate. One saccharide moiety is attached to residue 13 of ovotransferrin fragment BCd, as shown in Fig. 4. The corresponding residue in human transferrin (present in CN-10) is not glycosylated, confirming the observation by Graham and Williams (4) that there is no homology around the glycosylation sites of these two proteins.

A second striking feature is the assymmetric distribution of half-cystines between the two domains of the transferrin molecule. In the regions of sequence of the domains that we are able to compare, 16 half-cystines are found in the C-terminal region and only 8 in the N-terminal region. A lower conformational stability of the N-terminal region, indicated by a more rapid loss of iron from the N-terminal domain of ovotransferrin (8) and possibly human transferrin (16) at mildly acidic pH correlates well with this observation.

The necessity for conserving the metal-binding sites and other functional areas in the two homologous domains of the molecule would be expected to restrict the sequence variation around such areas. These areas should be apparent in the internal homology as regions of strongly conserved sequence, with the ligand-donating residues being invariant.

A number of invariant histidyl residues can be observed in the two domains, shown in Fig. 3. However, comparison of the sequences around the histidyl residues of human transferrin with some histidine-containing peptides from ovotransferrin (17) reveals that His^{245} is not conserved in ovotransferrin, so that the invariant histidyl residues at position 249 of Fig. 3 cannot be involved in iron-binding. Comparison of the human protein with the ovotransferrin peptides (17) also indicates that the sequences around His^{124} and His^{256} of Fig. 3 are very highly conserved in the two proteins. In every case, there is an arginyl residue 5 residues C-terminal to the histidyl residue. These arginines may be involved in anion binding (18).

It is also interesting to note that there are 2 tyrosine-rich regions of conserved sequence which are 51 and 57 residues C-terminal to the conserved histidyl/arginyl residues in the N- and C-terminal domains respectively. Thus, the N- and C-terminal domains can both be divided to produce two sub-domains (N-1, N-2, C-1, C-2 respectively), each containing an homologous histidyl residue, arginyl residue and one or two tyrosyl residues. Fig. 5 shows the alignment of the sequences of the four sub-domains around the conserved histidyl, arginyl and tyrosyl residues. Identical residues in corresponding positions are boxed only where the identities include residues

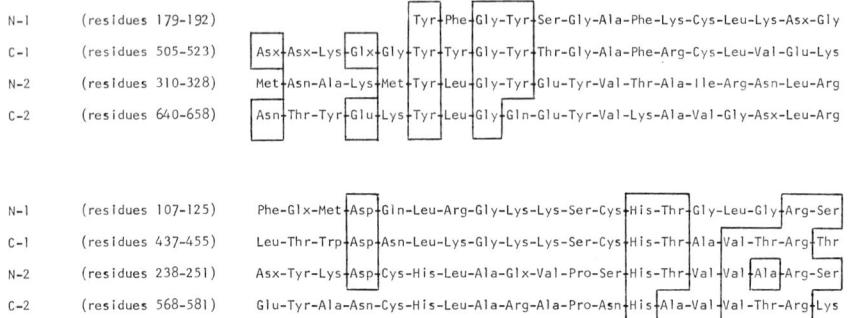

Fig. 5. Comparison of sequences around some conserved histidyl and tyrosyl residues of human transferrin. See text for discussion.

in sub-domains of the same half-molecule. Thus, there may be a weak four-fold homology in the sequence of human transferrin, suggesting that the molecule evolved from a protein of one quarter of its present size, as a result of two duplications of an ancestral structural gene.

The ancestral protein, of low molecular weight (20,000) may have been an iron-binding protein having a less avid iron-binding site, composed of a histidyl residue, one to two tyrosyl residues and an arginyl residue. A gene duplication could then give rise to a larger protein containing an iron-binding site with a very high affinity for ferric iron. The site would be a hybrid of the two weaker iron-binding sites and would be similar in structure to the metal-binding sites in present day transferrins, having two histidyl residues, three tyrosyl residues and two arginyl residues. A later gene duplication then gave rise to the protein of molecular weight 80,000, containing two sites with high affinities for iron.

The evidence for the second gene duplication is indisputable (see Figure 3). Although there is evidence that the more distant gene duplication also took place (Figure 5), the true test of this hypothesis must await the determination of the complete amino acid sequence of transferrin, and subsequent sequence comparisons within the whole polypeptide chain.

This work was supported by grant no. GM 21363 from NIH. KB is the recipient of a Research Career Development Award (K04GM00147) from NIH.

REFERENCES

1. Greene, F.C. and Feeney, R.E., Biochemistry 7, 1366-1371 (1968).
2. Mann, K.G., Fish, W.W., Cox, A.C. and Tanford, C., Biochemistry, 9, 1348-1354 (1970).
3. Sly, D.A. and Bezkorovainy, A., Physiol. Chem. and Physics, 6, 171-177 (1974).
4. Graham, I. and Williams, J., Biochem. J., 145, 263-279 (1975).
5. Guerin, G., Vreeman, H.J. and Nguyen, T.C., Eur. J. Biochem., 67, 433-446 (1976).
6. Jolles, J., Mazurier, J., Boutigue, M.H., Spik, G., Montreuil, J. and Jolles, P., F.E.B.S. Lett., 69, 27-31 (1976).
7. Williams, J., Biochem. J., 141, 745-752 (1974).
8. Williams, J., Biochem. J., 149, 237-244 (1975).
9. MacGillivray, R.T.A. and Brew, K., Science, 190, 1306-1307 (1975).
10. Sutton, M.R. and Brew, K., Biochem. J., 139, 163-168 (1974).
11. Sutton, M.R., MacGillivray, R.T.A. and Brew, K., Eur. J. Biochem., 51, 43-48 (1975).
12. Montreuil, J. and Spik, G., in R.R. Crichton (editor), "Proteins of Iron Storage and Transport in Biochemistry and Medicine", North-Holland, Amsterdam, pp. 27-38 (1975).
13. Sutton, M.R. and Brew, K., F.E.B.S. Lett., 40, 146-148 (1974).
14. MacGillivray, R.T.A., Ph.D. thesis, University of Miami, (1977).
15. Kingston, I.B. and Williams, J., Biochem. J., 147, 463-472 (1975).
16. Harris, D.C., Biochemistry, 16, 560-564 (1977).
17. Elleman, T.C. and Williams, J., Biochem. J., 116, 515-535 (1970).
18. Bates, G.W. and Schlabach, M.R., in R.R. Crichton (editor) "Proteins of Iron Storage and Transport in Biochemistry and Medicine", North-Holland, Amsterdam, pp. 51-58, (1975).

COMPARATIVE STRUCTURAL AND CONFORMATIONAL
STUDIES OF POLYPEPTIDE CHAIN,
CARBOHYDRATE MOIETY AND BINDING SITES OF
HUMAN SEROTRANSFERRIN AND LACTOTRANSFERRIN

GENEVIEVE SPIK
JOEL MAZURIER

Université des Sciences et Techniques de
Lille I ; France

I. INTRODUCTION

In 1975 (1) we have summarized our knowledges about the comparative studies of human serotransferrin (STF) and lactotransferrin (LTF). In this paper, we will describe the new results we have obtained during the last two years about the comparative structural and conformational studies of polypeptide chain, carbohydrate moiety and binding sites of human STF and LTF. These results reveal some similarities between the structure of the polypeptide chains and the carbohydrate groups of the two iron-binding proteins and some differences about the iron binding sites of STF and LTF.

II. CARBOHYDRATE STRUCTURE

A. Serotransferrin

Human serotransferrin contains essentially two identical glycans conjugated to the protein by a GlcNAc-Asn linkage(2,3). Structure of these glycans has been determined by the application of different techniques (4,5). Whereas, as discrepancies exist between the structure we had proposed and the structure given by Jamieson et al. (6) we have investigated in collaboration with H. Vliegenthart's dutch group the structure of the glycopeptides by high-resolution ^{1}H n.m.r. spectroscopy. The signals of the anomeric protons, the mannose-H-2 protons and the N-acetyl methyl groups were analyzed. The 360 MHz ^{1}H n.m.r spectrum of the asialo-glycan-Asn of STF is given in Fig. 1.

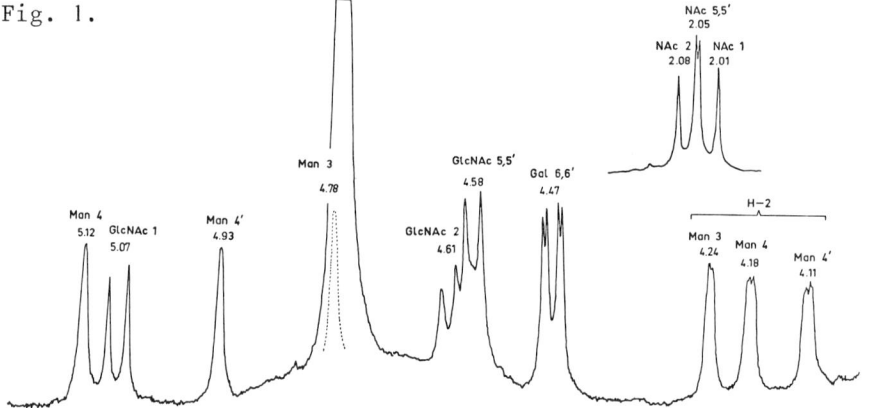

Fig. 1 Resonances in the anomeric region (4.4 - 5.2 ppm) of the 360 MHz ^{1}H n.m.r. spectrum of the asialo-glycan-Asn isolated from human STF (7). The monosaccharide units in this spectrum are numbered in correspondence to the numbering of the monosaccharide units of the Fig. 2.

The complete interpretation of this spectra is given in a recent paper (7). From this spectral data we can concluded that the chemical shifts and the coupling constants of the anomeric protons are characteristic for the nature of the monomer, the type and configuration of the glycosidic bond and the position in the carbohydrate chain. As indicated in Fig. 1, 3 mannose-H-2 signals with characteristic chemical shifts were obtained, giving the indication for mannotriosido-branching-region in the molecule and the confirmation of the structure of the glycan chains we had proposed (Fig. 2).

This "biantennary" structure of the glycan (8) which represents

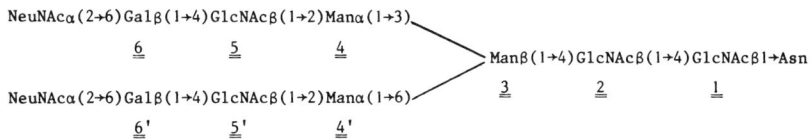

Fig. 2 *Structure of the carbohydrate units of human STF.*

most of the glycan structure found in human STF is able to bind with a Concanavalin A-Sepharose column (9) whereas the small proportion of a "triantennary" structure which results from the substitution of the manno-triosido di-N-acetylchitobiose core by 3 N-acetylneuraminyl-α(2 → 6)-N-acetyllactosamine residues (1,4,8) seems to be eluted from the Concanavalin-A-Sepharose column.

B. Lactotransferrin

The glycopeptides isolated from human LTF (4,10) were also analyzed by 360 MHz proton magnetic resonance spectroscopy (11). The results we obtained confirm that LTF contains 2 "biantennary" glycans with a structure identical with the STF "biantennary" glycans (8). The only difference consists in the presence of additional fucose residues linked by α-1,6 or (and) α-1,3 to N-acetylglucosamine residues. The variability of the fucose residue linkages makes the n.m.r study more difficult to assess without standards containing well known fucose residue linkages.

C. Conclusion

On the basis of the results obtained by 360 MHz ^{1}H n.m.r. analysis of human STF glycopeptides (7) we can confirm the structure we had proposed by the application of chemical techniques (4,5). These results make possible the use of high-resolution ^{1}H n.m.r. spectroscopy as a "fingerprint" method for the analysis of carbohydrate chains of glycoproteins.

III. PROTEIN STRUCTURE

A. Serotransferrin

The primary sequence of human STF has been undertaken by Sutton and Brew (12) and by Charet (13). 9 peptides were isolated by cyanogen bromide cleavage and the structure of 2 of these (CN-7 and CN-9) were published by our laboratory (14). Our results were confirmed by Sutton *et al.* (15) who reported also the sequence of CN-8. Various discrepancies exist between the structure proposed by Sutton and Brew (16) and our results (17) about the N-terminal sequence. The primary structure of this N-terminal fragment is given in Fig. 3.

B. Lactotransferrin

Cyanogen bromide cleavage of human LTF has been described by Mazurier *et al.* (18). An improvement of the purification procedure leads at that time to the isolation of 8 CNBr fragments instead of the 7 previously described. The structure of the supplementary fragment called F-V' (19) which corresponds to the amino terminal region of LTF is given in Fig. 3. This result confirms the presence of glycine as the N-terminal amino acid of human LTF (20).

C. Conclusion

The N-terminal amino-acid sequences of human STF and LTF reported in Fig. 3 have been compared to the amino terminal sequence of hen Ovo-TF (21). The alignement proposed in Fig. 3 demonstrates the extensive similarity between the N-terminal regions of the 3 proteins. However, we cannot conclude that all the polypeptide chains of the transferrins present homologies because the sequences of the polypeptides chains of chymotryptic and tryptic glycopeptides from STF (22,23), LTF (24) and Ovo-TF (25) reveal important differences.

```
           1                                  10                          20                 25
LTF        Gly-Arg-Arg-Arg-Arg-Ser-Val-Gln-Trp-Cys-Ala-Val-Ser-Gln-Pro-Glu-Ala-Thr-Lys-Cys-Phe-Gln-Trp-Gln-   -

STF        Val-Pro-Asp-Lys- -Θ- -Thr-Val-Arg-Trp-Cys-Ala-Val-Ser-Glu-His-Glu-Ala-Thr-Lys-Cys-Gln-Ser-Phe-Arg-Asp-

Ovo-TF     Ala-Pro-Pro-Lys- -Θ- -Ser-Val Arg-Trp-Cys-Thr-Ile-Ser-Ser-Pro-Glu-Gln-Lys-Lys-Cys-Asn-Asn-Leu-Arg-Asp-
                                        \   /
                                         Ile
```

Fig. 3 Homology of the amino terminal regions of human LTF (19), human STF (17) and hen Ovo-TF (21)

IV. METAL BINDING SITES

The nature and the number of amino-acids which participate to the 2 iron-binding sites of human STF and LTF have not yet been completely characterized. It is generally accepted that 2 or 3 Tyr are involved in the complex formation with each Fe^{3+} ion in STF (26-28) and LTF (29-30). By chemical and physical techniques we have investigated the role of His and Trp in the iron-binding sites of STF and LTF.

A. Involvement of Histidine Residues

Diethylpyrocarbonate (D.E.P.) was used as specific reagent for His modification and the number of modified His was determined by difference spectra measurements. The results obtained (31) show that at least 1 and 3 His are involved in the iron binding sites in LTF and in STF respectively. We suppose that these His are located only in one of the iron binding site because after alkylation by D.E.P. of the apoderivatives and resaturation with iron, only one atom of iron is still able to bind to these molecules.

B. Possible Involvement of Tryptophan in the Iron Binding Sites

The circular dichroism studies realized on apoderivatives and on iron-saturated forms of human STF and LTF show that the specific iron binding induces in both proteins a new dichroic band at 305 nm. This band is probably due to the Trp residues which have a disturbed electronic environment (32). The possible participation of Trp has been shown, also, by ultraviolet difference spectral studies performed at 292 nm in the presence of guanidine chloride solution (33). Indeed, the molar absorption differences at 292 nm of iron-saturated forms versus iron-free forms of human STF and LTF are - 15,000 ± 1,000 and - 14,000 ± 775 respectively. These important modifications may be attributed to the involvement of Trp residues in the iron-binding sites of the two proteins. Possible participation of Trp in the iron binding sites has been suggested by Tan and Woodworth about Ovo-TF (34). Recently, difference spectra studies realized by Tomimatsu and Donovan (35) on Ovo-TF and human serum STF saturated with Ga^{3+} and Al^{3+} demonstrated perturbation of Trp and Tyr residues.

C. Conformational Modification of the Proteins upon Iron Binding.

Conformational modifications of the proteins upon iron binding were analyzed by circular dichroism and by ultraviolet difference spectral studies. The obtained data show that α-helical content of apo and iron-saturated LTF is similar (28 and 26 %, respectively) whereas some differences appear in apo and iron-saturated STF (17 and 23 %, respectively)(32). In the presence of increasing molarity of guanidine chloride, the secondary structure of apo-STF appears to be more resistant whereas iron-LTF complex appears to be more stable upon denaturation. These results have been confirmed by UV difference spectra (33) since the guanidine introduces more important unfolding between LTF and apo-LTF than between STF and apo-STF.

D. Conclusion

Circular dichroism and UV difference spectral studies of human STF and LTF suggest that the protein conformation of STF is quite different from that of LTF. The behaviour of one of the iron-binding site of STF from which one iron atom can be removed without alteration of the secondary structure differs from the behaviour of the 2 iron binding sites of LTF from which the two iron atoms are removed simultaneously and with a concomitant unfolding of the protein. The differences between STF and LTF can be explained if we consider that the iron binding sites are located on or near the surface of the STF molecule but more deeply in the LTF molecule.

ACKNOWLEDGEMENT. This work was supported in part by the C.N.R.S. (Laboratoire Associé n° 217) and the D.G.R.S.T. (contrats 75-7-1334 and 75-7-0414).

REFERENCES

1. Montreuil, J., Spik, G., in "Proteins of Iron Storage and Transport in Biochemistry and Medicine" (Crichton, R.R.ed.), p. 27-38. North-Holland, American Elsevier, New York, 1975.
2. Spik, G., Thesis Fac. Sci. Lille (1968)
3. Spik, G., Monsigny, M., Montreuil, J., C.R. Acad. Sci. Paris 260, 4282-4284 (1965)
4. Spik, G., Vandersyppe, R., Fournet, B., Bayard, B., Charet, P., Bouquelet, S., Strecker, G., Montreuil, J., in Montreuil, J. "Actes du Colloque International n° 221 du Centre National de la Recherche Scientifique, Villeneuve d'Ascq, 20-27 juin 1973" (Ed. CNRS), vol. I, pp. 483-499, 1974.
5. Spik, G., Bayard, B., Fournet, B., Strecker, G., Bouquelet, S., Montreuil, J., FEBS-Letters 50, 296-299 (1975)
6. Jamieson, G.A., Jett, M., Debernardo, S.L., J. Biol. Chem. 246, 3686-3693 (1971)
7. Dorland, L., Haverkamp, J., Schut, B.L., Vliengenthart, J.F.G. and Spik, G., Strecker, G., Fournet, B., Montreuil, J., FEBS-Letters (in press)
8. Montreuil, J., Pure Appl. Chem. 42, 431-377 (1975)
9. Krusius, T., Finne, J., Rauvala, H., FEBS-Letters 71, 117-120 (1976)

10. Spik, G., Fournet, B., Bayard, B., Vandersyppe, R., Strecker, G., Bouquelet, S., Charet, P., Montreuil, J., Arch. Inter. Physiol. Biochim. 82, 791 (1974)

11. Spik, G., Debray, R., Fournet, B., Strecker, G., Dorland, L., Vliegenthart, J.F.G., Montreuil, J., Biochimie (in press).

12. Sutton, M.R., Brew, K., Biochem. J. 139, 163-168 (1974)

13. Charet, P., C.R. Acad. Sci. Paris 280D, 2049-2052 (1975)

14. Jollès, J., Charet, P., Jollès, P., Montreuil, J., FEBS-Letters 46, 276-280 (1974)

15. Sutton, M.R., MacGillivray, R.T.A., Brew, K., Eur. J. Biochem. 51, 43-48 (1975)

16. Sutton, M.R., Brew, K., FEBS-Letters 40, 146-148 (1974)

17. Boutigue, M.H., Jollès, J., Charet, P., Montreuil, J., Jollès, P., Biochimie 58, 891-892 (1976)

18. Mazurier, J., Spik, G., Montreuil, J., FEBS-Letters 48, 262-265 (1974)

19. Jollès, J., Mazurier, J., Boutigue, M.H., Spik, G., Montreuil, J., Jollès, P., FEBS-Letters 69, 27-31 (1976)

20. Bluard-Deconinck, J.M., Masson, P.L., Osinski, P.A., Heremans, J.F. Biochim. Biophys. Acta 365, 311-317 (1974)

21. Williams, J., in Bluard-Deconinck, J.M., Osinski, P., Querinjean, P., Masson, P., Heremans, J.F., Proc. 1st Int. Conf. "Solid Phase Methods in Protein Sequence Analysis" (Laursen ed), p. 203-209, Boston (1975)

22. Charet, P., Montreuil, J., C.R. Acad. Sci. Paris, 273D, 533-536 (1971)

23. Graham, I., Williams, J., Biochem. J. 145, 263-279 (1975)

24. Spik, G., Vandersyppe, R., Tetaert, D., Han, K.K., Montreuil, J., FEBS-Letters 38, 213-216 (1974)

25. Elleman, T.C., Williams, J., Biochem. J. 116, 515-535 (1970)

26. Line, W.F., Grohlich, D., Bezkorovainy, A., Biochemistry 6, 3393-3402 (1967)

27. Komatsu, K., Feeney, R.E., Biochemistry 6, 1136-1141 (1967)

28. Aasa, R., Aisen, P., J. Biol. Chem. 243, 2399-2404 (1968)

29. Teuwissen, B., Masson, P.L., Osinski, P., Heremans, J.F., Eur. J. Biochem. 31, 239-245 (1973)

30. Aisen, P., Leibman, A., Biochim. Biophys. Acta 257, 314-323 (1972)
31. Krysteva, M.A., Mazurier, J., Spik, G., Montreuil, J., FEBS-Letters 56, 337-340 (1975)
32. Mazurier, J., Aubert, J.P., Loucheux-Lefevre, M.H., Spik, G., FEBS-Letters 66, 238-242 (1976)
33. Krysteva, M.A., Mazurier, J., Spik, G., Biochim. Biophys. Acta 453, 484-493 (1976)
34. Tan, A.T., Woodworth, R.C., Biochemistry 8, 3711-3716 (1969)
35. Tomimatsu, Y., Donovan, J.W., FEBS-Letters 71, 299-302 (1976)

IRON-BINDING FRAGMENTS OBTAINED BY PROTEOLYSIS OF BOVINE TRANSFERRIN AND LACTOFERRIN

J.H. Brock, Fanny R. Arzabe,*N.E. Richardson, and *E.V. Deverson.

Fundación F. Cuenca Villoro, Zaragoza, Spain.
*Institute of Animal Physiology, Cambridge, England.

I. INTRODUCTION

The structure and function of transferrin and, in particular, of its iron-binding sites have been subjected to much recent investigation, stimulated largely by the proposal of Fletcher and Huehns (1) that these two sites might differ functionally. A major problem in these investigations is the difficulty of studying the properties of one site without interference by the other, and it would clearly be desirable to study each site in isolation. In the past limited proteolysis of multifunctional proteins has been used to advantage to provide monofunctional fragments - notably in the case of immunoglobulins - but this technique has not been widely used for transferrins. This is probably due to the fact that early studies showed iron-saturated transferrins to be largely resistant to proteolysis, whereas the iron-free (apo) proteins were rapidly degraded to inactive material (2). A recent advance has, however, been made by Williams (3,4), who, working with ovotransferrin, showed that under certain conditions it was possible to digest one half of the ovotransferrin molecule yielding a half-molecule fragment containing one iron-binding site.

During studies aimed at investigating the relative susceptibilities of bovine transferrin and lactoferrin to proteolysis we observed (5) that both these proteins, when treated with trypsin, yielded iron-binding fragments. As there appeared to be no previous report of such fragments from a mammalian transferrin or lactoferrin, we decided to undertake their isolation and characterisation.

II. METHODS

Full details are published elsewhere of trypsin digestion (5), the isolation and characterisation of fragments from transferrin (6, J.H. Brock, F.R. Arzabe, N.E. Richardson, E.V. Deverson & A. Feinstein, manuscript in preparation), and chromatography of lactoferrin digests (7). Uptake of iron from transferrin and its fragments by reticulocytes was carried out by a method, described in detail elsewhere (Brock et al. manuscript in preparation), based on those of Workman et al. (8) and of Zapolski and Princiotto (9).

III. RESULTS

A. Fragments from Iron-Saturated Bovine Transferrin.

When bovine Fe_2-transferrin was incubated with trypsin (50:1 w/w) little change in the E_{470} occurred even after 24h incubation. Subsequent analysis of the digest by gel filtration on acrylamide-agarose AcA 34 and electrophoresis in sodium dodecyl sulphate (SDS) polyacrylamide gel revealed, however, that some 40% of the transferrin had been converted to two fragments with molecular weights of approximately 32 000 and 38 500 respectively (5). These fragments were isolated by chromatography on Sephadex G 100 and DEAE Sephadex A 50 (6). When examined by cellulose-acetate electrophoresis in 0.1M Tris-0.3M glycine, pH 8.7, the smaller fragment (F) moved ahead of intact transferrin whereas the larger fragment (S) had a slower mobility than transferrin (Fig. 1). In this buffer system both intact transferrin and fragment S showed

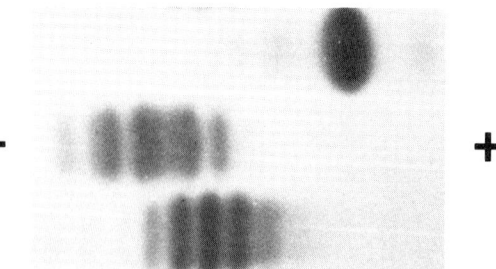

Fig. 1 Cellulose-acetate electrophoresis of bovine transferrin (lower), and fragments F (upper) and S (centre) in 0.1M Tris-0.3M glycine, pH 8.7 for 90 min at 200V. (Reproduced from ref. 6 with permission).

multiple banding. This effect was originally attributed to genetic variants (5), but subsequent studies (J.H. Brock and I. Esparza, unpublished) with purified variants A and D1 of bovine transferrin[1] have shown that the phenomenon is more complex, involving genetic variation, differences in sialic acid content and also a further unexplained heterogeneity (10). Staining of the electrophoresis strips with a specific reagent for iron showed that both fragments contained iron (5), and subsequent studies on the isolated fragments indicated that each fragment bound one ferric ion per mole (6).

The visible absorption spectra of the two fragments differed somewhat, S having a more pronounced maximum at 470 nm than F (Fig. 2a). An equimolar mixture of the two fragments gave, however, a spectrum almost indistinguishable from that of intact transferrin (Fig. 2b).

Ouchterlony gel-diffusion analysis of the fragments

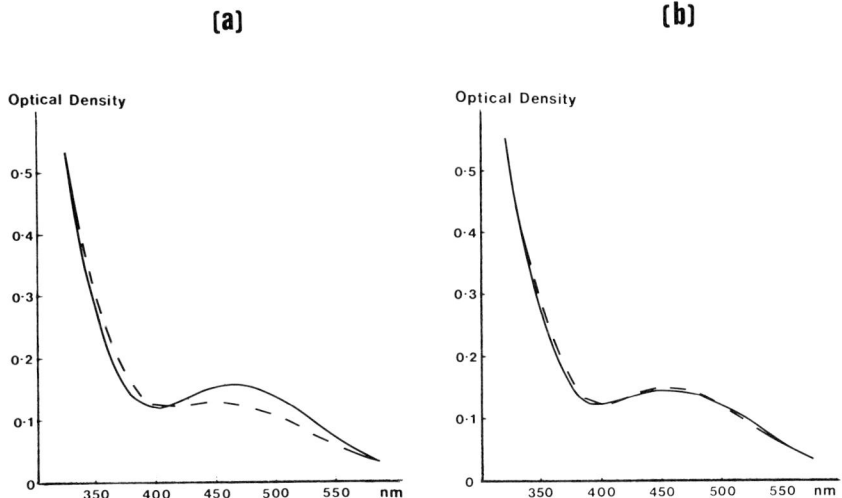

Fig. 2 Visible absorption spectra of (a) fragments F (-----) and S (_____) and (b) of an equimolar mixture of F and S (-----) and intact transferrin (_____). Samples were 2.2 mg/ml in 0.01M Tris-0.01% $NaHCO_3$, pH 7.8. (Reproduced from ref. 6 with permission).

---

1. These homozygous transferrins were kindly provided by Dr. R.L. Spooner, Animal Breeding Research Organisation, Edinburgh.

Proteins of Iron Metabolism

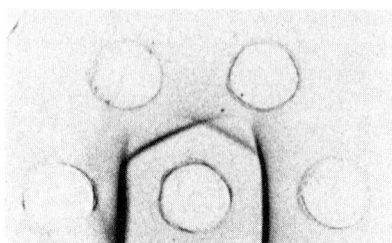

Fig. 3 Gel diffusion pattern of fragment F (upper left), fragment S (upper right), and undigested bovine transferrin (lower left and right) against antiserum to bovine transferrin. (Reproduced from ref. 6 with permission).

using anti-transferrin antiserum showed that the two fragments were immunologically distinct, but each gave a line of partial identity with transferrin (Fig. 3). Absorption of the antiserum with one of the fragments did not affect its reactivity with the other, but absorption with both fragments caused a total loss of activity against the fragments and transferrin. This is similar to the behaviour reported by Williams (4) of fragments derived from the N- and C-terminal halves of ovotransferrin, and suggested thet F and S derived from different halves of the transferrin molecule.

To determine whether fragments F and S did indeed each contain one of the transferrin iron-binding sites, an asymmetrical labelling of the sites with ^{59}Fe was carried out (6), making use of the fact that one site binds iron more readily than the other in the pH range 5.0-5.8 (11). Iron bound at low pH was always associated predominantly with F, thus confirming that each fragment originated from a different iron-binding region, and indicating in addition that fragment F contained the more acid-stable site.

The amino-acid compositions of the two fragments were basically similar (Brock et al., manuscript in preparation) but some differences were observed, S being richer in lysine and threonine, and F richer in proline and phenylalanine. S contained appreciable amounts of carbohydrate (5.8, 3.4 and 4.2 moles of neutral sugars, hexosamines and sialic acid respectively per mole of protein) but F contained only very small amounts of carbohydrate (1 mole per mole for each component) which were probably artifacts, since only S showed modification of electrophoretic mobility after treatment with neuraminidase.

Peptide maps of the fragments and of transferrin revealed little similarity between the fragments. Only 4 out of the 20-25 peptides from F and the 40-45 from S were common to the two fragments (Brock et al., manuscript in preparation), whereas nearly all of the total of 60-70 peptides corresponded to peptides obtained in maps of transferrin.

A comparison of SDS-polyacrylamide gel electrophoresis patterns of the fragments with and without prior reduction by 2-mercaptoethanol showed that fragment F was essentially a single polypeptide chain. Fragment S, however, was heterogeneous, some but not all the material running as faster bands after reduction, indicating that some internal cleavage of the polypeptide chain had occurred.

B.   Fragments from Bovine Apotransferrin.

Bovine apotransferrin was much more susceptible to proteolysis by trypsin than Fe_2-transferrin (5), in agreement with previous findings for human transferrin and ovotransferrin (2,3). However, it was nevertheless possible, after 3h digestion, to detect a fragment with electrophoretic characteristics similar to S, as well as some undegraded transferrin. No fragment corresponding to F could be detected.

The S-like fragment from apotransferrin (hereafter referred to as S_{apo}) has been isolated by chromatography on Sephadex G 100. No differences between this fragment and fragment S (from Fe_2-transferrin) were found in molecular weight, carbohydrate content or immunological properties. However, reduction of S_{apo} by 2-mercaptoethanol followed by SDS-polyacrylamide gel electrophoresis showed that this fragment did not show any evidence of internal cleavages of the polypeptide chain, in contrast to S.

C.   Uptake of Iron by Reticulocytes.

Bovine transferrin was an effective donor of iron to rabbit reticulocytes (Table 1). However, only very slight uptake occurred from fragment S (or S_{apo}), and uptake from F was essentially nil. An equimolar mixture of F and S was also ineffective. Thus cleavage of transferrin causes a loss of ability to interact with reticulocytes even though the iron-binding sites remain intact.

D.   Fragments from Bovine Lactoferrin.

Incubation of bovine Fe_2- and apolactoferrin with trypsin followed by SDS-polyacrylamide gel electrophoresis of the digest revealed, as with transferrin, the formation of large

Table 1   Uptake of ^{59}Fe from Bovine Transferrin and its Tryptic Fragments by Rabbit Reticulocytes

Iron donor	Percentage of ^{59}Fe in cell fraction after 90min at 37°
Transferrin	43.8
Fragment F	1.0
Fragment S	4.8
Fragment S_{apo}	7.0

The cell suspension contained approx. 20% reticulocytes.

fragments. Cleavage of Fe_2-lactoferrin into fragments occurred more readily than with Fe_2-transferrin and, as with transferrin, the apoprotein was much more readily degraded than the iron-saturated form. Although up to 5 different fragments were detected in the lactoferrin digests, only 2 were observed in significant proportions, these having molecular weights of approximately 31 800 and 52 700 (5). The latter fragment tended to disappear and be replaced by a slightly smaller fragment (MW 47 300) as the digestion time was increased. Brief digestion of apolactoferrin gave a qualitatively similar spectrum of fragments, though the fragment of MW 31 800 was relatively less abundant than in Fe_2-lactoferrin digests.

Attempts to isolate these fragments have been hindered by strong non-covalent interactions which cause the fragments to coelute from gel filtration and ion-exchange media, and it has so far been possible to achieve a partial separation only after denaturation and carboxymethylation of the digest (7). This, together with the fact that bovine lactoferrin and some, at least, of its fragments interact with cellulose acetate electrophoresis strips means that it has not been possible to demonstrate directly that the fragments can bind iron. This property can, however, be inferred from the fact that the E_{470} of Fe_2-lactoferrin is not significantly affected by trypsin digestion.

IV.   DISCUSSION

Fragments containing one or other of the two iron-binding sites of transferrins in an intact state have two potential uses: they may enable the properties of each site to be studied in isolation without interference from the other site, and they may be of use in relating iron-binding properties with protein structure. Bovine transferrin appears to be particularly suited to this type of study, since it is possible to obtain simultaneously two monoferric fragments, each containing one of the iron-binding sites, by tryptic cleavage of the iron-saturated protein. In contrast, iron-saturated ovotransferrin and human transferrin are apparently resistant to trypsin (2,3). It thus seems certain, at least for bovine transferrin, that the two iron-binding sites are located in separate regions with no part of the peptide chain of one region being involved in iron-binding by the other, and that no disulphide bridges exist between the two regions.

Since digestion of apotransferrin yields only an S-type fragment, it would appear that the structural destabilisation occuring when iron is lost is greater for the region contained by fragment F than for that contained by S. It may be relevant that S contains most, if not all, of the carbohydrate moiety of transferrin and that F has a higher proline content, which would give proportionally less helical structure and more exposed, protease-susceptible regions.

The failure of either fragment to effectively donate iron to rabbit reticulocytes, even though intact bovine transferrin is an efficient donor, shows that the structural features of the transferrin molecule which are directly involved in iron-binding are not in themselves sufficient to mediate interaction with reticulocytes. Since both F and (when derived from apo - transferrin) S appear to consist of intact polypeptide chains, it is possible that the cleavage of transferrin into two fragments occurs in a region involved in interaction with reticulocytes. The central part of the transferrin polypeptide chain may therefore be a fruitful region to explore for localisation of the site of interaction with reticulocyte receptors.

Regarding bovine lactoferrin, it seems that this protein is certainly no less susceptible to tryptic digestion than transferrin, which is perhaps surprising in view of the fact that its secretory and intracellular localisation must frequently bring it into contact with active proteases. Isolation of tryptic fragments is impeded by their strong non-covalent interaction with each other. Recently the production of a single monoferric fragment from human Fe_2-lactoferrin by partial degradation with pepsin has been reported (12). It seems likely that isolation of biologically active fragments from bovine lactoferrin will also require procedures in which only one iron-binding site remains intact.

V. REFERENCES

1. Fletcher, J., Huehns, E.R., Nature 218, 1211 (1968).

2. Azari, P.R., Feeney, R.E., J. Biol. Chem. 232, 293 (1958).

3. Williams, J., Biochem. J. 141, 745 (1974).

4. Williams, J., Biochem. J. 149, 237 (1975).

5. Brock, J.H., Arzabe, F., Lampreave, F., Piñeiro, A., Biochim. Biophys. Acta 446, 214 (1976).

6. Brock, J.H., Arzabe, F.R., FEBS Lett. 69, 63 (1976).

7. Piñeiro, A., Lampreave, F., Brock, J.H., Arzabe, F.R., Carboné, R., Rev. Españ. Fisiol., submitted.

8. Workman, E.F., Graham, G., Bates, G.W., Biochim. Biophys. Acta 399, 254 (1975).

9. Zapolski, E.J., Princiotto, J.V., Biochim. Biophys. Acta 421, 80 (1976).

10. Richardson, N.E., Buttress, N., Feinstein, A., Stratil, A., Spooner, R.L., Biochem. J. 135, 87 (1973).

11. Princiotto, J.V., Zapolski, E.J., Nature 255, 87 (1975).

12. Line, W.F., Sly, D.A., Bezkorovainy, A., Int. J. Biochem. 7, 203 (1976).

CHEMICAL MODIFICATION OF BASIC AMINO ACID RESIDUES IN THE TRANSFERRINS

T. B. Rogers, R. A. Gold, R. E. Feeney

University of California at Davis

I. INTRODUCTION

Spectroscopic studies of transferrin-metal complexes have shown that at least one and perhaps as many as three nitrogen ligands are furnished by the protein to chelate metals in the binding sites (1,2,3). EPR studies on copper-transferrin complexes indicate that there are as many as four nitrogens available for coordination in the iron binding site (2,3). A number of chemical modification studies have been pursued utilizing reagents specific for nitrogen-containing amino acid side chains in order to determine the nature of the amino acids involved in iron binding. Reductive alkylation and trinitrophenylation of human serum transferrin indicate that the amino groups of lysine are not directly essential for activity (4,5,6). Tryptophan is considered nonessential from the chemical modification studies of human serum transferrin with 2-hydroxy-5-nitrobenzyl bromide (7). However, histidines have been implicated as essential for activity from chemical modification studies using bromoacetate and ethoxyformic anhydride (8,9). Modification of human serum transferrin with bromoacetate suggested two histidines were located in each iron binding site, but ethoxyformylation of human serum transferrin and lactoferrin gave variable results, indicating 1.5 and 0.5 histidines per iron site respectively.

Studies involving arginine are not reported in the literature. One recent report on the sequences of CNBr fragments of human serum transferrin, however, suggests they may play an essential role in transferrin activity because of some invariancy of arginines in the sequence (10). The purpose of this study has been to determine whether histidines (11) and arginines (12) are essential for iron binding.

II. METHODS

The materials and methods of this study are reported in more detail elsewhere (11,12).

Ethoxyformic anhydride reactions with either ovotransferrin or human serum transferrin were run at pH 7.35 in 50 m$\underline{M}$ phosphate. Typically, aliquots of reagent were added to a 3.90 µ$\underline{M}$ protein solution to a final concentration of 0.5 to 3.0 µl per ml. The extent of ethoxyformyl-histidine formation was determined by an increase in A_{242} of the reaction solution ($E_{242} = 3.2 \times 10^3 \ \underline{M}^{-1} cm^{-1}$).

Ovotransferrin was photooxidized in the presence of methylene blue using a Gilson Differential Respirometer as the light source. In a typical experiment, 4.0 ml of a solution containing 6.50 µ$\underline{M}$ ovotransferrin, in 0.05 $\underline{M}$ phosphate buffer, pH 7.35, and 0.05% methylene blue was irradiated at 22°C. The reaction was completed by separating the dye from the protein by gel filtration. Losses in amino acid residues during the course of the reaction were followed by amino acid analysis after hydrolysis in 6 $\underline{N}$ HCl.

The modification of ovotransferrin by phenylglyoxal was accomplished in 0.10 $\underline{M}$ NaHCO$_3$ pH 7.8. Usually, a 0.14 m$\underline{M}$ protein solution was exposed to 17 m$\underline{M}$ phenylglyoxal. Aliquots were removed and the protein was separated from low molecular weight reactants by gel permeation chromatography. Losses in arginine, lysine and histidine were followed by amino acid analysis. The iron-binding activity was determined by a spectrophotometric titration at 470 nm using Fe-NTA solution as the titrant.

III. RESULTS

Initial experiments showed that all of the histidines in both apo-ovotransferrin and apo-human serum transferrin could be modified using 6 m$\underline{M}$ to 18 m$\underline{M}$ ethoxyformic anhydride. Maximum histidine values of 13.3 and 19.5 were obtained for ovotransferrin and human serum transferrin respectively. Figure 1 shows the rate of modification for apo-human serum transferrin.

When 40-50% of the histidines were modified, little iron binding activity remained. The diferric-transferrins were modified under identical conditions as the apo-protein reac-

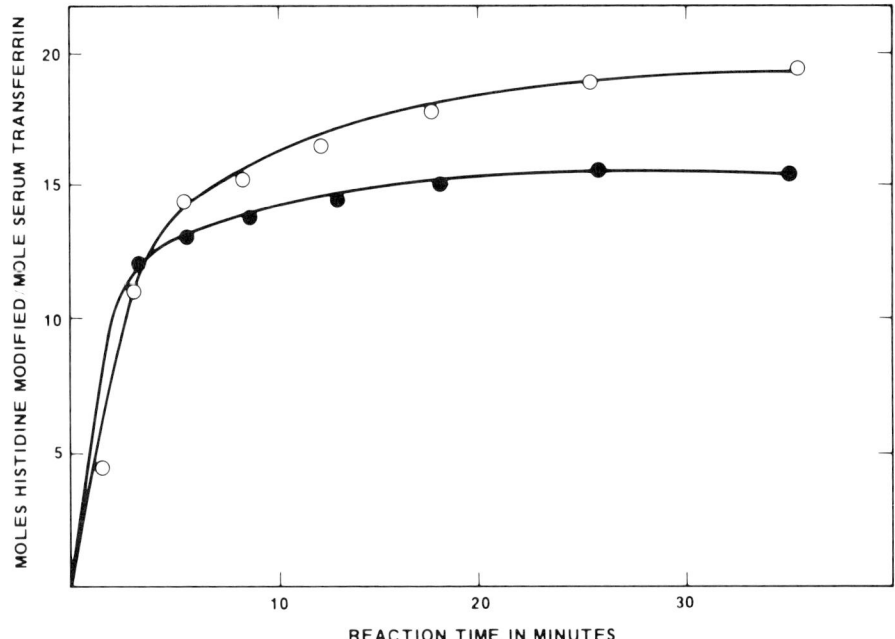

Fig. 1. Relative reactivities of apo-human serum transferrin (○, 3.90 μ$\underline{M}$) and diferric-human serum transferrin (●, 3.90 μ$\underline{M}$) toward ethoxyformic anhydride (2.17 m$\underline{M}$).

tions. The total histidines modified were 9.3 and 15.6 for ovotransferrin and serum transferrin respectively (Fig. 1). In both proteins the binding of each iron protected two histidines from modification.

Modified apo-ovotransferrin samples were exposed to 0.10 $\underline{M}$ NH_2OH in order to regenerate free histidines and iron binding activity. In one case, the activity increased from 76% to 94% incubation with NH_2OH. After longer reaction times only part of the activity is regenerated, increasing from 44% to 67% after NH_2OH treatment.

When apo-ovotransferrin was irradiated in the presence of methylene blue, a first order loss in the amount of histi-

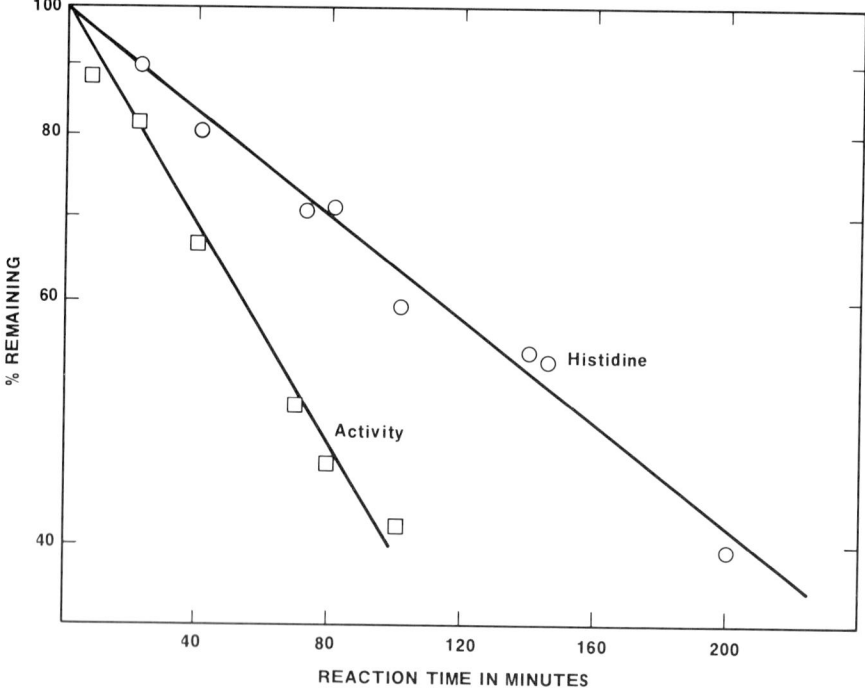

Fig. 2. Semilogarithmic plot of the losses of histidine and iron binding activity upon photooxidation of apo-ovotransferrin. The reaction conditions are described in the text.

dine in the protein was observed (Fig. 2). The rate of loss of tyrosine was very low or essentially zero. When diferric-ovotransferrin was subjected to the same oxidative conditions, the rate of loss of histidine and of tyrosine for the first 60 minutes was essentially the same as for apo-ovotransferrin. There is a first order loss in activity upon irradiation of apo-ovotransferrin as shown in Figure 2. After 100 minutes, approximately 40% of the iron binding activity remained. When the first order rate constants for the loss in amino acid residues and activity for apo-ovotransferrin are calculated from the slopes of the semilog plots in Figure 2 (13), the ratio of the rate of loss of histidine to the rate of loss in activity is 2.2. Analysis of this data suggests that there are two essential histidines per binding site.

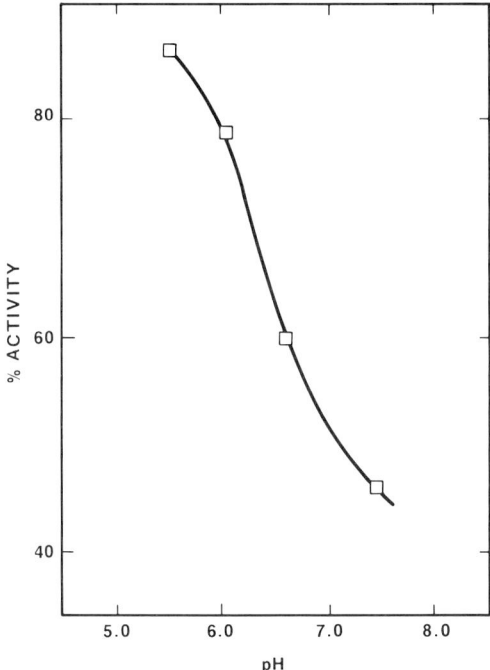

Fig. 3. Effect of pH on photoinactivation of apo-ovotransferrin. Each sample was irradiated for 100 minutes. Acetate buffers (0.05 M) were used from pH 4.5 to 6.5 and phosphate buffers (0.05 M) from pH 6.5 to 8.0.

The effect of pH on the photoinactivation of ovotransferrin was also investigated in order to further associate histidines with the inactivation process. The photoinactivation of apo-ovotransferrin showed a profound pH dependence as shown in Figure 3. This plot has an inflection point around pH 6.4, close to the pK associated with imidazole ring nitrogens of histidine.

When apo-ovotransferrin is exposed to phenylglyoxal, a gradual loss in arginine and activity is observed (Table 1). An eight-fold excess of reagent to arginine was used in these experiments. When 100 molar excess of phenylglyoxal was used, 90% of the arginines were lost in 45 minutes.

TABLE 1

Reactivity of Apo-ovotransferrin toward Phenylglyoxal

Time of Reaction (min.)	Number of Arginines Remaining	% Activity
0	33.0	100.0
5	28.3	93.5
10	24.7	86.5
15	22.5	75.5
30	16.5	49.5
45	13.2	33.0

IV. DISCUSSION

When ovotransferrin was subjected to either ethoxyformylation or dye sensitized photooxidation, a gradual loss in histidine and iron binding activity was observed. The binding of Fe^{3+} protected some of the histidines from reaction with ethoxyformic anhydride, and correspondingly the binding activity was unchanged. When ovotransferrin was irradiated in the presence of methylene blue, iron binding had no effect on the rate of loss of histidine. This is discussed in more detail elsewhere (11).

All of the histidines in apo-ovotransferrin and apo-human serum transferrin reacted with ethoxyformic anhydride. Protection experiments showed that the binding of each atom of iron protected two histidines from ethoxyformylation. In order to demonstrate that the loss in activity could be correlated with a loss in histidine, reversal of the ethoxyformylation of histidines in ovotransferrin was accomplished by incubation with hydroxylamine. These regeneration experiments showed that the removal of the ethoxyformyl group from histidine restored some of the iron binding activity.

Upon irradiation of ovotransferrin, only losses in histidine were observed. Since tyrosines are involved in iron chelation to the protein, the possibility of this side reaction occurring had to be excluded before interpretations on the importance of histidine could be made. The pH profile for the photoinactivation implicates the destruction of histidine in the inactivation process.

A primary objective of this work was the quantitation of the number of essential histidines at the active site of ovotransferrin. The binding of each iron protected two histidines from ethoxyformylation. This represents a maximum number of essential histidines, since the binding of iron may cause a conformational change, thereby protecting histidines some distance from the active center. The analysis of the kinetic data of the photooxidation reaction indicates that two histidines are essential. This was determined by the observation that the rate constant of inactivation reaction was 2.2 times the first order rate constant for the destruction of histidine. This value of two essential histidines obtained from kinetic data represents a minimum number of essential residues. There may be other residues of the same type which are nonreactive or very slowly reacting which will not contribute to the kinetic data for inactivation. Thus, by utilizing two different chemical modification techniques and two distinct methods of analysis of the reaction data, it is shown that two essential histidines are found in each of the iron binding sites of ovotransferrin.

Preliminary results for the modification of arginines in transferrin are reported here. Gradual losses in activity and arginines are seen upon reaction of apo-ovotransferrin with phenylglyoxal. Studies are now in progress in order to further probe the possible essential role of arginines.

V. REFERENCES

1. Spartalian, K., Oosterhuis, W. T., Window, B., Mossbauer Effect Methodology 8, 137 (1973).
2. Windle, J. J., Wiersema, A. K., Clark, J. R., Feeney, R. E., Biochemistry 2, 1341 (1963).
3. Aasa, R., Aisen, P., J. Biol. Chem. 243, 2399 (1968).
4. Means, G. E., Feeney, R. E., Biochemistry 7, 2192 (1968).
5. Zschocke, R. H., Chiao, M. T., Bezkorovainy, A., Eur. J. Biochem. 27, 145 (1972).
6. Buttkus, H., Clark, J. R., Feeney, R. E., Biochemistry 4, 998 (1965).
7. Ford-Hutchinson, A. W., Perkins, D. J., Eur. J. Biochem. 25, 415 (1972).
8. Line, W. F., Grohlich, D., Bezkorovainy, A., Biochemistry 6, 3393 (1967).
9. Krysteva, M. A., Mazurier, J., Spik, G., Montreuil, J., FEBS Lett. 56, 337 (1975).
10. Brew, K., MacGillivray, R. T. A., Science 190, 1306 (1975).
11. Rogers, T. B., Gold, R. A., Feeney, R. E., Biochemistry. In press (1977).
12. Rogers, T. B., Feeney, R. E. In preparation.
13. Ray, W. J., Jr., Koshland, D. E., Jr., J. Biol. Chem. 237, 2493 (1962).

THE IRON BINDING PROPERTY OF OVOTRANSFERRIN

J. Williams and R.W. Evans
University of Bristol

I INTRODUCTION

Recent studies on a variety of transferrins show that the polypeptide chain folds up into two compact regions, each of which has one iron binding site (1-3). These regions have been isolated by limited proteolysis and, in some cases, have been identified as the N-terminal and C-terminal halves of the protein. The two halves may be homologous to one another in an evolutionary sense but they are not identical and many differences in their structure and iron binding properties exist.
The relative strengths of iron binding by the two sites has been studied often and conflicting theories have been proposed. In two early studies (4,5) association constants were determined by equilibrium dialysis and indicated large differences between the first and second iron atoms. Indeed, in the first case it was suggested that the iron atoms bind to the protein in pairs. However, true equilibrium was probably not reached in either study and later work with this method (6-8) showed approximately equal association constants for the two iron atoms. If the sites are equivalent and independent, as suggested, it would be expected that the iron atoms will be distributed at random. These experiments involved high concentrations of iron chelating agents and the conclusion might not hold for different conditions since exchange of bound iron can be brought about by citrate (9). However, isoelectric focusing (10,11) also suggests that iron is randomly distributed, although redistribution can occur during the run (10,12) possibly because of exposure of the protein to low pH. In a chromaotgraphic study (13) which showed an apparently random distribution of iron the metal was added as ferric chloride and allowed to bind to the transferrin as the pH was raised from 5.0 to 7.9. Despite these difficulties it is still widely believed that iron is bound randomly to the two sites of transferrin (14).
Other data, however, are inconsistent with either pairwise or random binding. Two electrophoresis experiments (15,16)

showed absence of diferric transferrin at iron levels below 50%, although in both cases the authors wrongly claimed that their data supported random binding of iron. Calorimetry (17) also showed that when iron, as ferric nitrilotriacetate (FeNTA), is added to an alkaline solution of transferrin the diferric complex is not seen below 50% saturation with iron. Finally, the quenching of protein fluorescence on adding iron to ovotransferrin showed two linear phases instead of the smooth curve expected for random binding (18). In a similar study of serum transferrin (19) the data were interpreted in terms of random binding, but probably could be fitted equally well to two linear curves. Thus, these studies suggest that the first iron atom is bound more readily than the second, due either to strong negative interaction between the sites or to a large difference between the intrinsic association constants of independent sites. Non-random binding of iron to serum transferrin has also been observed under acidic conditions (20).

However, none of the theories described above explain the results of experiments in which the N-terminal half of ovotransferrin was isolated (1). Ovotransferrin in bicarbonate buffer was partially saturated with FeNTA and that protein not protected by iron was digested away with proteolytic enzymes. The maximum yield of the N-terminal fragment was obtained at 30% saturation and no detectable C-terminal fragment was found. It was, therefore, proposed that the first iron atom binds specifically to the N-terminal site. An inexplicable feature of these experiments was the appearance of a large amount of diferric ovotransferrin even at low levels of iron saturation.

In the present paper we have made use of the availability of iron binding fragments to measure the equilibrium distribution of iron in partially saturated ovotransferrin. From these and other results we will propose a new theory of iron binding which will explain the formation of diferric transferrin and reconcile some of the other discrepancies. In particular it will become apparent that iron binding is markedly affected by the nature of the experimental conditions.

II EXPERIMENTAL DATA

In order to determine the distribution of iron a double isotope labelling method was used. Iron-59, in various forms, was added to solutions of ovotransferrin to give different levels of saturation. After 24h. the E-470nm was measured and iron 55 was added to give a full saturation. After a further

24h. unbound iron was removed by passage through sephadex G-100. The C-terminal fragment was then prepared by subtilisin digestion (1). Radioactivity in the whole protein and in the fragment was measured by liquid scintillation counting. Iron-55 was counted in the H-3 channel and iron-59 in the H-3 and C-14 channels. Appropriate standards of diferric labelled ovotransferrin were also counted and the ratio Fe-59/Fe-55 was determined. The isotope ratios were used to calculate the distribution of iron-59 between the N-terminal and C-terminal sites, as follows:

$$\frac{\text{Fe-59 in N-terminal site}}{\text{Fe-59 in C-terminal site}} = \frac{2R + Rr - r}{Rr + r},$$

where R and r represent the ratio Fe-59/Fe-55 in the whole protein and in the C-terminal fragment, respectively. If iron-59 is distributed equally between the two sites the distribution ratio will have the value 1.0 and in the case of unequal distribution it will approach 1.0 as the level of iron-59 approaches 100%.

Unequal distributions were observed under a variety of conditions (Fig. 1). In alkaline solutions iron preferentially binds to the N-terminal site, the inequality being greatest when FeNTA was added to the protein in 0.1M-Tris HCl pH 8.1. In 0.02M-ammonium bicarbonate or 0.1M-sodium bicarbonate the inequality was smaller. N-terminal preference was also seen with Fe(II) ascorbate although it was less marked than with FeNTA.

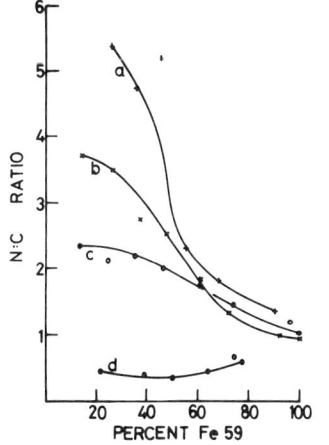

Fig.1. The distribution ratio N:C at different levels of saturation under the following conditions:
(a) FeNTA, 0.1M-Tris HCl pH8.1
(b) FeNTA, 0.02M-NH$_4$HCO$_3$ pH8.4
(c) Fe(II), 0.02M-NH$_4$HCO$_3$
(d) FeNTA, 0.05M-MES 0.02M-NaHCO$_3$ pH6.0.

Iron (FeNTA) was also added to ovotransferrin dissolved in 0.05M-MES [2-N morpholino ethane sulphonic acid] 0.02M-NaHCO$_3$ pH 6.0. At all levels of FeNTA used free and bound iron existed in equilibrium and the unbound complex was removed by passage through sephadex G-100 in pH 6.0 buffer. Further dissociation of iron from the protein was prevented by rapidly raising the pH to 8.1 with ammonium bicarbonate, before completing iron saturation with iron-55. As shown in Fig. 1 the iron bound at pH 6.0 is found mainly in the C-terminal site, which confirms the earlier finding (1) that acid treatment causes preferential loss of iron from the N-terminal site.

However, the isotope ratios do not show the proportion of iron-59 present in monoferric and diferric complexes. Therefore, enzyme digestion was carried out on 30% saturated ovotransferrin and the products separated by gel filtration. Fragments were identified by double diffusion tests using specific rabbit antisera against the N-terminal and C-terminal fragments (1). When FeNTA was added to the protein in 0.1M sodium bicarbonate there was a large peak of undigested diferric transferrin. This peak was much smaller when the buffer was 0.1M-Tris HCl pH 8.1 (Fig.2). Conversely the yield of fragment was much greater in Tris buffer than in bicarbonate.

In order to decide whether the formation of diferric transferrin is facilitated by the presence of bicarbonate or inhibited by the presence of Tris the experiment was performed with 0.1M-Tris HCl 0.1M-sodium bicarbonate pH 8.2 as buffer. After digestion of the 30% saturated protein a large peak of diferric transferrin was seen. Thus, bicarbonate favours the formation of the diferric transferrin. In all these cases the fragment peak consisted of material with only N-terminal specificity.

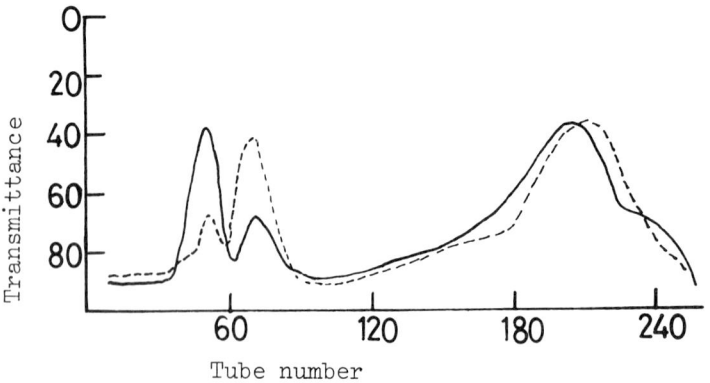

Fig.2. *Gel filtration patterns of enzyme digests of 30% saturated ovotransferrin.* ___ *0.1M-sodium bicarbonate*, ___ *0.1M-Tris HCl pH 8.1*.

In the experiment shown in Fig. 3 solutions of ovotransferrin in bicarbonate buffer received different amounts of FeNTA before being digested with chymotrypsin. Polyacrylamide gel electrophoresis in SDS containing buffer showed that the enzyme-resistant Fe_2OT begins to appear even at the lowest concentrations of iron.

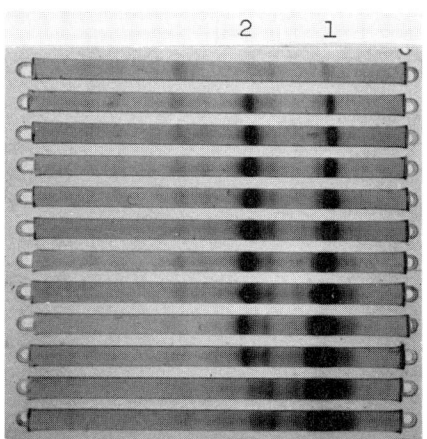

Fig. 3. *Polyacrylamide gel electrophoresis of chymotrypsin digests of ovotransferrin protected by different amounts of FeNTA. From top to bottom the gels represent the following levels of iron: 0, 10%, 20%, 30%, 40%, 50%, 60%, 70%, 80%, 90%, 100% and 200%. Band 1 represents Fe_2OT and band 2 the N-terminal fragment derived from digestion of FeOT.*

In health the blood transferrin is incompletely saturated with iron, the level in the cockerel being about 70% (21). To see whether the distribution of iron in the serum transferrin is similar to that found in the experiments described above the transferrin in a sample of cockerel blood was purified in a single step procedure by affinity chromatography. The adsorbent was sheep anti-ovotransferrin linked to sepharose 4B. As judged by optical density at 280 nm and 470 nm the purified protein was 65-70% saturated. To one sample iron-55 (FeNTA) was added to full saturation and the specific radioactivity of the whole protein and of the C-terminal fragment after subtilisin digestion was determined. The distribution ratio N:C calculated from these figures was 1.47 and the original level of saturation, from the specific radioactivity, was 66%. Digestion of a second sample of the protein gave a small peak of fragment material which contained only the N-terminal fragment. The distribution of iron in the blood in vivo thus appears to be similar to that observed experimentally when

FeNTA is added to transferrin in 0.02M-bicarbonate buffer.
When Fe(II) ascorbate was used to 30% saturate ovotransferrin in bicarbonate buffer the elution pattern after digestion showed a large peak of diferric protein and a smaller fragment peak. The fragment peak, however, possessed both N-terminal and C-terminal antigenic specificities which suggests that in addition to the diferric protein both the monoferric complexes (FeOT and OTFe) are formed. In a single observation with ferric chloride at 30% saturation a distribution ratio close to 1.0 was obtained.

Preliminary experiments in which FeNTA was added to human transferrin in bicarbonate buffer showed markedly unequal distribution of the iron between the sites and it is tentatively suggested that preferential binding occurs at the N-terminal site, as in hen transferrin. However, further work is required on this.

III INTERPRETATION

Consider first the addition of FeNTA to ovotransferrin in bicarbonate buffer. The absence of a C-terminal fragment in the trypsin digestion experiments implies that the only monoferric species present at 30% saturation is FeOT. Nevertheless, the distribution ratio shows that iron is present at the C-terminal site even at low levels of saturation and this iron must be present as diferric transferrin complexes. Thus, the first iron atom binds to the N-terminal site and then makes possible the binding of the second iron at the C-terminal site. Presumably a conformational change occurs in the C-terminal half. We have estimated the relative affinities of the two sites by comparing the observed distribution ratios with the expected values if K_2 was equal to $0.5K_1$, K_1 and $2K_1$ assuming in all cases that the first iron atom must occupy the N-terminal site. From Fig.4, it is seen that K_1 and K_2 must be approximately equal.

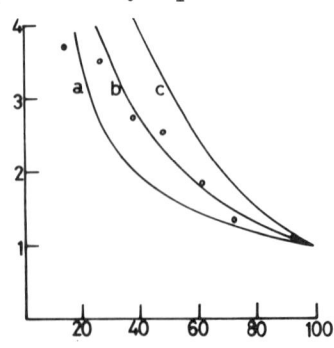

Fig.4. Distribution of FeNTA on ovotransferrin in bicarbonate. Circles represent observed values Theoretical curves represent (a) $K_1 = 0.5K_2$, (b) $K_1 = K_2$ (c) $K_1 = 2K_2$.

In the absence of added bicarbonate the distribution ratio is larger and, therefore, K_2 is smaller than K_1. The iron binding process will then appear to show negative cooperativity although in reality only positive cooperativity is involved since effectively iron does not bind to the C-terminal site until binding has occurred at the N-terminal site.

When ferrous iron is bound to ovotransferrin it prefers the N-terminal site but the pattern of binding is clearly different from FeNTA. Fig.5 shows the agreement between the distribution ratios observed and those expected if iron binds independently to the two sites, with the intrinsic binding affinity of the N-terminal site 2.3 times larger than that of the C-terminal site.

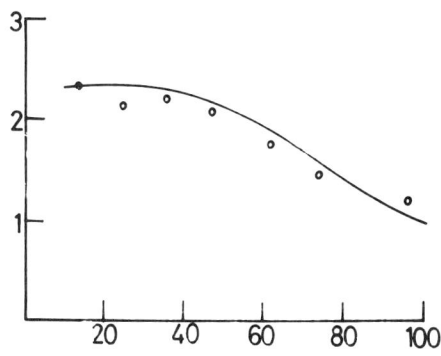

Fig.5. Distribution of Fe(II) on ovotransferrin. Circles are observed values. Theoretical curve represents binding to independent sites with $K_n = 2.3 K_c$.

IV CONCLUSIONS

In a recent review of iron binding by transferrin (14) it was stated that "simple addition of radioiron results in random labelling of the two sites - a fact agreed on by all observers". Our observations emphasise the importance of the experimental conditions in determining the mode of iron binding. Only specially selected conditions give rise to random distribution of iron.

The theory of sequential iron binding which we are proposing can be represented as a two step process:

$$Tf + Fe \underset{K_1}{\rightleftharpoons} FeTf + Fe \underset{K_2}{\rightleftharpoons} FeTfFe$$

and it explains (i) the presence of diferric transferrin at low levels of iron saturation (1) and (ii) the phenomenon of "negative cooperativity" or "anti-cooperativity" described by several authors who employed conditions of low bicarbonate concentration.

In addition to the level of bicarbonate it is clear that pH and the nature of the iron presented to the protein also play critical roles. Although the precise conditions of iron binding in vivo are not known it is likely, at least in the fowl, that the sites of serum transferrin are occupied unequally just as they are when FeNTA is added to ovotransferrin in bicarbonate buffer.

The theory of sequential binding, as given here, disagrees with the theory proposed by Donovan (17) to explain the results of differential scanning calorimetry. According to his theory the binding of FeNTA to ovotransferrin is highly "anti-cooperative" but indiscriminate with respect to N-terminal and C-terminal sites. The observed anti-cooperativity may be explained by the absence of added bicarbonate from the buffer. The claim that iron binds indiscriminately to the two sites depends only on the presence of two intermediate endotherms at sub-saturating levels of iron. These were speculatively identified as being due to the two monoferric species FeOT and OTFe. However, this identification appears doubtful because (a) the immunological evidence given here indicates absence of OTFe during the binding of FeNTA at alkaline pH and (b) iron-free human transferrin also gives two endotherms (Donovan - personal communication) whose significance is not yet understood.

V ACKNOWLEDGEMENTS

We are grateful to the Medical Research Council for financial support in this work and to Mrs. K. Moreton for her excellent assistance.

VI REFERENCES

1. Williams, J., in "Proteins of Iron Storage and Transport in Biochemistry and Medicine" (Crichton, R.R. ed.), p.81 North-Holland, Amsterdam, 1975.

2. Brock, J.H., Arzabe, F., Lampreave, F., Pineiro, A., Biochim. Biophys. Acta 446, 214 (1976).

3. Line, W.F., Sly, D.A., Bezkorovainy, A., Int. J. Biochem. 7, 203 (1976).

4. Warner, R.C., Weber, I., J. Amer. Chem. Soc. 75, 5094 (1953).

5. Davis, B., Saltman, P., Benson, S., Biochem. Biophys. Res. Commun. 8, 56 (1962).

6. Aasa, R., Malmstrom, B.G., Saltman, P., Vanngard, T., Biochim. Biophys. Acta 75, 203 (1963).

7. Aisen, P. Leibman, A., Biochem. Biophys. Res. Commun. 30, 407 (1968)

8. Aisen, P., Leibman, A., Biochim. Biophys. Acta 257, 314 (1972).

9. Aisen, P., Leibman, A., Biochem. Biophys. Res. Commun. 32, 220 (1968)

10. Wenn, R.V., Williams, J., Biochem. J. 108, 69 (1968)

11. Hovanessian, A.G., Awdeh, Z.L., Eur. J. Biochem. 68, 333 (1976).

12. Van Eyk, H.G., Vermaat, R.T., Leijnse, B., FEBS Lett. 3, 193 (1969).

13. Lane, R.S., Brit. J. Haematol. 29, 511 (1975).

14. Aisen, P., Brown, E.B., Seminars in Haematology 14, 31 (1977).

15. Aisen, P., Leibman, A., Reich, H.A., J. Biol. Chem. 241, 1666 (1966).

16. Williams, J., Phelps, C.F., Lowe, J.M., Nature, Lond. 226, 858 (1970).

17. Donovan, J.W., Beardslee, R.A., Ross, K.D., Biochem. J. 153, 631 (1976).

18. Evans, R.W., Holbrook, J.J., Biochem. J. 145, 201 (1975).

19. Lehrer, S.S., J. Biol. Chem. 244, 3613 (1969).

20. Princiotto, J.V., Zapolski, E.J., Nature, Lond. 255, 87 (1975).

21. Planas, J., De Castro, S., Recio, J.M., Nature, Lond. 189, 668 (1961).

DIFFERENCES BETWEEN OVOTRANSFERRIN AND HUMAN SERUM TRANSFERRIN IN STRUCTURAL AND METAL-BINDING COOPERATIVITY

J. W. Donovan

Western Regional Research Center, ARS, USDA

Because a native globular protein is only marginally stable, binding of ligands can significantly alter its denaturation temperature and enthalpy of denaturation. Differential scanning calorimetry (DSC) of transferrins partially saturated with iron show a number of endotherms which have been interpreted as showing the presence in solution, not only of the apo- and diferric-transferrins, but also of species which have a single iron bound (1, 2). These experiments indicate that chicken ovotransferrin (OT) and human serum transferrin (HST) differ. The difference can be interpreted in terms of cooperativity between the two structural units (the "domains") which each contain one binding site (3).

I. CHARACTERISTICS OF THE CALORIMETRIC METHOD

The calorimeter, calibration, experimental procedures and OT preparation have been described (1). The HST, a commercial sample, was used as received, except as noted. The experimental output of the DSC (the "thermogram") shows "peaks" of heat absorption (represented downward for endotherms) at temperatures at which protein species are heat-denatured. A homogeneous protein normally gives a single endotherm. A mixture of non-interacting proteins normally gives the same number of endotherms as the number of species present.

II. DOMAINS OF THE TRANSFERRINS

Covalently-linked domains of some proteins may undergo heat denaturation separately (4). Accordingly, the two endotherms of apo-HST (Fig. 1) may result from denaturation of two domains which have different thermal stabilities as a function

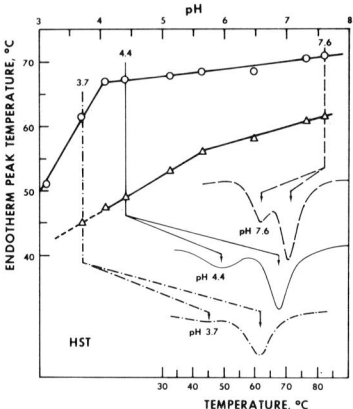

Fig. 1. DSC thermograms of HST and endotherm temperatures of HST as a function of pH. Heating rate, 10°/min.

of pH. Alternatively, the two endotherms could result from denaturation of entirely different molecules. But gross heterogeneity of this HST sample appears unlikely. Not only are impurities (notably hemopexin) present in very small concentrations, but, when this HST is fully saturated with iron, only a single endotherm is observed. However, Mr. J. G. Davis of this laboratory has fractionated this HST (both in the apo- and in the Fe_2-form) into at least four distinct components on DEAE-cellulose at pH 6.5. Each component gives two endotherms for the apo form, and a single endotherm for the Fe_2-form. Other endotherms obtained at partial saturation with iron varied slightly in temperature from fraction to fraction (J. G. Davis, unpublished). Accordingly, although the HST sample shows some heterogeneity, the two endotherms found for the apo forms are those of a homogeneous protein, and thus appear to result from separate heat denaturation of two domains. These domains and endotherms are referred to as the 62° and 72° domains and endotherms, even though the denaturation temperatures are a function of pH and ionic strength. At pH 7.5, heat denaturation of the 62° domain is reversible; the endotherm has the same area when a sample is reheated. This is not true for the 72° domain.

A. Relationships between Binding Sites and Domains

1. *Binding of Aluminum to HST and OT*

The changes observed when metals are added to HST indicate that the two endotherms represent domains containing

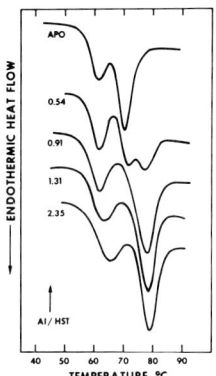

Fig. 2. DSC thermograms for binding of Aluminum to HST in 0.5M tris, pH 8.8. Heating rate, 10°/min.

separate binding sites. At pH 8.8, binding of one Al causes the 72° endotherm to be replaced by one at 78°, while the 62° endotherm is unchanged (Fig. 2). When a second Al is added, the 62° endotherm shifts to 66°, and at the same time the 78° endotherm shifts to 79°. These observations suggest that binding of Al is sequential, first to the 72° domain, then to the 62° domain, and that there is some effect of the binding of the second Al on the thermal stability of the 72° domain. At pH 7.5-8.3, addition of one Al shifts the 72° endotherm to 77°. Addition of a second Al shifts the 62° domain perhaps 1°, depending on pH. In contrast, at pH 7.5, addition of Al raises the single endotherm of apo-OT 5°. Addition of a second Al has no discernible effect, although two Al are bound at the metal-binding sites (5).

Binding of Ga to OT and HST is similar to Al (6). For these two metals, little interaction between domains is apparent for HST. That is, stabilization of one domain does not produce marked stabilization of the other. In contrast, addition of only one Al or Ga to OT stabilizes both domains, and only one endotherm is observed (5, 6).

2. Binding of Iron to HST and OT

Stepwise titrations of OT and HST with FeNTA (NTA: nitrilotriacetate) are shown in Fig. 3. A titration of OT at lower protein concentration has been presented earlier (7). The most marked difference between these sets of thermograms is the presence of two endotherms for apo-HST, and only one for OT. When fully saturated with iron, both OT and HST give one endotherm at 85°-90°. When only one equivalent of Fe is added, both the OT and the HST thermograms show two main

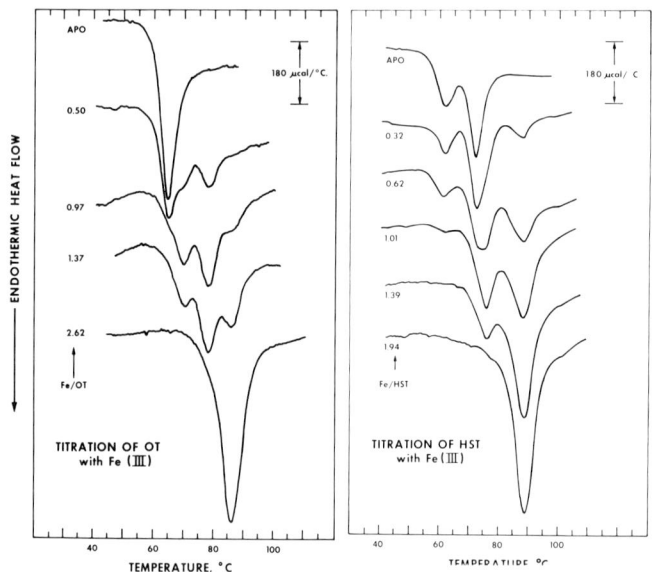

Fig. 3. DSC thermograms for binding of FeNTA to OT and HST, at pH 7.5 in 0.5 M tris, 0.02M CO_2, at a heating rate of 10°/min. OT concentration, 28 mg/ml; HST, 31 mg/ml.

endotherms, characteristic of the monoferric transferrins. However, there are significant differences between the two proteins. The intermediate endotherms observed for OT can occur in any ratio, depending on whether iron is added as Fe^{++}, $FeCl_3$, or at various times after mixing apo- and Fe_2-OT in the presence of NTA (7). These observations, together with a careful material balance for iron and protein (1), indicate that each of these two intermediate endotherms results from heat denaturation of a complete (two-domain) molecule which has one Fe bound to it. Since there are two of these intermediate endotherms, and the rate of conversion of one to the other is increased in the presence of a chelate (7), they appear to result from the denaturation of two different monoferric forms of OT which have iron bound to one or to the other of the two binding sites/domains.

Binding of one Fe to HST alters the denaturation temperatures of both domains, and new major endotherms at 76° and 88° result. One is at the same temperature as the endotherm for Fe_2-HST. However, the thermogram obtained at 1 Fe/HST is not that of a mixture of Fe_2-HST and a single intermediate monoferric species with endotherm at 76°, since the material balance for iron will not allow this interpretation. Accordingly, as with the apo-HST, the two endotherms result from the

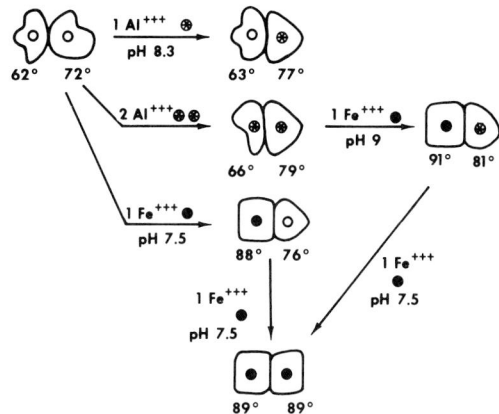

Fig. 4. Model of HST, showing its metal-binding and interdomain interactions. Domains, drawn as different shapes to represent different conformations, have the denaturation temperatures shown. Open circles are empty binding sites. Only one of two possible unsymmetrical species is shown.

separate denaturation of two domains, one of which has an iron bound to it. By analogy with the increased stabilization observed for both members of a protein--protein complex (8), the higher temperature endotherm must result from the domain with attached iron. If this domain were denatured first (at 76°), the other domain, no longer stabilized by interaction with a stabilized domain, would denature simultaneously: its denaturation temperature must be either 62° or 72°. It seems probable that there is mainly only one monoferric HST, otherwise *four* endotherms would be observed for a mixture of two different monoferric species.

In certain cases, two different metals can be bound (5, 9). Displacement by Fe of one of two Al bound to HST gives a thermogram similar to that for monoferric HST. A summary of these experiments for HST is presented in Fig. 4.

B. Decreasing the Interaction between the Domains of OT

Since the binding sites of the transferrins are located on separate domains (3), cooperativity (or anticooperativity) of metal binding requires interaction between the sites and their respective domains (perhaps an artificial distinction, since a binding site is a part of its domain), and interaction between domains. Accordingly, binding cooperativity between sites and structural cooperativity between domains cannot be

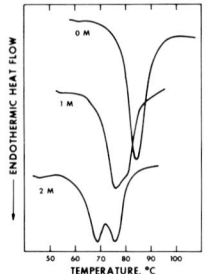

Fig. 5. *Effect of $NaClO_4$ on Fe_2-OT at pH 7.*

separated -- they are two aspects of interdomain interaction.

A fundamental difference between OT and HST is that the interaction between the domains of OT is so great that when one domain is heat-denatured, it "triggers" denaturation of the other, so that the two domains denature simultaneously, like complexes between proteases and their protein inhibitors (8). Although the domains of OT are denatured simultaneously, they must differ, since the two monoferric OT species are denatured at different temperatures. Realizing that ClO_4^- abolishes the difference observed by EPR between the iron-binding sites of OT (3, 10), Dr. R. W. Evans (personal communication) suggested repeating the DSC experiments in perchlorate to observe its effect on the monoferric ovotransferrins. 1M perchlorate lowered the denaturation temperature of all the species, but did not change the relative positions or sizes of the endotherms corresponding to the two Fe-OT species. However, 2M perchlorate "split" the Fe_2-OT endotherm into two endotherms of approximately equal size (Fig. 5). Thus, high concentrations of ClO_4^- appear to decrease the interaction between the two domains so that they are denatured separately.

III. DISCUSSION

Because DSC produces separate endotherms for the different species present in solution, the amounts of the various species (but not the amount of free metal) can be determined from the areas of their respective endotherms. Accordingly. more information about binding can be obtained by DSC than by equilibrium dialysis, for example, as long as *relative* values of the binding constants are adequate. Some cautionary statements: The amount of each species is measured only at its denaturation temperature; it must be assumed (or tested by operating at different heating rates) that the effect of

temperature on the binding equilibrium is minor, or occurs so
slowly relative to the heating process, that the sizes of the
endotherms are a reasonably accurate measure of the concentrations of species present at room temperature. It must be
determined whether an endotherm results from denaturation of
an entire molecule or merely of a domain. Generally, tests
in which the amount of protein, added metal, and areas of
endotherms are all carefully measured and compared to separately determined enthalpies of denaturation of the species
present are required for quantitative measurement of binding
constants.

Relative values of the four *site* binding constants of OT
for iron have been determined (7), and the binding shown to be
anticooperative, but not sequential, with the ratio of the
stoichiometric binding constants, $K_1/K_2 \simeq 40$. Sequential
binding would not allow significant amounts of two Fe-OT species (represented by the two main endotherms observed at Fe/OT
= 1 in Fig. 3) to be present simultaneously. Measurement of
quenching of fluorescence of OT upon binding of iron indicates
$K_1/K_2 > 30$ (11). For HST, binding of iron may be both anti-
cooperative (since very small amounts of apo- and Fe_2-HST are
present at Fe/HST = 1) and sequential (since the two endotherms observed at Fe/HST = 1 appear to be produced by the two
domains of a single species). Relative values of site binding constants for HST have not been calculated because the
enthalpies of denaturation of the two domains of HST are not
known. However, from visual inspection of Fig. 3, K_1/K_2 must
be of the order of 10^3. This is not in agreement with one
series of equilibrium dialysis experiments (12), but does
agree with a study in which EDTA was present (13).

The different behavior of OT and HST is consistent with
a model of two domains which, if separated, would have different thermal stabilities, and site binding constants for iron
which differ by a factor of, say, 10^3 to 10^4. Covalent
connection of domains and some structural cooperativity produced by proximity of these domains could result in "HST behavior": a decreased difference in site binding constants and
some interdomain stabilization, but not enough so that the
domains are heat-denatured simultaneously. "OT behavior"
could result from greater interdomain cooperativity, such that
the thermal stability of the connected domains is like that of
a single protein. This greater structural cooperativity
could also produce a greater degree of cooperativity of iron
binding. Accordingly, the site binding constants might differ by as little as a factor of 10, as observed for OT (7).
The cooperative/anticooperative aspects of metal binding to
transferrins thus result not only from the intrinsic structure
of the domains and their binding sites, but also from the

degree of interdomain cooperativity expressed in each transferrin. Determination of the thermal stabilities and binding constants of the separated domains is desirable; the degree of interdomain cooperativity could thus be established.

Acknowledgments. A number of colleagues, particularly K. D. Ross and R. A. Beardslee, contributed to this work. I take this opportunity to thank J. G. Davis and C. J. Mapes for some of the previously unpublished work described here, and J. A. Garibaldi and Y. Tomimatsu for helpful discussions.

IV. REFERENCES

1. Donovan, J. W., Ross, K. D., *J. Biol. Chem. 250*, 6026 (1975).
2. Donovan, J. W., Ross, K. D., *Fed. Proc. 35*, 1464 (1976).
3. Williams, J., *in* "Proteins of Iron Storage and Transport in Biochemistry and Medicine" (R. R. Crighton, Ed.), p. 81. North-Holland, Amsterdam, 1975.
4. Donovan, J. W., Mihalyi, E., *Proc. Natl. Acad. Sci. USA 71*, 4125 (1974).
5. Donovan, J. W., Ross, K. D., *J. Biol. Chem. 250*, 6022 (1975).
6. Tomimatsu, Y., Donovan, J. W., *this volume*, p.
7. Donovan, J. W., Beardslee, R. A., Ross, K. D., *Biochem. J. 153*, 631 (1976).
8. Donovan, J. W., Beardslee, R. A., *J. Biol. Chem. 250*, 1966 (1975).
9. Aisen, P., Lang, G., Woodworth, R. C., *J. Biol. Chem. 248*, 649 (1973).
10. Price, E. M., Gibson, J. F., *J. Biol. Chem. 247*, 8031 (1972).
11. Evans, R. W., Holbrook, J. J., *Biochem. J. 145*, 201 (1975).
12. Aasa, R., Malmström, B. G., Saltman, P., Vänngård, T., *Biochim. Biophys. Acta 75*, 203 (1963).
13. Davis, B., Saltman, P., Benson, S., *Biochem. Biophys. Res. Commun. 8*, 56 (1962).

VANADYL(IV) LABELLED TRANSFERRIN:

AN OVERVIEW[1]

N. Dennis Chasteen
Robert C. Campbell
Lawrence K. White
J. David Casey

*University of New Hampshire
Durham, N.H., USA*

I. INTRODUCTION

The question of the equivalency of the two metal sites in transferrin has been of considerable interest in recent years. Various spectroscopic and chemical studies have yielded conflicting results (see references 1 and 2 for a review of the literature). Functionally, the sites appear to be equivalent or non-equivalent depending on the system under investigation. For example, it has been found that the two sites of rabbit serum transferrin (RST) are homogeneous with respect to the release of iron to rabbit reticulocycles (3). The same is true of human serum transferrin (HST) when human reticulocytes are employed (3). However, in the mixed human-rabbit system, site inequivalency with respect to iron donation from HST to rabbit reticulocytes is observed (4).

The question of metal site homogeneity is related to the role, if any, of anion in the release of the iron. More specifically, the structural relationship between the metal and anion binding sites has not been clearly established although some models have been proposed (5,6). It has been well established that an anion is required for the metal to bind (7) and that the spectroscopic properties, visible and

---

[1] Supported by the National Institute of General Medical Sciences Grant GM20194-05.

EPR, of the Fe(III) depend markedly on the identity of the anion, e.g., carbonate, oxalate, glycolate, etc. (6,8).

We have undertaken an extensive investigation of the transferrins through the use of the vanadyl(IV) ion, VO(IV), as an electron paramagnetic resonance (EPR) probe of the metal binding sites. We present here a survey of our key findings which provide some insight into the nature of the metal and anion binding sites.

II. EXPERIMENTAL

Conalbumin (CA) (9,10) and RST (11) were isolated according to established procedures. HST (99% pure) was purchased from Behring Diagnostics and used without further purification. Substantially bicarbonate free protein solutions were prepared by the method of Bates and Schlabach (12) except that argon was used instead of nitrogen and the protein was acidified to only pH 4.0 instead of 3.2. Procedures for handling the proteins and introducing the vanadyl ion have been detailed elsewhere (2).

III. RESULTS AND DISCUSSION

A. Conformation States

Figure 1 shows the X-band (9.2 gHz) EPR spectra of frozen solutions of divanadyl HST, RST, and conalbumin. In all cases, the spectrum is a superposition of resonances arising from two or more chemical environments of the metal. The effect of pH on the intensity of the A and B components of the HST spectrum has been examined in detail. The ratio of EPR intensities, I_A/I_B, (measured from the low field parallel line) as a function of pH is shown in Figure 2 (14). At pH 7.5 to 9.0, $I_B/I_A \sim 1$, corresponding to one VO(IV) ion in each environment. However, above pH 9.0, the B resonances rapidly gain intensity at the expense of the A resonances. Finally, at pH 10.2, the spectrum consists nearly completely of B-type resonances. Below pH 7.0 the B signals fall off while the A signals remain relatively constant. At pH 6.0 only one VO(IV) remains bound to the protein (2); it is in the A environment.

The A to B conversion above pH 9.0 can best be understood in terms of interconvertible conformational states of the metal sites of the protein. We have shown that throughout the pH range 7.0 - 10.0, two VO(IV) remain bound (2,13, 14). When the data above pH 9.0 in Figure 2 is cast in the

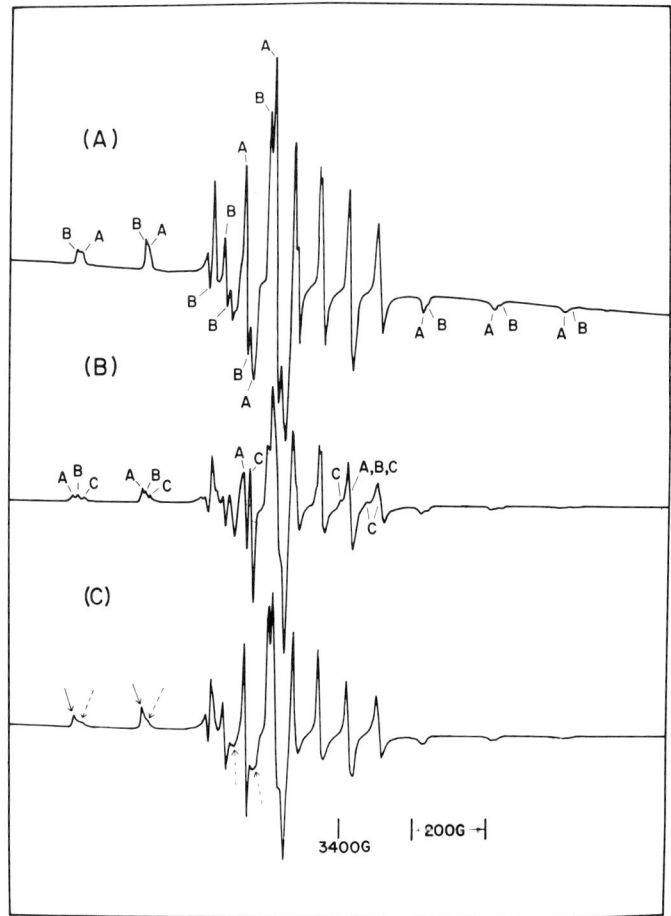

Fig. 1 X-band (9.2 gHz) EPR spectrum of divanadyl (A) HST, pH 8.0, (B) CA, pH 9.0, (C) RST, pH 8.1. Protein concentration nominally 0.4 mM in 0.01 M $NaHCO_3$, 77 K.

form of a Henderson-Hasselbalch plot, one obtains a straight line relationship with a slope of -1.27 and a corresponding pK_a = 10.0 ± 0.1 (14). We conclude that the A to B conformational change is associated with the ionization of a single functional group with an apparent pK_a = 10.0, a value consistent with tyrosine or lysine.

Subsequent experiments at Q-band (35 gHz) have afforded better resolution of the spectrum and reveal the presence of at least three conformations of the metal sites, designated as A, B_1, and B_2 (Figure 3) (15). The resonances from B_1 and B_2 largely overlap at X-band, being designated collectively as B

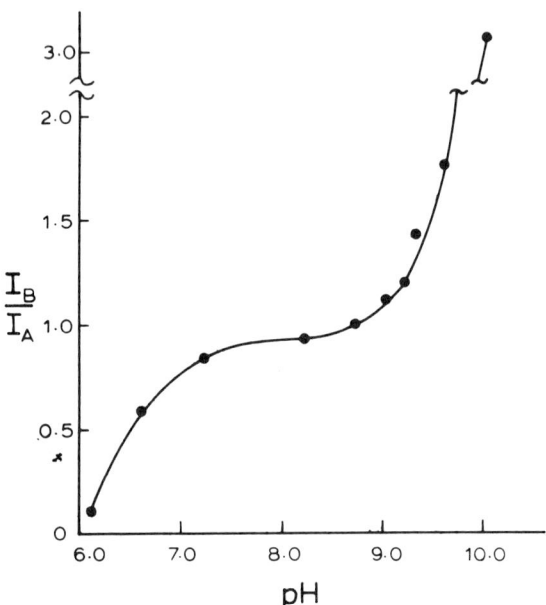

Fig. 2 Plot of the ratio of the first derivative EPR intensities, I_B/I_A, of the $M_I = -7/2$ low-field parallel lines of the A and B conformations as a function of pH. Each point represents a different sample preparation. Divanadyl transferrin concentration typically 0.5 mM in 0.01 M $NaHCO_3$. (Reprinted from reference 14 with permission.)

at that frequency (Figure 1). B_1 and B_2 have the same pH dependence and exist in near equal amounts throughout the pH range 7.5 to 10.0. B_1 and B_2 could possibly arise from two variants of the protein.

Three conformations (designated as A, B, and C in Figure 1) likewise exist for divanadyl CA with resonances resolvable at both X-band and Q-band frequencies (16). However, the relative populations of the A, B, and C conformations change with pH in a manner more complicated than for HST. At pH > 10.7, the X-band spectrum consists of mostly B resonances, while the Q-band spectrum indicates that small amounts of A and C resonances are also present. The pH behavior cannot be described by a simple model involving only one ionizable functional group.

---

[2] We are indebted to Dr. Richard Deming of the University of Vermont for this suggestion.

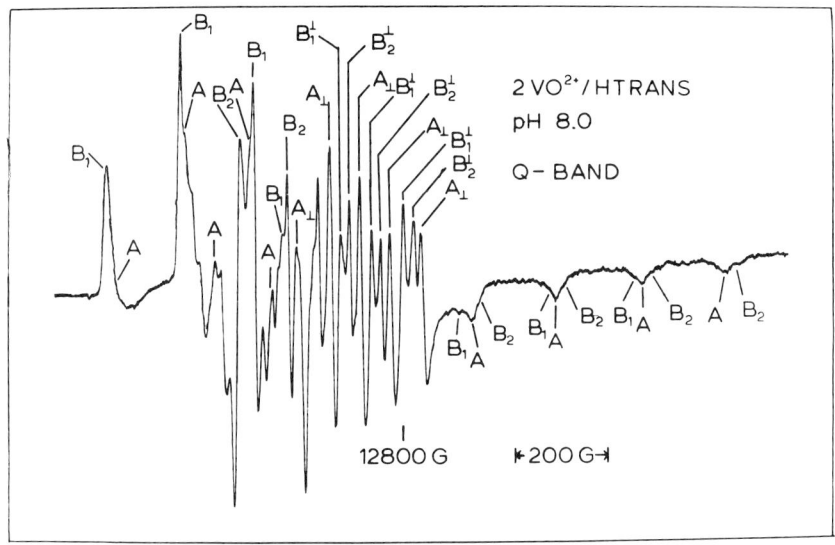

Fig. 3 Q-band (35 gHz) EPR spectrum of divanadyl HST. Conditions as in Figure 1 (A), 100 K.

The overlapping sharp and broad resonances (designated by solid and dashed arrows respectively in Figure 1) of RST are also probably due to different conformational states for this protein; however, they have not been examined in detail. The Q-band spectrum of RST (not shown) suggests possibly three environments of the VO(IV) ion.

Price and Gibson (17) pointed out the possible significance of metal state conformational states in their study of the effect of perchlorate, a chaotropic agent, on the EPR spectrum of iron HST and iron CA. Subsequent studies have shown similar perchlorate effects on the iron binding fragments of CA (18). We have examined the effect of perchlorate on the EPR spectrum of vanadyl HST and CA (2,19). Reported X-band measurements on HST suggested that perchlorate primarily causes a shifting in the positions of the B resonances (2); however, recently obtained Q-band spectra indicate that the situation is more complicated than that (19,20). Concentrations of $NaClO_4$ in excess of 0.4 M cause no further change in the VO(IV) spectrum of HST while the spectrum of CA continues to change up to 1.7 M. This behavior parallels that of iron HST and iron CA (17).

Perchlorate effects on the EPR spectrum of HST is not limited to Fe(III) and VO(IV). We have observed changes in the EPR spectra of Cu(II) and Gd(III) HST attending addition of perchlorate (20). Thus, this phenomena appears to be largely associated with structural changes in the protein which are reflected in changes the metal binding site independent of the identity of the metal. The sensitivity of the metal site toward external factors is further illustrated by the fact that 1M NaCl, which is not normally considered a chaotropic agent, perturbs the EPR spectrum of divanadyl CA (19). Cl^- does, however, bind to the surface of proteins. Moreover, it has been observed that N-2-hydroxyethylpiperazine-N'-2-ethanesulfonic acid (Hepes) buffer markedly sharpens the EPR spectrum of iron (20), copper (20), and vanadyl (21) HST.

It is well known that the transferrins can accomodate a wide variety of metal ions, including many of the lanthanides (1). This fact, plus the observation of several metal site conformational states suggests that there is considerable flexibility in the geometry and/or ligands of the metal site. External agents such as $NaClO_4$ and NaCl presumably act in part by affecting the H bonding, charge, and conformation of the surface of the protein, an effect which is readily transmitted to the metal site. The fact that the metal sites are spectroscopically equivalent or inequivalent depending on the circumstances suggests that the species variation in homogeneity of iron release (see Introduction) could reside in different conformational states of the metal sites. This could be dictated by the configuration of the transferrin binding site on the reticulocyte membrane of the particular species, i.e, rabbit or man, and the way in which it binds HST or RST. Accordingly, it is possible that inequivalence of iron release is only induced when a mixed species system is studied, i.e, rabbit-human, as opposed to human-human or rabbit-rabbit (see Introduction). In this connection, it would be interesting to examine iron donation from RST to human reticulocytes.

B. Anion Binding

Attempts to remove bicarbonate from the HST solution always resulted in an EPR spectrum of much reduced intensity (15% or less of divanadyl HST) after introducing the VO(IV) ion and adjusting the pH with gaseous ammonia (12,22). EPR spectrometric titrations indicate that 0.96 equivalents of $NaHCO_3$ are required per VO(IV) bound to the protein (14).

When other anions were employed in place of carbonate, the EPR spectrum changed dramatically. In order for an anion to be considered synergistic it had to 1) promote VO(VI) binding in excess of that producing the residual signal due to low background bicarbonate and 2) exhibit an anisotropic room-temperature solution spectrum indicative of VO(IV) binding to the protein as opposed to merely binding to the free anion.

Anions were found to fall in three groups: I. Non-synergistic Anions, II. Synergistic Anions, B-Conformation Only, and III. Synergistic Anions, A and B-Conformations. The non-synergistic anions (Group I) included: acetate, ethylenediamine tetraacetate, succinate, and glycine. The Group II anions enhanced only B-signals (designated arbitrarily) and included: lactate, thiosalicylate salicylate, phenyllactate, glycolate, thioglycolate, and glyoxalate. One common feature of Group II anions is that they exist as mononegative species at pH 7-9, consisting of a carboxylate group and a second non-ionized proximal functional group. The resonance positions of the B signals varied with anion. Group III anions enhanced both A and B signals with only the positions of the B resonances showing any anion dependence. Group III consisted of maleate (which was barely synergistic and showed only weak A signals), nitrilotriacetate, malonate, and oxalate. These anions are all dinegative at pH 7-9 due to the presence of two ionized carboxylic acid groups.

Of particular interest is the nature of the interaction between the anion and metal binding sites. To correlate the EPR data with the structure of the anion, we prepared a series of 2:1 ligand/metal model compounds employing ligands of the type $L-CO_2^-$. The variation in the EPR parameter, $A_{\shortparallel}$, for the model compounds largely reflects changes in L since the other coordinating functional is a carboxylate group in all cases. If the anion in the metal-protein-anion ternary complex is directly ligated to the metal through the L group as proposed by Schlabach and Bates (6), then one would expect the variation in $A_{\shortparallel}$ for the ternary complexes to correlate with the variation observed with the model compounds. Indeed, a good correlation is observed for the B-resonances with a correlation coefficient of 0.978 (Figure 4).

We have examined the pH dependence of the loss of VO(IV) binding below pH 7 in some detail (Figure 2). The loss of binding of one VO(IV) (B-conformation) with carbonate as the anion is associated with the protonation of a single functional group with an apparent $pK_a = 6.6 \pm 0.1$ (14). However, when oxalate serves as the anion, two VO(IV) ions bind at pH 6.0 in the A and B-conformations (22). This result suggests that the functional group with the $pK_a = 6.6$ for the carbonate system is quite possibly the anion itself. (The two pK_a's of oxalic acid are much lower than those of carbonic acid, 1.2 and 4.2 versus 6.4 and 10.3.) This idea is attractive in view of the proposal from several groups (see reference 1 for literature citations) that the anion is first attacked, possibly by a protonation step, prior to the release of the iron.

It is perhaps significant that the pK_a of the functional group is elevated considerably from 6.6 in H_2O to 7.8 in 90%

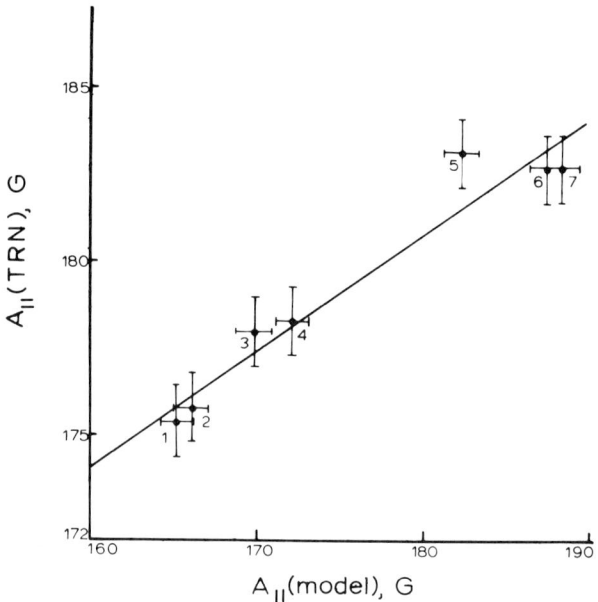

Fig. 4 Correlation of the parallel hyperfine splittings, $\underline{A}_{\prime\prime}$, between VO^{2+}-transferrin-anion complexes and simple model complexes employing the anion as a ligand. Data for the VO^{2+}-transferrin-anion complexes are for the $\underline{B}$ conformation. Linear regression line with a correlation coefficient of 0.978 is shown. 1. thioglycolate 2. thiosalicylate 3. phenyllactate 4. lactate 5. salicylate 6. oxalate 7. malonate.

D_2O. Invariably pK_a's of functional group are elevated by D_2O, but only to the extent of 0.33 to 0.80, units (see reference 14). The much larger than expected D_2O effect for the protein suggests that more is involved in the release of the metal than simply protonation of a functional group. Other structural changes probably occur, such as changes in stability of protein conformation induced by D_2O, which help to trigger the release of the metal. This is consistent with the view proposed in the previous section in which the release of the metal is a function of the protein conformation.

REFERENCES

1. Chasteen, N.D., Coord. Chem. Rev., (1977) in press.

2. Cannon, J.C. and Chasteen, N.D., Biochemistry, 14, 4573 (1975).
3. Harris, D.C. and Aisen, P., Biochemistry, 14, 262 (1975).
4. Harris, D.C. and Aisen, P., Nature, 257, 823 (1975).
5. Harris, D.C. and Aisen, P., in "Proteins of Iron Storage and Transport in Biochemistry and Medicine", (R.R. Crichton, Ed.) p. 59, North Holland Press, Amsterdam, 1975.
6. Schlabach, M.R. and Bates, G.W., J. Biol. Chem., 250, 2182 (1975).
7. Price, E.M. and Gibson, J.F., Biochem. Biophys. Res. Commun., 46, 646 (1972).
8. Aisen, P., Pinkowitz, R.A. and Leibman, A., Ann. N.Y. Acad. Sci., 222, 337 (1973).
9. Woodworth, R.C. and Schade, A.L., Arch. Biochem. Biophys. 82, 78 (1959).
10. Aisen, P., Lang, G., and Woodworth, R.C., J. Biol. Chem., 248, 649 (1973).
11. Baker, E., Shaw, D.C., and Morgan, E.G., Biochemistry, 7, 1371 (1968).
12. Bates, G.W. and Schlabach, M.R., J. Biol. Chem., 250, 2177 (1975).
13. Cannon, J.C. and Chasteen, N.D., in "Proteins of Ion Storage and Transport in Biology and Medicine", (R.R. Crichton, Ed.) p. 67, North Holland Press, Amsterdam, 1975.
14. Chasteen, N.D., White, L.K., and Campbell, R.F., Biochemistry, 16, 363 (1977).
15. White, L.K. and Chasteen, N.D., manuscript in preparation.
16. Casey, J.D. and Chasteen, N.D., manuscript in preparation.
17. Price, E.M. and Gibson, J.F., J. Biol. Chem., 247, 8031 (1972).
18. Butterworth, R.M., Gibson, J.F., and Williams, J., Biochem. J., 149, 559 (1975).
19. Casey, J.D. and Chasteen, N.D., work in progress.
20. White, L.K., and Chasteen, N.D., unpublished observations.
21. Harris, D.C., Biochemistry, 16, 560 (1977).
22. Campbell, R.C. and Chasteen, N.D., submitted for publication.

DIFFERENT PHYSICAL PROPERTIES AND
SIMILAR FUNCTIONAL PROPERTIES OF THE
TWO SITES OF HUMAN TRANSFERRIN

Daniel C. Harris
University of California, Davis

I. INTRODUCTION

The two sites of transferrin are so similar in their metal-binding properties (1) that it is of great interest to know whether or not the sites share the same physiologic function. The suggestion of Fletcher and Huehns (2,3) that the two sites fulfil different roles has inspired a flood of research leading to highly contradictory conclusions (4-8). Recently it was discovered that the pH dependence of iron binding by transferrin is different for the two sites (9,10). One site, designated in this paper as A, binds iron above a pH of approximately five. Site B, however, binds iron only above pH six. The electron paramagnetic resonance (EPR) spectrum of Fe^{3+} bound to transferrin under physiologic conditions does not show any distinction between metals at each site. However, the EPR spectra of vanadyl (VO^{2+}) transferrin (11) and of chromium transferrin (1) are both interpretable as a superposition of the spectra of metals at two distinct locations.

The research described here had three objectives: (1) To confirm the pH dependence of metal binding and to label transferrin at each site with a different isotope of iron. (2) To obtain spectroscopic labels for sites A and B by the correlation of the binding of Fe^{3+}, Cr^{3+}, and VO^{2+}. (3) To test for a difference in function of iron at site A or site B. Complete experimental details have been published elsewhere (12,13).

II. SELECTIVE LABELLING OF THE SITES WITH ^{55}Fe AND ^{59}Fe.

Spectrophotometric titration (monitoring A_{470}) of apotransferrin at pH 6.0 with Fe-NTA (iron-nitrilotriacetate) (7) shows a clear end point after just one Fe^{3+} is bound. To make doubly labelled protein, therefore, one equivalent of ^{59}Fe-NTA was added to transferrin at pH 6.0 in succinate buffer. After 30 min the pH was raised to 7.5 with Hepes

TABLE 1. Results of Double Label Experiments

	Percent of Fe Lost Upon Lowering pH to 6.2-6.3		
^{59}Fe added at pH =	^{55}Fe	^{59}Fe	$\dfrac{^{55}\text{Fe lost}}{^{59}\text{Fe lost}}$
6.0	66	11	6.0
6.0	46	4.1	11
7.5	54	8.6	6.3
7.5	38	3.0	13
7.6*	46	5.0	9.2
7.5**	20	20	1.0

* In this case, ^{59}Fe-nitrilotriacetate was added to a buffer at pH 7.6 prior to adding it to the apoprotein.
** In this control experiment, ^{59}Fe and ^{55}Fe were mixed together before addition to the transferrin,

(N-2-hydroxyethylpiperazine-N'-2-ethanesulfonic acid) buffer and one equivalent of ^{55}Fe-NTA was added. After another 30 min, the solution was passed through a column of Bio-Rad AG1-X4 anion exchange resin at pH 7.5 to remove any unbound Fe-NTA. This procedure was intended to place ^{59}Fe at site A and ^{55}Fe at site B.

To confirm that each site was selectively labelled we sought to remove ^{55}Fe from site B by gently lowering the pH to 6.3 with succinate buffer (pH 6.0). The resulting solution was immediately passed through a small column of Sephadex G-25 to remove dissociated iron. The transferrin, analyzed by liquid scintillation counting, lost approximately half of its ^{55}Fe but only a small fraction of its ^{59}Fe (Table 1, entries 1 and 2).

When the pH was lowered to 6.0 instead of 6.3, and the solution was left for 30 min prior to gel filtration, the two isotopes were completely randomized. It was therefore essential to use the mildest possible conditions to remove the iron. Since some scrambling of the isotopes might occur even under mild conditions, the selectivity observed in our experiments represents a minimal specificity of labelling of the two sites.

We repeated the double labelling experiment, but added both isotopes sequentially at pH 7.5. Much to our surprise, the first isotope went to site A with about the same specificity observed at pH 6.0 (Table 1, entries 3-4). It seemed possible that the first iron went to site A because the Fe-NTA solution has a pH of 4.0. If the droplet of Fe-NTA lowered the pH of the protein prior to iron binding, then only site A would be occupied. To exclude this possibility, ^{59}Fe-NTA was first mixed with buffer at pH 7.6 and then added to the protein. This ^{59}Fe still went mostly to site A (Table 1, entry 5). As a control, both isotopes of Fe-NTA were mixed before addition to the protein. In this case, they were taken up randomly (Table 1, entry 6).

It had been anticipated that Fe-NTA added to transferrin at pH 7.5 would bind to both sites equally. This expectation was based on equilibrium dialysis studies which indicated that the two binding constants were equal (within a large experimental uncertainty) (14). That Fe-NTA added to transferrin went first to site A might be a kinetic effect rather than an equilibrium effect. To test for this possibility, a sample of ^{59}Fe-transferrin in 0.1 M NaClO$_4$-0.01 M Hepes (pH 7.5) left at 4° for 5 months was treated with 1.0 equiv of ^{55}Fe-NTA and any unbound iron was removed by anion exchange. The pH was then lowered to 6.2 and unbound iron was removed. In two different experiments, the ratio ^{55}Fe lost/^{59}Fe lost was 2. Attempts to achieve true equilibrium of iron between the two sites in Fe$_1$-transferrin using citrate as a scrambling agent (15) have so far been unsuccessful.

The conclusion to be drawn from these experiments is that iron goes first to site A at pH 7.5, but only for kinetic reasons. We have yet to establish the true ratio of binding constants at pH 7.5, but it apparently is not greater than about 2:1.

III. SPECTROSCOPIC IDENTIFICATION OF THE SITES

The EPR spectrum of divanadyl transferrin at pH 7.5 in Fig. 1 has been interpreted by Cannon and Chasteen (11) in terms of VO^{2+} ion at two different sites. We will argue below that the labels A and B in Fig. 1 correspond to the same sites A and B already defined for iron binding. At pH 6.0, vanadyl-saturated transferrin gives the mostly A-site spectrum also shown in Fig. 1. (Excess unbound VO^{2+} aggregates and gives no signal in frozen solution.) It has been shown that only one VO^{2+} is bound to transferrin at pH 6.1 (11). We were unable to demonstrate binding of VO^{2+} to monoferric transferrin at pH 6.0, suggesting that Fe^{3+} occupies the same site to which VO^{2+} binds at this pH. When chelate-free (16) monoferric transferrin prepared at pH 7.5 was treated with VO^{2+}, the site B vandyl spectrum in Fig. 2 was observed. Since we showed by isotopic labelling that the Fe^{3+} is mainly at site A in the monoferric transferrin, site B for VO^{2+} must be the same as site B for Fe^{3+}.

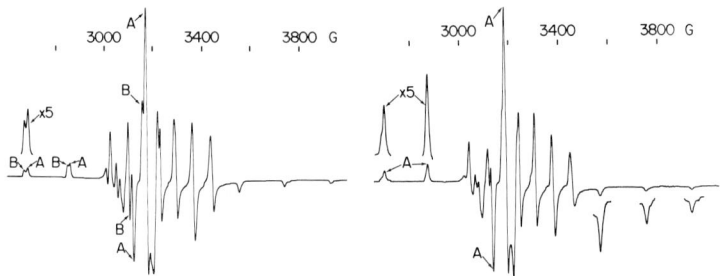

Fig. 1. EPR spectrum of transferrin with a VO^{2+}/protein molar ratio of 2.0. Left: pH 7.5 in 0.1 M Hepes buffer. Right: pH 6.0 in bicarbonate buffer. The site B component is nearly completely absent in a preparation in which VO^{2+}/protein = 1.0.

Dichromium transferrin was prepared at pH 7.7 and the excess Cr^{3+} was removed by gel filtration. The resulting preparation exhibited the signals of type 1 and type 2 Cr^{3+} designated previously (Fig 3a) (17). When the same procedure was carried out at pH 5.9, only the type 2 spectrum was observed (Fig 3b). This implies that only one Cr^{3+} is bound to transferrin at pH 5.9. When VO^{2+} was added to this monochromium transferrin at pH 6.1, the site A VO^{2+} spectrum was observed. Type 2 Cr^{3+} is therefore at site B of the protein. This assignment is supported by the observation that Fe-NTA added to monochromium transferrin at pH 6.0 was readily bound. It was also observed some time ago (17,1) that one equivalent of Fe-NTA added to dichromium transferrin at pH 7.5 selectively displaces type 1 Cr^{3+} from the protein. Since we have

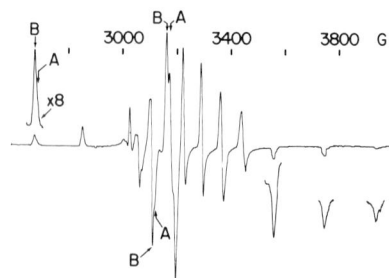

Fig. 2. EPR spectrum of chelate-free monoferric transferrin to which 1.0 mol VO^{2+} was added at pH 7.5.

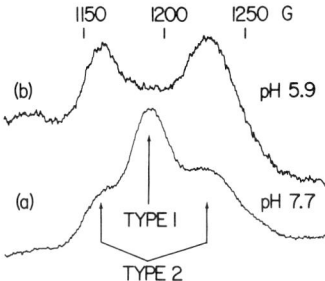

Fig. 3. EPR spectrum of Cr^{3+}-transferrin prepared at (a) pH 7.7 and (b) pH 5.9.

shown that Fe-NTA selectively binds to site A, type 1 Cr^{3+} is probably at site A. Interestingly, Co^{3+} appears to selectively displace type 1 Cr^{3+} at pH 7.5 also (17).

In summary, it appears that Fe^{3+} and VO^{2+} bind selectively to site A at pH 6.0. Cr^{3+} appears to bind at site B under the same conditions.

IV. FUCTIONAL EQUIVALENCE OF THE SITES

In 1967 Fletcher and Huehns presented evidence to show that one site of transferrin is a better donor of iron than the other toward immature red blood cells (2). They suggested that the occupation of each site by metal may serve to regulate iron metabolism. This attractive hypothesis has now been tested by many investigators, with conflicting results. In the most recent approach, transferrin saturated with ^{59}Fe was incubated with reticulocytes until about half of the iron was consumed. If ^{59}Fe had been preferentially removed from one site, then that site would be largely vacant. ^{55}Fe was then used to saturate the protein. This doubly labelled protein was fed to fresh reticulocytes and the uptake of each isotope measured. Experiments with human or rabbit cells and the homologous transferrin failed to discern selective uptake of iron from one site (7,8). In contrast, experiments conducted *in vivo* with rats showed highly selective uptake (4-6).

We labelled human transferrin with ^{59}Fe at site A at pH 6.0 and with ^{55}Fe at site B at pH 7.5. Incubation of this transferrin with reticulocytes from three different patients resulted in random uptake of the two isotopes (Table II). Once again, we failed to observe any difference in function of the two sites with respect to reticulocytes.

Table 2. *Iron Uptake by Human Reticulocytes*

time (min)	Experiment 1		Experiment 2		Experiment 3	
	Total Fe consumed	%^{55}Fe consumed / %^{59}Fe consumed	Total Fe consumed	%^{55}Fe consumed / %^{59}Fe consumed	Total Fe consumed	%^{55}Fe consumed / %^{59}Fe consumed
30	30%	1.02	1.3%	1.03	14%	1.09
60	58%	1.02	2.4%	1.10	26%	1.09
90	78%	1.01	3.8%	1.10	38%	1.09
120	92%	1.00	4.7%	1.12	47%	1.10
150	91%	1.02	6.4%	1.11	55%	1.08
Patient's condition	hemolytic anemia (autoimmune)		hemolytic anemia (autoimmune)		hemolytic anemia (sickle cell disease)	
Percent reticulocytes	44%		6%		9%	

V. DISCUSSION

Perhaps our most interesting finding is that Fe-NTA added to transferrin at pH 7.5 goes first to site A with a high degree of selectivity ($\sim$ 10:1). When such preparations of monoferric transferrin were allowed to stand for 5 months at 4° or were incubated with citrate for 2 days at 37°, the iron was redistributed. We observed as little as a 2:1 ratio of iron in sites A and B. The selective binding to site A therefore appears to be kinetic in nature.

In a recent gel electrophoresis study (18) two monoferric components of transferrin were observed in approximately equal concentrations at pH 8.4 over a wide range or iron saturation. The method of preparing the iron-transferrin was not stated, but the gels contained 50 mM EDTA which might have caused the iron to equilibrate between the sites. Another investigation (19) showed that monoferric and diferric transferrin were formed in the ratio expected on the basis of equal binding to the two sites. In this case iron was added to apoprotein as Fe-NTA and the monoferric and diferric transferrin were separated by isoelectric focussing on slab gels. We feel that isoelectric focussing must be interpreted with caution, because the isoelectric points of the different species are in the range 5.2-5.6. Of considerable interest is the finding (19) that the ratio of diferric transferrin to monoferric transferrin in normal human sera is about seven times *greater* than expected on the basis of random binding to the two sites. This could be a result of a physiologic difference between monoferric and diferric transferrin *in vivo*. It appears that we have much to learn about the kinetics, thermodynamics and physiology of iron binding by transferrin.

VI. SUMMARY

1. One ion of Fe^{3+} or VO^{2+} binds to transferrin at site A at pH 6.0. One ion of Cr^{3+} binds to site B at pH 6.0.

2. When one equivalent of Fe-NTA is added to transferrin at pH 7.5, the iron is bound almost entirely at site A. This is a kinetic effect, for the iron is eventually found at both sites.

3. Human transferrin was labelled with ^{59}Fe at site A at pH 6.0 and with ^{55}Fe at site B at pH 7.5. Both isotopes were equally available for consumption by human reticuloytes.

Acknowledgement. This work was supported in part by the Petroleum Research Fund and by grant HL 20141 from the National Institutes of Health.

VII. REFERENCES

1. Aisen, P. and Brown, E.B., *Prog. Hematology*, 9, 25 (1975).
2. Fletcher, J., and Huehns, E.R., *Nature*, 218, 584 (1967); 218, 1211 (1968).
3. Fletcher, J., *Clin. Sci.*, 37, 273 (1969).
4. Awai, M., Chipman, B., and Brown, E.B., *J. Lab. Clin. Med.*, 85, 769, 785 (1975).
5. Brown, E.B., Okada, S., Awai, M. and Chipman, B., *J. Lab. Clin. Med.*, 86, 576 (1975).
6. Beamish, M.R., Keay, L., Okigaki, T., and Brown, E.B., *Brit. J. Haem.*, 31, 479 (1975).
7. Harris, D.C. and Aisen, P., *Biochem.*, 14, 262 (1975).
8. Harris, D.C. and Aisen, P., *Nature*, 257, 821 (1975).
9. Princiotto, J.V. and Zapolski, E.J., *Nature*, 255, 87 (1975).
10. Lestas, A.N., *Brit. J. Haem.*, 32, 341 (1976).
11. Cannon, J.C. and Chasteen, N.D., *Biochem.*, 14, 4573 (1975).
12. Harris, D.C., *Biochem.*, 16, 560 (1977).
13. Harris, D.C., *Biochim. Biophys. Acta*, (1977), in press.
14. Aasa, R., Malmström, B.G., Saltman, P., and Vänngård, T., *Biochim. Biophys. Acta*, 75, 203 (1963).
15. Aisen, P., and Leibman, A., *Biochem. Biophys. Res. Comm.*, 32, 220 (1968).
16. Bates, G.W., and Schlabach, M.R., *J. Biol. Chem.*, 248, 3228 (1973).
17. Aisen, P., Aasa, R., and Redfield, A.G., *J. Biol. Chem.*, 244, 4628 (1969).
18. Makey, D.G. and Seal, U.S., *Biochim. Biophys. Acta*, 453, 250 (1976).
19. Hovanessian, A.G. and Awdeh, Z.L., *Eur. J. Biochem.*, 68, 333 (1976).

TRANSFERRIN IRON-BINDING: OBSERVATION ON NONRANDOM
(HETEROGENEIC) BINDING

E.J. Zapolski and J.V. Princiotto

Department of Physiology and Biophysics
Georgetown University Schools of Med. & Dent.
Washington, D.C. U.S.A.

1. INTRODUCTION

The two iron-binding sites of human transferrin are described as equivalent and non-interacting, but there is at least one distinct difference between iron-binding at each site. One retains its affinity for iron at acid pH (1,2) but the other, which preferentially surrenders its ferric ion to rabbit reticulocytes (3) cannot.

On the basis of equilibrium dialysis (4), fluorescence quenching (5) and reticulocyte iron uptake (6) data, it was concluded transferrin binds ferric ions randomly, rather than pairwise or by filling one site before the other. The reaction between a ferric complex and transferrin, whereby metal is transferred from complex to protein ligands is not yet completely understood. The chemical nature of the iron complex markedly alters the reaction rate (7). Even for ferric dinitrilotriacetate ($FeNTA_2$), under ideal conditions a rapid iron donor, an equilibrium period of ½ hour is recommended to assure complete occupancy of human transferrin iron-binding sites (8). Recent evidence that both ovotransferrin iron-binding sites are nonequivalent, one site binds ferric ion more rapidly than the other (9), prompted us to investigate iron-binding by human tranferrin using the different pH dependent iron dissociation properties of each site to identify the locus of bound iron.

2. MATERIALS AND METHODS

Human apotransferrin (Behring Diagnostics) was further purified before use (10). Ferric chloride (6) was mixed with a solution of recrystalized nitrilotriacetic acid (NTA) or citric acid and the pH was carefully adjusted to 5.5 - 6.0

with 0.1 M NaOH to prepare 1.0×10^{-4}M iron chelate solutions (1 mole Fe: 2 moles of citrate or NTA). Ferrous-ascorbate (11) was prepared by mixing an acid solution of Mohr's salt (ferrous ammonium sulphate) with 3 molar equivalents of L-ascorbic acid in the same manner. Solutions were tracer labeled by mixing ^{59}FeCl$_3$ with nonradioactive iron solution.

The concentration of transferrin was precisely determined by measurement of ^{59}Fe-iron-binding capacity (1). In our experiments, nontransferrin bound iron was removed from reaction mixtures by passing the solutions through small columns of anion exchange resin (Biorad AG 1x4). Iron specifically bound to transferrin was washed from the columns into eluate while nonspecifically transferrin bound iron, excess iron chelate or ferric ion dissociated from transferrin, was retained by resin.

Fractionally ^{59}Fe-labeled diferric transferrin solutions were prepared for study of iron dissociation at pH 5.8. Enough iron complex was mixed with apotransferrin to achieve 25, 50 or 75% transferrin iron saturation with either ^{59}Fe-labeled or unlabeled iron complex in HEPES buffered-bicarbonate-saline media (pH 7.4). After incubation for 1 hour, nonspecifically bound or unreacted iron was removed on equilibrated resin columns, bicarbonate was replaced and a second increment of iron chelate was added to provide sufficient alternate iron isotope necessary to achieve 110% iron saturation. After equilibration, solutions were passed through resin columns and diluted with bicarbonate media to yield 2.5 uM transferrin solutions. Because of the slow reactivity of ferric dicitrate (7), the second iron increment was added as FeNTA$_2$ with 1 hour incubation or if ferric dicitrate was again employed, the solutions were equilibrated overnight. For each source of iron, four such fractionally ^{59}Fe-labeled diferric transferrin solutions were prepared (A-D). Solutions A and D contained 0.5 mole ^{59}Fe and 1.5 mole Fe/mole of transferrin, differing only in the order of addition of radioiron. In solutions B and C, half the total diferric transferrin bound iron was labeled with ^{59}Fe. (A: 25% ^{59}Fe, 75% Fe; B: 50% ^{59}Fe, 50% Fe; C: 50% Fe, 50% ^{59}Fe; D: 75% Fe, 25% ^{59}Fe). Control, uniformly ^{59}Fe-labeled diferric transferrin solutions, were prepared using 10% excess iron (Solution E).

3. Results

Specific binding of iron to transferrin at pH 7.4 as measured by ^{59}Fe activity recovered in eluate, is illustrated in Fig. 1. Transferrin initially bound nearly 1 mole Fe/mole transferrin from FeNTA$_2$ within 1-2 min (open circles) and then slowly acquired its full complement of metal (2Fe/Tr) by 1 hour. Considerably more iron was initially bound from ferrous-ascorbate (solid circles). At lower temperatures the reactions were somewhat slower. (Not illustrated in Fig. 1). Sample eluates collected from the FeNTA$_2$ iron-binding study (1-5 mins) were saturated with unlabeled iron and these fractionally ^{59}Fe-labeled diferric transferrin solutions were studied by partial dissociation at pH 5.8. From fully ^{59}Fe-labeled transferrin (1 hour sample) 44% of radioactivity was recovered in eluate but between 74% (1 min sample) and 67% (5 min sample) was recovered from test eluates, suggesting that iron from FeNTA$_2$ was initially bound at the site which retains affinity for iron at acid pH.

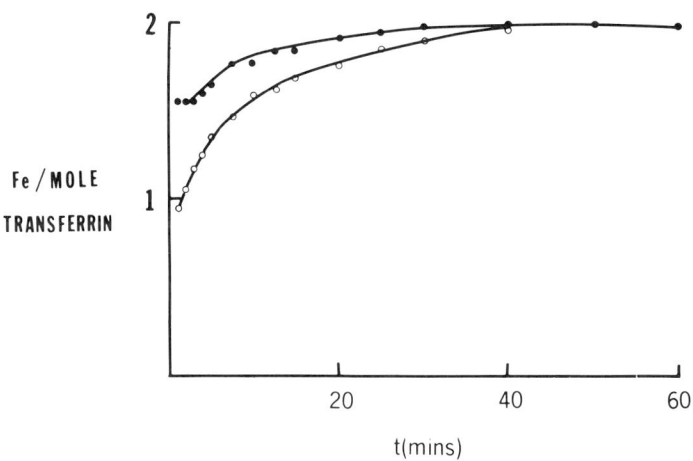

Fig. 1 Iron bound to transferrin measured by elution from anion exchange resin columns. (●): transferrin binding of ferrous ascorbate iron; (o): transferrin binding of FeNTA$_2$ iron. (concentration of reactants : transferrin, 1 x 10^{-5}M; Fe: 2.2 x 10^{-5}M)

The results obtained after partial dissociation of the fractionally ^{59}Fe-labeled diferric transferrins (A-D) and control uniformly labeled diferric transferrin (E) are listed in Table 1 where the average percent ^{59}Fe recovered in eluate is tabulated. Data was tested for significance by analysis of variance (N=6/sample). When ferrous-ascorbate was the iron donating complex, the same eluate percent ^{59}Fe was measured from all fractionally labeled preparations (A-D) and from control (E). Data obtained when ferric dicitrate was the iron source, indicate a greater percentage of radioiron (compared to control E) was still bound to transferrin after partial dissociation at pH 5.8 when ^{59}Fe occupied sites remaining vacant after initial binding of unlabeled citrate iron (C & D). The reverse was observed when FeNTA$_2$ was the iron source (A & B).

TABLE 1. Percent ^{59}Fe activity in eluates after partial dissociation at pH 5.8 of fractionally labeled diferric transferrin solutions

Iron Complex	Diferric transferrin solutions				
	A	B	C	D	E
Ascorbate	43.8a	43.8a	43.0a	43.0a	44.5a
Citrate	29.4	31.4	56.7	60.8	43.2a
NTA	81.9	75.8	15.7	5.1	42.8a

a No significant difference (p= 0.01 or 0.05)

The diferric transferrin preparations (A-E) were stoppered, stored at 4° for 7 weeks, and retested. There were no significant differences between the data obtained initially (Table 1) or 7 weeks later, indicating that no site-site exchange of the bound ferric ions had occurred during this equilibrium period.

4. DISCUSSION

The finding that one human transferrin iron-binding site loses its affinity for iron below pH 6 has been successfully employed to identify this site as the preferential rabbit reticulocyte iron donator (3). There was no a priori reason to suspect human transferrin iron-binding occurred other than randomly. The present study of transferrin iron-binding was initially undertaken simply to determine minimal equilibration time required to prepare iron transferrin complexes. The observation that human transferrin rapidly bound nearly one ferric ion from iron supplied as FeNTA$_2$ suggested that perhaps like ovotransferrin (9), there was a difference in rate of iron-binding by each site. This supposition was strengthened when we studied dissociation of ^{59}Fe from partially labeled diferric transferrin of samples collected during the reaction. Substantially more than 44% of the ^{59}Fe activity remained bound to transferrin in these samples after dissociation at pH 5.8.
We have measured transferrin iron binding by removal of

all nonspecific transferrin bound iron with anion exchange resin. Spectrophotometric measurements of transferrin FeNTA-iron-binding indicate iron-binding is complete within a few seconds after mixing (7,12). This product may be a mixture of transferrins bearing NTA and bicarbonate anions (8). Since absorbance at 470 is greater for pure Fe-NTA-transferrin complex than for bicarbonate complex, (12) the constant absorption at 470 may reflect incomplete occupancy of iron-binding sites and concomittant replacement of anion NTA by HCO_3^- as vacant sites bind iron so that total occupancy of the transferrin iron sites measured by light absorption may not be completed within a few seconds.

For random binding, iron must be equally distributed between the iron-binding sites (6). This must be so, regardless of whether radioisotope was added initially or if it occupied sites that were left vacant after initial binding of unlabeled iron. In samples A and D of our study, 1/4 of the transferrin iron-binding sites were occupied by ^{59}Fe and in B and C, half bore radioiron. These pairs differed only in the order of addition of isotope. Samples prepared in this manner were previously tested by allowing reticulocytes to partially remove iron, and it was concluded that transferrin iron-binding from ferrous ascorbate was random (6). It is clear from the data of the present study (Table 1) that there was no difference in percent eluate ^{59}Fe collected from samples A-D (prepared from ferrous-ascorbate and then dissociated at pH 5.8). Iron from ferrous ascorbate was randomly distributed between sites. Iron from ferric citrate was directed to the site which loses its ability to bind iron at low pH. Samples A and B initially bound ^{59}Fe-dicitrate iron. After partial dissociation, only 30% of the ^{59}Fe activity remained bound to transferrin, whereas 60% was bound in samples C and D. An even more marked polarization of radioactivity was noted when $FeNTA_2$ was the iron source. In this case, iron was directed to the alternate site which retains affinity for iron at low pH. This finding was completely unexpected and is contrary to Lehrer's report that FeNTA iron was randomly bound (5). However, his study gave no direct information as to specific site occupancy, considering only quenching due to one and two iron transferrin species.

Ligands of citrate include possible involvement of three carboxylate and an oxygen with the octahedrally oriented ferric ion. NTA also has three carboxylate ligands but has a nitrogen group instead of oxygen. Possibly the difference in these ligand groups, or their geometry, influences iron exchange from chelate to the protein's ligand groups which may differ at each site (6). Another difference between these ferric complexes is the rapid iron donating rate and ability to form a protein ternary complex that has been observed for NTA(7,12). However, iron-binding rate differences could not account for our observations unless each site exhibited different iron-binding rates for NTA or citrate bound iron. The possibility that our observations are due to NTA or citrate occupancy of the anion site can be ruled out in light of the data obtained from these solutions after 7 weeks equilibration (when media HCO_3^- should displace other bound anion). Citrate is ineffective as a syngergistic anion (13). Our finding that different ferric complexes can direct iron to specific transferrin iron-binding sites is a basis for heterogeneic iron donation *in vivo*.

REFERENCES

1. Princiotto, J.V., Zapolski, E.J., Nature 255, 87 (1975).
2. Lestas, A.N., Brit, J. Haemat. 32, 341 (1976).
3. Princiotto, J.V., Zapolski, E.J., Biochim. Biophys. Acta 428, 766 (1976).
4. Aasa, R., Malmstrom, B.G., Saltman, P., Vannguard, T., Biochim. Biophys. Acta 75, 202 (1963).
5. Lehrer, S.S., J. Biol. Chem. 244, 3613 (1969).
6. Zapolski, E.J., Ganz, R., Princiotto, J.V., Amer. J. Physiol. 226, 334 (1974).
7. Bates, G.W., Billups, C., Saltman, P., J. Biol Chem. 242, 2810 (1967).
8. Bates, G.W., Schlabach, M.R., J. Biol. Chem. 248, 3228 (1973).
9. Evans, R.W., Holbrook, J.J., Biochem, J. 145, 201 (1975).
10. Zapolski, E.J., Princiotto, J.V., Biochim. Biophys. Acta 421, 80 (1976).
11. Masson, P.L., Heremans, J.F., European J. Biochem. 6, 579 (1968).
12. Bates, G.W., Wernicke, J., J. Biol. Chem. 246, 3679 (1971).
13. Schlabach, M.R., Bates, G.W., J. Biol. Chem. 250, 2182 (1975).

HISTIDYL RESIDUES OF TRANSFERRIN AND CONALBUMIN AS NMR[1] REPORTER GROUPS FOR THE BINDING OF HYDROGEN IONS, METAL IONS AND ANIONS

Robert C. Woodworth[2]
Robert J.P. Williams
Basim M. Alsaadi

Department of Biochemistry, University of Vermont
and
Inorganic Chemistry Laboratory, University of Oxford

I. INTRODUCTION

We have previously reported the general effects of the binding of trivalent metal ions on the aromatic region of the proton magnetic resonance spectra of the siderophilins, transferrin from human plasma and conalbumin from egg white (1). In this report we consider the implication of histidyl sidechains in specific metal ion binding (2,3) and anion binding to these proteins and find these residues to be suitable nmr reporter groups for the binding of hydrogen ions, metal ions and anions.

II. MATERIALS AND METHODS

Transferrin (4) and conalbumin (5) were isolated as previously described. Glass distilled water or reagent grade D_2O (99.7 atom % D) was used for making all solutions and dilutions. Chemicals were reagent grade and were used without further purification. Protein samples were exchanged

---

[1] Abbreviations used: pmr - proton magnetic resonance, nmr - nuclear magnetic resonance, TF - transferrin, CA - conalbumin, - chemical shift, ppm - parts per million relative to tetramethylsilane, dipic - 2,6-dipicolinate

[2] Supported in part by Department of HEW Fogarty International Senior Fellowship FOG-00050.

into D_2O by concentration to minimum volume on an Amicon PM-10 membrane followed by dilution with D_2O. This process was repeated two or three times.

Spectra were obtained on the Oxford Enzyme Group's 270 MHz nmr spectrometer operating in the Fourier transform mode. For each spectrum 2048 scans were accumulated and stored on magnetic disc. Transformation of the stored free induction decays, phase correction and plotting of spectra were carried out with a Nicolet 1085 computer and an x-y plotter. Chemical shifts were calculated relative to tetramethylsilane, but with acetone as the internal standard.

Reported pD values are from protein samples in D_2O with a glass electrode standardized with buffers in H_2O at pH 9.2, 6.5 and 4.0.

III. RESULTS

Whereas we previously considered the complete pmr spectra of the siderophilins in question, we confine ourselves here to the chemical shift region of 7.5 to 8.7 ppm, the region of absorption by the C(2)-H's of the imidazole sidechain of histidine (6). For transferrin with 16His, 30Phe, 27Tyr (7) and 12Trp (8) these C(2)-H's comprise 4.6% of the total aromatic hydrogens of the protein. For a sample of transferrin at pD4, all the C(2)-H resonances appeared in a single band centred at 8.56 ppm. The area of this peak was 6.4% relative to the area of the entire aromatic region. The discrepancy from the theoretical value can arise from variations in the relaxation time, T1, among the aromatic hydrogen atoms.
Fig. 1a shows a series of spectra for the C(2)-H histidyl region of transferrin as a function of pD. When the relative areas of this region, corrected for baseline, were calculated for nine different pD values the mean (± std dev) was 6.6± 0.9%, in good agreement with that for the pD4 sample.

The histidyl region for transferrin at pD5.45 (Fig.1a) can be divided into three subgroups, A (8.45 to 8.65 ppm), B (8.25 to 8.45 ppm) and C (8.00 to 8.25 ppm). These groups account for 4, 6 and 6 of the C(2)-H's respectively. As the pD increases these three groups move upfield and nearly coalesce at high pD values. Each group titrates according to pK_a of 6.5±0.1. The residues in group A show a normal pattern of chemical shifts whilst the residues of groups B and C display a compressed range of chemical shifts on titration to low pD. At high pD all C(2)-H's center around a normal value of 7.7 ppm.

In contrast to transferrin, $Ga_2TF(C_2O_4)_2$ and $Fe_2TF(C_2O_4)_2$

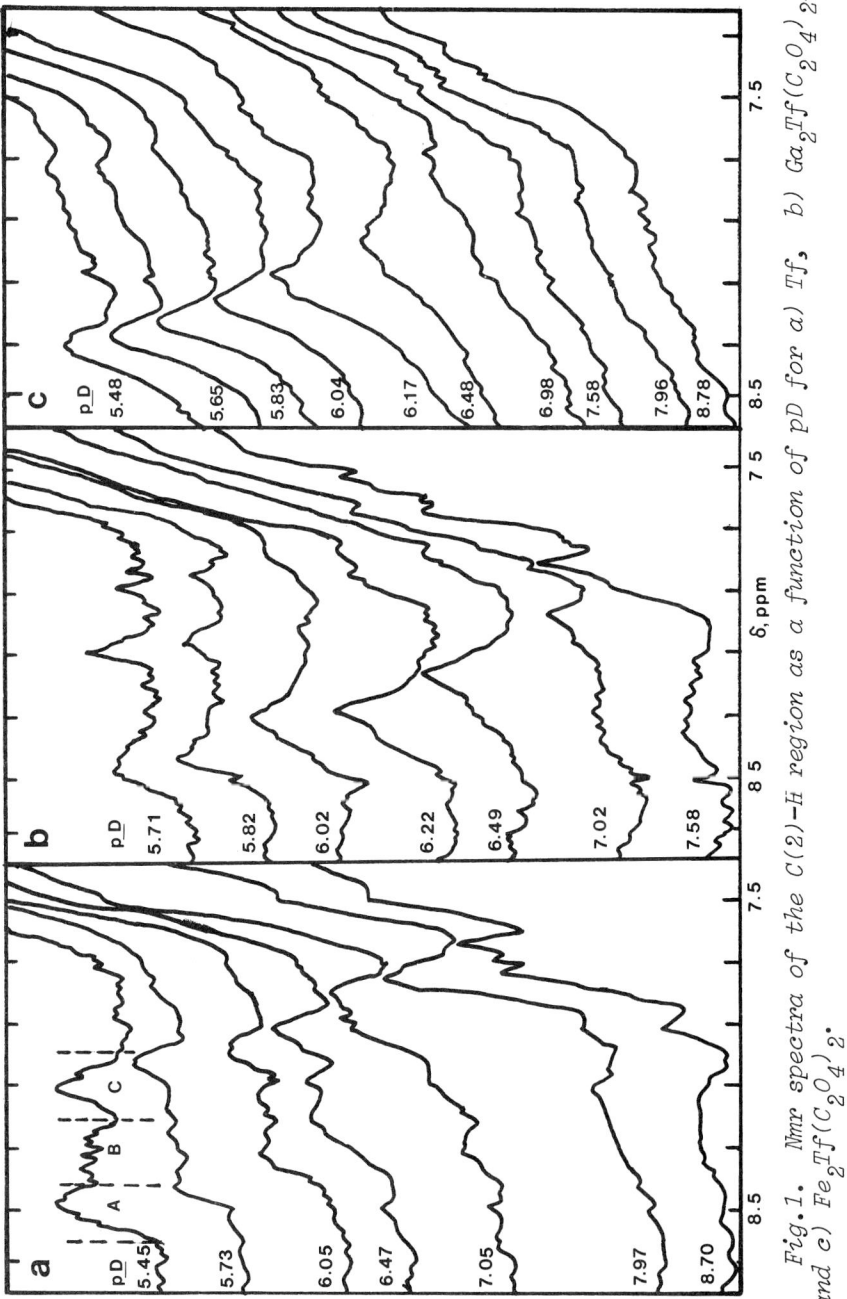

Fig.1. Nmr spectra of the C(2)-H region as a function of pD for a) Tf, b) $Ga_2Tf(C_2O_4)_2$, and c) $Fe_2Tf(C_2O_4)_2$.

gave a single unresolved absorption band in the mid-titration range, pD 6 to 6.5, (Fig. 1b, c) which integrates for 10±1 (C2)-H's. Group C reappears at pD lower than 6 for Ga_2TF $(C_2O_4)_2$ as a single gallium is released. With time at pD5.5 the spectrum of this sample reverted to that of transferrin.

In $Ga_2TF(C_2O_4)_2$ at pD7.58 the C(2)-H resonances have split relative to TF, some appearing at lower field, 7.82 ppm, and others at higher field in the slope of the major aromatic absorption envelope. In $Fe_2TF(C_2O_4)_2$ the C(2)-H resonances are very poorly resolved at high pD. Note also that the baseline for these spectra (Fig.1c) has a pronounced positive slope relative to those for Figs. 1a, b. This slope arose from a machine artifact, not from the sample.

Addition of oxalate to a final concentration of 10mM in a sample of TF at pD6.3 caused a shift of certain C(2)-H resonances to lower field. Similar low field shifts were observed on addition of 2,6-dipicolinate to CA, at a mole ratio of 2:1, in 50mM trideuteroacetate, pD5.7 (Fig.2a,b). The new band at 7.77 ppm arises from free dipic. Addition of diamagnetic Ln·dipic to CA, at a mole ratio of 2:1, resulted in a broadened band for Ln·dipic at 8.12 ppm and shifts in certain C(2)-H resonances. Addition of the shift reagent Pr·dipic to CA, at mole ratios of 1:1 and 2:1, results in similar shifts in the C(2)-H resonances as in the Ln·dipic case plus the appearance of a broad resonance at 12.2 ppm (Fig.2c, d). This latter band is characteristic of that seen for the Pr·(dipic)$_3$ complex. Further addition of Pr·dipic to Pr_2CA (dipic)$_2$ gave rise to resonances at 13.8 and 15.0 ppm characteristic of the p- and m-protons in the free Pr·dipic complex.

IV. DISCUSSION

The nmr titration of transferrin revealed three distinguishable groups of histidine C(2)-H resonances. Group A contains 4 His residues which titrate in a normal manner with a pK_a of 6.5±0.1 and a chemical shift on protonation of ca +0.8 ppm. These histidines probably lie on the surface of the protein. Group B represents 6 His residues with a pK_a of 6.5±0.1 and a somewhat reduced chemical shift of +0.64 ppm. In group C another 6 His residues show a pK_a similar to those of groups A and B, but a chemical shift restricted to 0.47ppm. We conclude that the histidines of groups B and C experience ring current shifts and therefore lie in proximity to other aromatic residues.

Titration of $Ga_2TF(C_2O_4)_2$ and $Fe_2TF(C_2O_4)_2$ complexes reveals that some six His residues fail to titrate in groups B

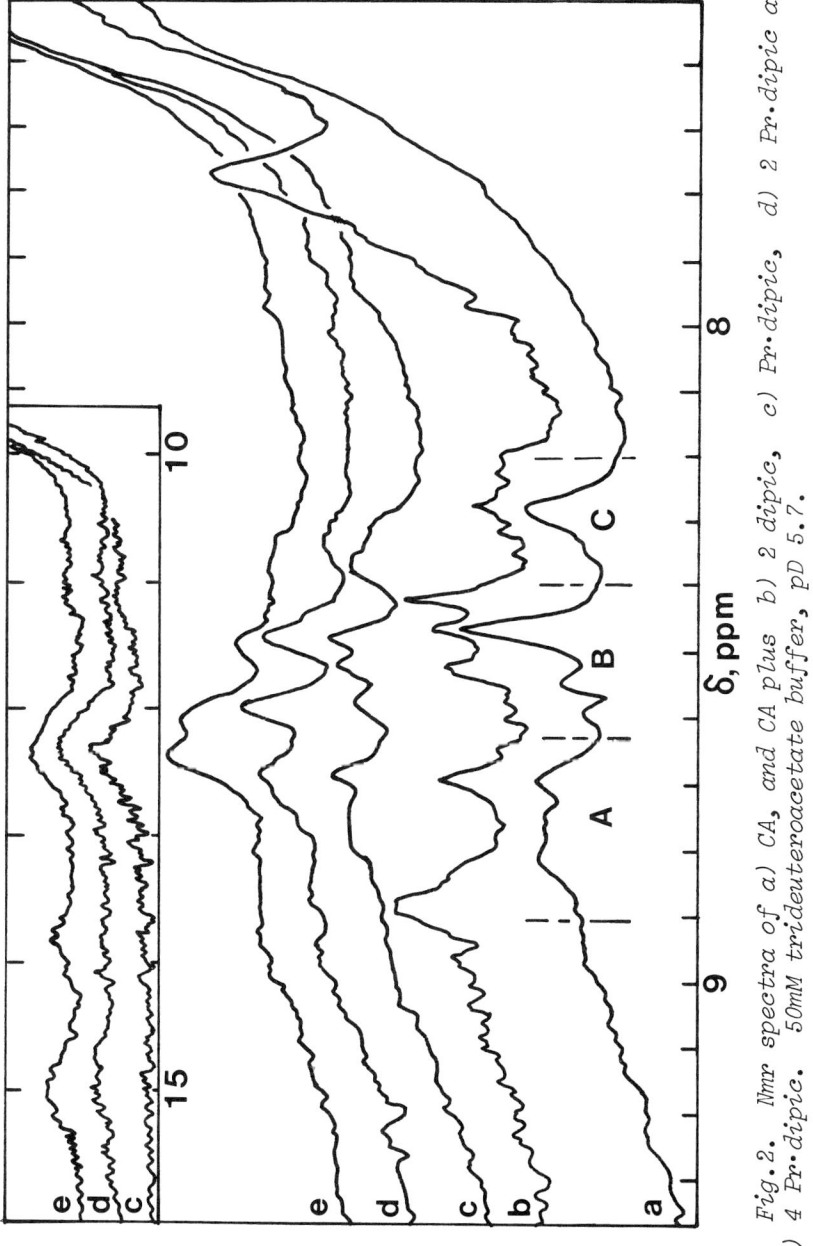

Fig. 2. Nmr spectra of a) CA, and CA plus b) 2 dipic, c) Pr·dipic, d) 2 Pr·dipic and e) 4 Pr·dipic. 50mM trideuteroacetate buffer, pD 5.7.

and C. Because these residues show a ring current shift in the apoprotein and because tyrosyl residues are known to participate in metal ion binding to siderophilins (1,9), we conclude that the six His residues missing from the nmr spectra of the metal complexes are located at the metal binding sites of the proteins, possibly three residues per site.

In $Ga_2Tf(C_2O_4)_2$ the reappearance at low pD of one set of ring-current-shifted C(2)-H resonances before the other is consistent with assignment of histidines of group B to the acid stable metal binding site and of group C to the acid labile metal binding site. In the iron complex coordinating histidine resonances would be expected to be broadened or shifted beyond recognition as is borne out by the nmr spectra (Fig.1c).

The shift to higher field of certain C(2)-H resonances on addition of oxalate to Tf or dipic to CA (Fig.2a,b) at constant pD is consistent with protonation of the affected His residues and with the known simultaneous binding of anions and hydrogen ions to these proteins (10). The addition of a stoichiometric amount of Ln·dipic to CA results in the appearance of the expected resonance at 8.12 ppm for Ln·dipic. However, the band is considerably broadened, which indicates that the entire complex is bound to the protein. Addition of stoichiometric amounts of Pr·dipic to CA results in a resonance at 12.2 ppm, which is characteristic of the slowly exchanging Pr·(dipic)$_3$ complex (Fig.2c,d). Addition of excess Pr·dipic results in appearance of the expected resonances at 13.8 and 15.0 ppm for this complex in free solution (Fig.2e). We conclude that the Pr·dipic complex is bound to the proton as such and exhibits slow exchange behaviour because of additional ligands from the protein to the Pr. From Fig.2 one can also see that as with Tf the C(2)-H resonances of CA fall into three groups. As Pr·dipic is added the ring-current-shifted resonances in groups B and C decrease in relative intensity, thereby implicating them in the metal binding process.

On the basis of the findings reported herein we propose a model for the specific metal ion/anion binding sites of the sideophilins (Fig.3). We propose that in addition to two tyrosyl residues (9) each site contains three histidyl residues, one involved in anion binding, and two possibly coordinated to the metal ion. From the studies with Pr·dipic it appears that the disposition of anion to metal ion in the ternary complex with protein must be similar to that in free solution.

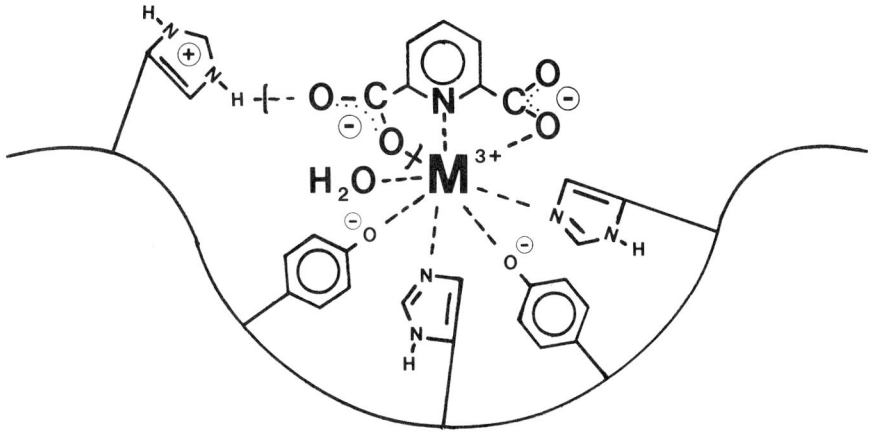

Fig.3. Model for metal ion and anion binding to siderophilins. The bracketed bonds from the left hand carboxyl group to the histidine and to the metal ion indicate variability in these bonds according to the pH and to the particular metal ion and anion involved in the ternary complex.

REFERENCES

1. Woodworth, R.C., Morallee, K.G., Williams, R.J.P., Biochemistry 9, 839 (1970).

2. Aasa, R., Malmström, B.G., Saltman, P., Vänngård, T., Biochim. Biophys. Acta 75, 203 (1963).

3. Line, W.F., Grohlich, D., Bezkorovainy, A., Biochemistry 6, 3393 (1967).

4. Woodworth, R.C., Balin, N.A., in "Proteins of Iron Storage and Transport in Biochemistry and Medicine" (Crichton, R.R. ed.), p.75, Fig.1, North-Holland, Amsterdam, 1975.

5. Woodworth, R.C., Schade, A.L., Arch. Biochem. Biophys. 82, 78 (1959).

6. Dwek, R.A., "Nuclear Magnetic Resonance in Biochemistry: Applications to Enzyme Systems," Ch.5, Oxford University Press, Oxford, 1975.

7. Montreuil, J., Spik, G., in "Proteins of Iron Storage and Transport in Biochemistry and Medicine," (Crichton, R.R. ed.), p.32, North-Holland, Amsterdam, 1975.

8. Woodbury, R.G., Ph.D. Thesis, University of Vermont, p.89 (1974).

9. Tan, A.T., Woodworth, R.C., Biochemistry 8, 3711 (1969).

10. Woodworth, R.C., Virkaitis, L.M., Woodbury, R.G., Fava, R.A., in "Proteins of Iron Storage and Transport in Biochemistry and Medicine" (Crichton, R.R. ed.), p.42, North-Holland, Amsterdam, 1975.

Section V

TRANSFERRIN FUNCTION

SPECTROPHOTOMETRIC AND DIFFERENTIAL SCANNING CALORIMETRIC MEASUREMENTS OF Zn(II), Al(III) AND Ga(III) BINDING TO OVO- AND HUMAN SERUM TRANSFERRIN

Yoshio Tomimatsu and John W. Donovan
Western Regional Research Center, USDA, Berkeley, California

I. INTRODUCTION

Recent work (1) indicates involvement of tryptophan (in addition to tyrosine, histidine and bicarbonate anion) in iron binding to transferrins. Our spectroscopic study (2) suggested that Al(III) and Ga(III) binding to ovotransferrin (OT) and human serum transferrin (HST) produce ultraviolet difference spectra which are characteristic of <u>perturbation</u> of tryptophan and tyrosine. But binding of Zn(II) gave difference spectra which are consistent with tyrosine <u>ionization</u>. To further elucidate the differences between Zn and Al or Ga binding to transferrins, which produce colorless complexes, we present comparisons between experimental and calculated difference spectra, spectrophotometric titration curves and differential scanning calorimetric (DSC) measurements of changes in denaturation temperature (T_d) produced by metal binding.

II. EXPERIMENTAL

The OT and HST samples and the techniques used to obtain uv difference spectra and DSC thermograms have been described (2,3). Difference spectra produced by ionization of N-acetyl tyrosine and by ethylene glycol perturbation of tryptophan and tyrosine (2) were used to calculate protein difference spectra. For ionization, integral numbers of moles of tyrosine were assumed. A factor was used to account for the apparent difference between the molar absorbance changes produced by metal binding to tyrosine and by ionization of tyrosine in alkaline solution. Tryptophan and tyrosine perturbation was assumed to result from transfer of those chromophores from aqueous solution to the interior of transferrin. So the 20% (v/v) ethylene glycol perturbation data were multiplied by

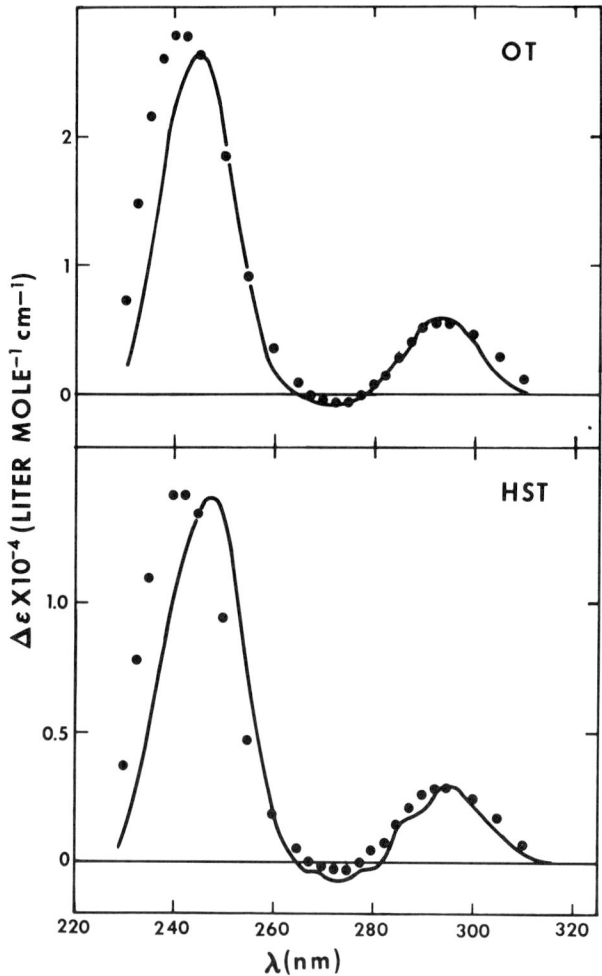

Fig. 1. Difference spectra produced by Zn binding to OT and HST at pH 8.4. Calculated Δε's (●). See text.

6 (4). Partial "burial" of tryptophan and tyrosine was assumed. In some cases, a small (2.5 nm) blue shift of the tryptophan perturbation difference spectra was introduced into the calculations.

III. RESULTS AND DISCUSSION

Difference spectra (obtained at full saturation of the metal-binding sites) produced by Zn binding to OT and HST are shown in fig. 1. Calculated Δε's were obtained from ioniza-

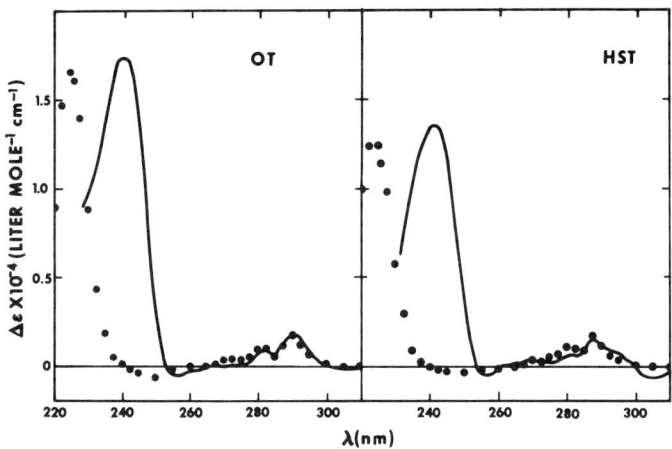

Fig. 2. Difference spectra produced by Ga binding to OT and HST at pH 8.4. Calculated Δε's (●). See text.

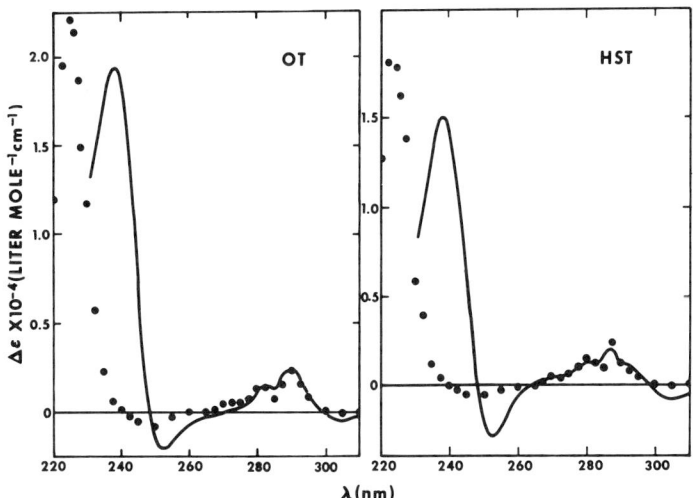

Fig. 3. Difference spectra produced by Al binding to OT and HST at pH 8.4. Calculated Δε's (●). See text.

tion data for 4 tyrosines by applying a factor of 0.7 for OT and 0.35 for HST. The good agreement between experimental and calculated difference data indicate that Zn binding to OT and HST produce difference spectra consistent with tyrosine ionization. Small irregularities in the difference spectra between 275-287 nm, particularly for Zn-HST, suggest that perturbation of tryptophan and tyrosine also occur.

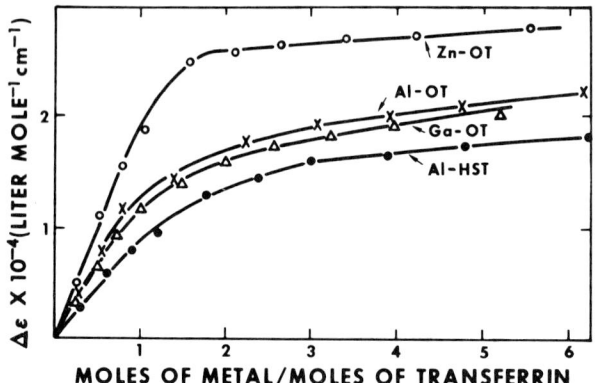

Fig. 4. Spectrophotometric titrations of OT with Zn (O-O-O) (245 nm), Al (X-X-X) (238 nm), and Ga (Δ-Δ-Δ) (241 nm) and HST with Al (●-●-●) (238.5 nm) at pH 8.4.

Fig. 2 shows difference spectra produced by Ga. For OT, calculated $\Delta\varepsilon$'s are for perturbation of 2 tryptophans which are 60% "buried". Data for one tryptophan were blue-shifted by 2.5 nm. For HST, calculated $\Delta\varepsilon$'s are for two tryptophans which are 40% "buried" and two tyrosines which are 50% "buried". Data for both tryptophans were blue-shifted by 2.5 nm. Fig. 3 shows Al difference spectra. Here, 80% (for OT) and 60% (for HST) "burial" of the tryptophans were assumed. The experimental difference curves in figs. 2 and 3, which are distinctly different from those in fig. 1, cannot be fit with tyrosine ionization data. The reasonable fit obtained assuming tryptophan and tyrosine perturbation suggests that tryptophan and tyrosine residues are in or near the metal-binding sites but they are not ligands in Ga and Al binding. Since the Ga- and Al-HST difference spectra also required tyrosine perturbation data for a best fit, this is further evidence for a difference in metal-binding between OT and HST (5).

Titration curves for Zn, Al and Ga binding are shown in fig. 4. Curves for Zn and Ga binding to HST (not shown) are almost identical in shape and magnitude with Al-HST. Linear changes in $\Delta\varepsilon$ occur up to a metal to transferrin molar ratio of about 1, followed by a non-linear change up to a ratio between 2-3. Similar results, previously reported for Mn(II) titration of OT (6), Zn titration of HST (7), and Tb(III), Eu(III), Er(III) and Ho(III) titration of HST (8), have been interpreted to indicate binding of two moles of metal. Non-equivalence of the two sites is well established (3,8-11). Evans and Holbrook (11) observed that the decrease in fluorescence on binding Fe(III) was biphasic. Thus, negative cooper-

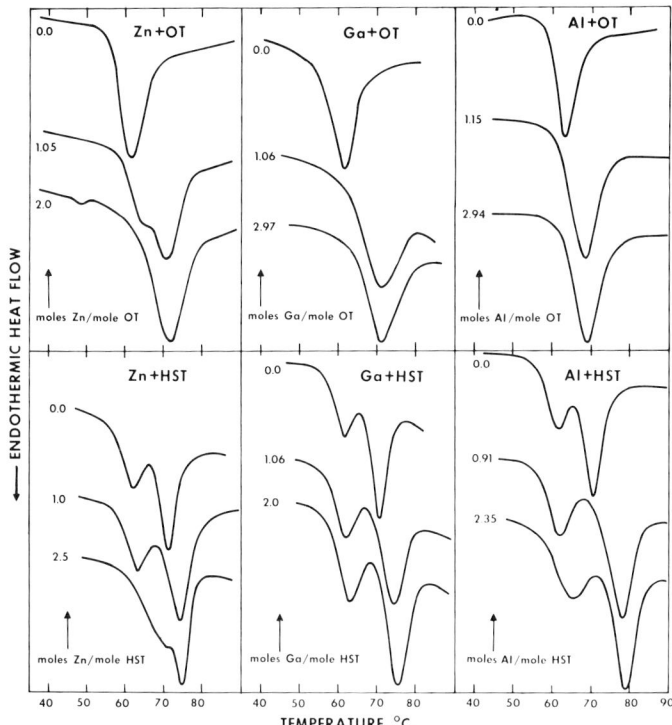

Fig. 5. Effects of Zn, Al and Ga on the heat stability of OT and HST. Heating rate, 10°/min. in 0.5 and 0.6 M TRIS buffer (except Zn-OT and Zn-HST, 0.025 M borate buffer, 0.025 M KCl) at pH 8.7-8.9 (except Al-OT, pH 7.5).

ativity exists between metal-binding sites. The initial linear change in Δε with metal added (fig. 4) could be due to occupation of a particular site. If this is so, then the subsequent non-linear change in Δε, which is then due to occupation of the other site, suggests that the second metal is more weakly bound than the first. For HST, and possibly for OT (see DSC results below), the Δε produced by binding the first metal is 1.3-2.7 times the Δε produced by binding of the second. Since the shapes of the difference curves remains the same throughout the titration measurements, the residues being ionized or perturbed are the same for the first and the second metal.

Fig. 5 shows the effects of Zn, Al and Ga binding on OT and HST heat stability. Apo-HST has two endotherms (62°, 71°) for the two domains (12), each of which contains one metal-binding site. The first metal is bound to the domain with the higher T_d, stabilizing it. The second metal, bound to the

other domain, produces a smaller increase in stability, except for Zn. Binding of the second Zn produces a large change (8°) in T_d compared to that produced by the first Zn (3°).
d_{Apo-OT} has a single endotherm (62°) which shifts to a higher temperature when one Ga (71°) or one Al (69°) is bound. Binding of a second Ga or Al has little further effect. Since the thermograms do not distinguish between the domains in OT, it is not clear whether the first Ga or Al enters a specific binding site. Mono-Zn-OT, however, behaves like apo-HST; it shows two endotherms (65°, 71°). These probably correspond to denaturation of domains with an empty and an occupied binding site, respectively. Binding of a second Zn produces di-Zn-OT which denatures as a single molecule.

The results of spectrophotometric and DSC measurements are consistent. They show i) sequential occupation of the binding sites for Zn, Al and Ga binding to HST, and possibly OT; ii) Zn binding to OT and HST produce difference spectra and DSC thermograms which are different from those produced by Al and Ga binding; and iii) Zn, Al and Ga binding to OT produce difference spectra and DSC thermograms different from those produced by binding of the corresponding metal to HST.

IV. REFERENCES

1. Krysteva, M.A., Mazurier, J., and Spik, G., Biochim. Biophys. Acta 453, 484 (1976).
2. Tomimatsu, Y., and Donovan, J.W., FEBS Lett. 71, 299 (1976).
3. Donovan, J.W., and Ross, K.D., J. Biol. Chem. 250, 6022, 6026 (1975).
4. Donovan, J.W., J. Biol. Chem. 244, 1961 (1969).
5. Tomimatsu, Y., Kint, S., and Scherer, J.R., Biochemistry 15, 4918 (1976).
6. Tan, A.T., and Woodworth, R.C., Biochemistry, 8, 3711 (1969).
7. Nagy, B., and Lehrer, S.S., Arch. Biochem. Biophys. 148, 27 (1972).
8. Luk, C.K., Biochemistry 10, 2838 (1971).
9. Price, E.M., and Gibson, J.F., J. Biol. Chem. 247, 8031 (1972).
10. Aisen, P., Lang, G., and Woodworth, R.C., J. Biol. Chem. 248, 649 (1973).
11. Evans, R.W., and Holbrook, J.J., Biochem. J. 145, 201 (1975).
12. Donovan, J.W., Fed. Proc. 35, 1464 (1976).

IRON RELEASE FROM TRANSFERRIN MEDIATED

BY ORGANIC PHOSPHATE COMPOUNDS

E. H. Morgan
University of Western Australia and
University of Washington

I. INTRODUCTION

The utilization of transferrin-bound iron by reticulocytes occurs in at least three stages, viz. (i) adsorption of transferrin to the cell membrane, largely at specific receptor sites, (ii) movement of transferrin into the cell by endocytosis, and (iii) release of iron from the transferrin (1, 2). The adsorption stage is insensitive to incubation temperature and appears to be independent of cellular metabolism (1). By contrast, the other stages are both temperature sensitive and are believed to be dependent on metabolism (1). However, although the rate of iron uptake is markedly diminished when the cells are incubated in the presence of a wide variety of inhibitors of energy metabolism, transferrin uptake is influenced to a lesser degree and only at high concentrations of the inhibitors (3). In addition, it has been shown that there is a close, linear relationship between the rate of iron uptake and the cellular concentration of ATP, while the rate of transferrin uptake is inhibited only when the ATP concentration is reduced to very low levels (4). This indicates that the rate of detachment of iron from transferrin is correlated with the cellular concentration of ATP. Such a correlation could occur if ATP or other metabolites which vary in concentration with that of ATP are able to release iron from transferrin. This possibility was therefore investigated. Two experimental systems were used, both of which included an acceptor of the iron released from transferrin. In one method the acceptor was apotransferrin and in the other it was desferrioxamine.

II. METHODS

Rabbit apotransferrin prepared as described previously (5) and human apotransferrin from Behringwerke were freed of any residual chelating agents by dialysis in turn against 0.1 M $NaClO_4$, 0.1 M $NaHCO_3$, and 0.15 M NaCl. Iron was added to the proteins in amounts equivalent to 95% of their calculated iron-binding capacities. The rabbit transferrin was labeled with ^{59}Fe added in the form of its complex with nitrilotriacetic acid (6), followed by dialysis against 0.1 M $NaHCO_3$ and 0.15 M NaCl. The nonradioactive iron complexes of both proteins were prepared by adding ferrous ammonium sulfate followed by $NaHCO_3$ (4.5 mM) to apotransferrin.

The first method used to measure iron release from transferrin was based on measurement of the transfer of ^{59}Fe between immunologically different types of transferrin which were separated by immunoprecipitation. Iron-59-labeled rabbit transferrin (100 µg/ml) was incubated with human apotransferrin (700 µg/ml) in 0.15 M NaCl buffered with 0.1 M HEPES. At timed intervals aliquots of the solution were removed, antiserum to human transferrin added, and the mixture incubated for 30 min at 37° C and overnight at 4° C. The precipitate was washed three times in ice-cold 0.15 M NaCl and counted for radioactivity. The system was tested by the use of ^{125}I-^{59}Fe-labeled rabbit transferrin. It was found that the rate of iron accumulation in the immunoprecipitates was not due to coprecipitation of rabbit transferrin. Hence, it represented iron transfer from rabbit to human transferrin. No measurable iron exchange between the two species of transferrin occurred during incubation at 37° C and pH 7.4 for the period studied (4-6 hrs) when phosphate compounds, cellular metabolites or iron chelators were not added to the incubation mixture.

In the second method the rate of iron release from transferrin was measured by the change in absorbance at 293 nM of iron-transferrin when read against a blank solution containing identical concentrations of apotransferrin and the other constituents of the solutions. The absorbance measurements were made in a Beckman Model DKII and a Varian Techtron Model 635 spectrophotometer, both of which were equipped with temperature controlled cell holders. The solutions contained 0.02 mM transferrin, 5 mM desferrioxamine methane sulphonate (Desferal, Ciba), and the substance under test dissolved in 0.15 M NaCl buffered with 0.05 M HEPES.

III. RESULTS

Several organic phosphates and a large number of cellular metabolites were tested for their ability to mediate iron exchange from rabbit transferrin to human transferrin (7). GTP, 2,3-diphosphoglycerate (DPG) and ATP, when used at concentrations present in immature erythroid cells (8), were all found to be active in this regard. The type of result obtained with these compounds is shown in Fig. 1. Mathematical analysis of the data showed that the curves closely fitted those which would be expected from a first order reaction which reached equilibrium when 80% iron exchange had occurred (7). Of the other substances tested only ADP and citrate, both of which induced iron exchange at about one-half the rate of DPG, were nearly as effective as the above three compounds. Pyrophosphate, which has been reported to cause the release of iron from transferrin (9) was ineffective in the two-transferrin system (Fig. 1). In other experiments it was shown that the rates of iron exchange induced by GTP, DPG, and ATP were sensitive to incubation temperature and pH of the incubation solution (7). The activation energy of the exchange reaction was calculated to be approximately 14 kcal mole^{-1}. A direct linear relationship was found between the exchange rate and H^+ concentration within the pH range tested (pH 6.8-8.0).

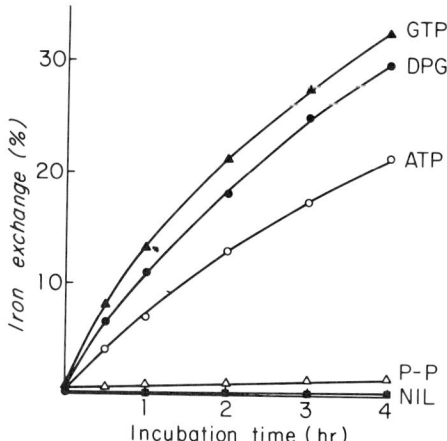

Fig. 1. Iron exchange between rabbit and human transferrin mediated by GTP (▲—▲), DPG (●—●), ATP (O—O), and pyrophosphate (△—△). The substances were present at concentrations of 3 mM. Incubation temperature was $37°$ C, pH 7.4. See text for details.

An experiment was performed to determine whether the cytosol of reticulocytes contained substances which were capable of mediating iron release from rabbit transferrin. Reticulocyte-rich blood cells from a rabbit with phenylhydrazine-induced anemia (65% reticulocytes) were washed with 0.15 M NaCl, hemolysed with an equal volume of water and the stroma-free supernatant solution obtained by centrifugation at 40,000 g for 1 hr. A similar preparation was obtained from mature erythrocytes from a normal rabbit (3% reticulocytes). The reticulocyte hemolysate, when diluted three-fold in the incubation mixture produced a rapid transfer of iron between the transferrin species (Fig. 2). The transfer was sensitive to incubation temperature and pH in a manner comparable with that found with solutions of GTP, DPG, and ATP. It was also found that the hemolysate from mature erythrocytes was equally as effective at producing iron transfer at 37° C and pH 7.4 as was that from reticulocytes. Ultrafiltrates were prepared from both types of hemolysate by the use of an Amincon PM10 filter with a stated exclusion limit of 10,000 daltons. The ultrafiltrate from reticulocytes and mature erythrocytes was found to be equal and approximately as effective as a 4 mM solution of ATP at mediating iron exchange.

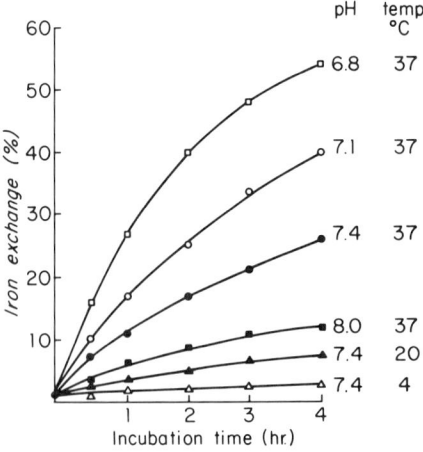

Fig. 2. Effect of incubation temperature and pH on iron exchange between rabbit and human transferrin mediated by an hemolysate from rabbit reticulocytes. The hemolysate was prepared as described in the text.

The method which employed desferrioxamine was developed in order to have a direct system of measuring iron release from transferrin which was not dependent on the ability of another molecule of transferrin to bind the iron which had been released. It was found that the absorbance of the solution decreased slowly when transferrin was incubated with desferrioxamine and the rate of change of absorbance was little affected by the pH of the solution. However, when ATP was added the absorbance decreased more rapidly. This effect became more pronounced as the pH was lowered (Fig. 3). Similar results were obtained when DPG was used in place of ATP. On the other hand, ATP (Fig. 4) or DPG produced very little change in absorbance in the absence of desferrioxamine. These results therefore confirmed those obtained with the double transferrin system that ATP and DPG can cause the detachment of iron from transferrin. However, in order for this to be demonstrated it is necessary that an additional iron-binding substance be present in the system.

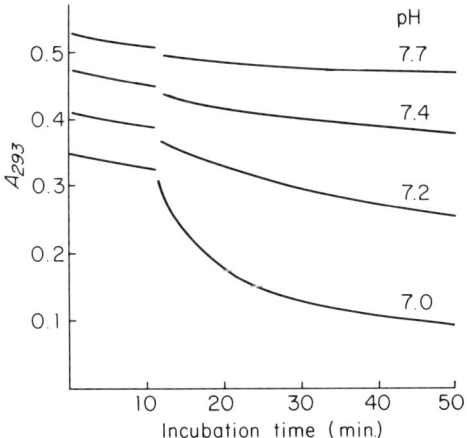

Fig. 3. Change in absorbance at 293 nm of a solution of diferric human transferrin read against a blank of apotransferrin. The solution contained 5 mM desferrioxamine and was incubated at 39° C at the pH values shown on the figure. At the time indicated by the discontinuity of the graphs 2.5 mM ATP was added to each solution. The bottom three curves have been displaced downward from the level of the top one in order to provide clearer presentation of the results.

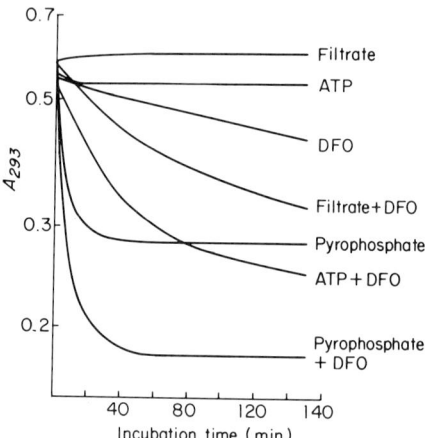

Fig. 4. Change in absorbance at 293 nm of diferric human transferrin read against apotransferrin. The solution was incubated at 37° C and pH 7.4. It contained ATP (7 mM), desferrioxamine (DFO, 5 mM), pyrophosphate (7 mM), an ultrafiltrate from rabbit reticulocytes or combinations of these materials as indicated.

Pyrophosphate and the ultrafiltrate prepared from the hemolysate of reticulocytes were studied in the system containing desferrioxamine (Fig. 4). It was found that the filtrate, alone, produced no release of iron from transferrin. However, when desferrioxamine was added to the incubation solution a progressive release of iron resulted. By contrast, pyrophosphate, alone, caused a rapid release of iron from transferrin and the rate of release increased little when desferrioxamine was added to the solution. It was also noted that semilogarithmic plots of iron release from human transferrin were curved in shape, suggesting the presence of two components in the release process (Fig. 4). In this regard the release of iron from human transferrin appeared to differ from the release from rabbit transferrin as observed with the two-transferrin system.

The desferrioxamine system was used to compare the release of iron from rabbit transferrin with that from human transferrin. At any given pH within the range studied (pH 6.6-7.6) and with both ATP and DPG it was found that rabbit transferrin released its iron more slowly than did human transferrin. Moreover, when the changes in absorbance were plotted semilogarithmically a straight line was obtained with rabbit transferrin. Human transferrin was again found to give a curved line which could be resolved graphically into two linear components each of which accounted for about 50% of the iron released (Fig. 5). This indicates that the two iron atoms bound by human transferrin were released at different rates while there was no detectable difference in the rates of release of iron from the two binding sites of rabbit transferrin.

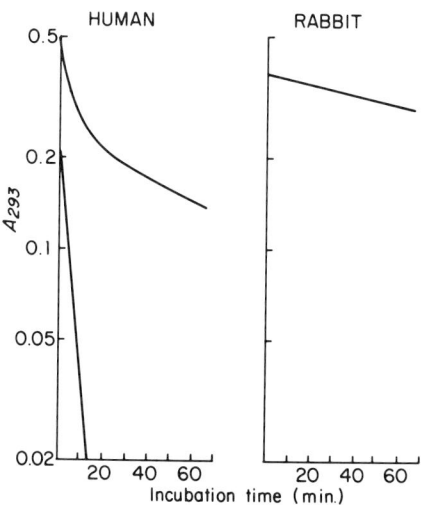

Fig. 5. Change in absorbance at 293 nm of diferric human and rabbit transferrin read against solutions of the corresponding apotransferrins. The solutions were incubated at 37° C, pH 7.0 and contained 5 mM desferrioxamine and 4 mM ATP. The lower line for human transferrin represents the fast component derived from the upper line.

IV. DISCUSSION

The results of these experiments show that GTP, DPG, and ATP can release iron from transferrin when used at concentrations similar to those present in reticulocytes (8). Stroma-free hemolysates and ultrafiltrates of the hemolysates from both reticulocytes and mature erythrocytes were also able to release the iron. The characteristics of the release were similar to those observed with the phosphate compounds with regard to the effects of temperature, pH, and the requirement for an iron acceptor such as desferrioxamine. Hence, it is possible that these substances have a direct role in the process of iron release from transferrin in immature erythroid cells and represent the link between cellular metabolism and iron uptake. Further work is required to show whether this is the case and, if so, to demonstrate whether the release reaction occurs within endocytotic vesicles, in the cell cytosol, or after binding of transferrin to the membranes of mitochondria (10). It will also be necessary to determine whether transferrin receptors play any part in the reaction.

The hemolysate derived from mature erythrocytes was able to cause the release of iron from transferrin even though these cells are unable to take up transferrin-bound iron (11). Failure of erythrocytes to assimilate transferrin-iron probably results from changes to the cell membrane which occur during maturation of the reticulocyte. These changes result in the loss of functional transferrin receptors and of the capacity for endocytosis (2). The loss of mitochondria during maturation could also have the same effect if the iron-release reaction occurs at the mitochondrial membrane as has been postulated (10).

The observation that iron release from rabbit transferrin could be described by a first-order plot but that release from human transferrin was bimodal indicates that the rates of iron release from the two binding sites are similar for the rabbit protein but different for human transferrin. This is compatible with results which showed that there are differences in the utilization of the two irons of human transferrin by rabbit reticulocytes but that no differences could be demonstrated with rabbit transferrin (12). It is also in accordance with the finding that the two bicarbonates are released from human transferrin at different rates (13).

Previous investigations have failed to demonstrate iron release from transferrin in the presence of ATP (9) or of rabbit reticulocyte hemolysates (14). The reason for this can probably be accounted for by the absence of a suitable iron acceptor in the systems used. It is likely that, although ATP and some components of the hemolysate can release iron from transferrin, they have relatively low affinities for iron so that the iron is rebound by transferrin and at equilibrium nearly all of the iron is present as the complex with transferrin. In the present work apotransferrin and desferrioxamine, which has an even higher affinity for iron than transferrin (15), were used at concentrations equivalent to 7 and 125 times the iron-binding capacity of the iron-donor transferrin. As a result iron released from the latter transferrin would have a high probability of being rebound by the apotransferrin or desferrioxamine and the net reaction would be iron transfer to these compounds. By contrast to GTP, DPG, and ATP, pyrophosphate probably has an affinity for iron much closer to that of transferrin so that at equilibrium much of the iron is present as the complex with pyrophosphate. This would account for the low rate of iron exchange between transferrin molecules which was observed in the presence of pyrophosphate (Fig. 1) and the rapid release of iron from transferrin produced by pyrophosphate even in the absence of desferrioxamine (Fig. 4 and ref. 9).

The mechanism by which phosphate compounds release iron from transferrin is uncertain. They may react with the transferrin molecule to produce a conformational change in the protein which weakens the iron-transferrin bond. Possibly, iron release results from prior release of bicarbonate as has been postulated to occur during iron uptake by reticulocytes (9, 16, 17). It seems unlikely that the mechanism involves a replacement of bicarbonate in the ternary complex which it forms with iron and transferrin because Egyed was unable to find any evidence that ATP or pyrophosphate could form such a ternary complex (9).

V. REFERENCES

1. Morgan, E.H., in "Iron in Biochemistry and Medicine" (Jacobs, A., and M. Worwood, eds), p. 29. Academic Press, London, 1974.

2. Hemmaplardh, D., and Morgan, E.H. Brit. J. Haem. 36, 85 (1977).

3. Morgan, E.H., and Baker, E. Biochim. Biophys. Acta 184, 442 (1969).

4. Kailis, S.G., and Morgan, E.H. Biochim. Biophys. Acta 464, 389 (1977).

5. Baker, E., Shaw, D.C., and Morgan, E.H. Biochemistry 7, 1371 (1968).

6. Hemmaplardh, D., and Morgan, E.H. Biochim. Biophys. Acta 373, 84 (1974).

7. Morgan, E.H. Biochim. Biophys. Acta (submitted for publication).

8. Bartlett, G.R. Biochem. Biophys. Res. Comm. 70, 1055 (1976).

9. Egyed, A. Biochim. Biophys. Acta 411, 349 (1975).

10. Neuwirt, J., Borová, J., and Ponka, P., in "Proteins of Iron Storage and Transport in Biochemistry and Medicine" (Crichton, R.R., ed.), p. 161. North Holland, Amsterdam, 1975.

11. Jandl, J.H., Inman, J.K., Simmons, R.L., and Allen, D.W. J. Clin. Invest. 38, 161 (1959).

12. Harris, D.C., and Aisen P. Biochemistry 14, 262 (1975).

13. Aisen, P., Leibman, A., Pinkowitz, R.A., and Pollack, S. Biochemistry 12, 3679 (1973).

14. Morgan, E.H. Biochem. J. 158, 489 (1976).

15. Neilands, J.B., in "Inorganic Biochemistry" (Eichhorn, G.L., ed), p. 167. Elsevier, Amsterdam, 1973.

16. Aisen, P., and Leibman, A. Biochim. Biophys. Acta 304, 797 (1973).

17. Schulman, H.M., Martinez-Medellin, J., and Sidloi, R. Biochim. Biophys. Acta 343, 529 (1974).

EVIDENCE FOR THE DIRECT INVOLVEMENT OF ATP IN THE
IRON UPTAKE BY RETICULOCYTES

A. Egyed
National Institute of Haematology
Budapest, Hungary

I. INTRODUCTION

Recently the transferrin-reticulocyte interaction has been a subject of extended research. Several researchers suggested that the removal of iron from transferrin must be preceded by the displacement of transferrin-bound carbonate (1-4). The mechanism of this process is not yet understood. The reticulocytes have very active metabolism and the ATP content of these cells is significantly higher than that of mature erythrocytes (5,6). The involvement of ATP in the iron uptake process was suggested earlier (7-9). As to the function of ATP two hypotheses were proposed. Mazur and Carleton (8) suggested that after the reduction of transferrin-bound ferric iron ATP would remove ferrous iron from the protein and transfer it to intracellular ferritin. In contrast to the above hypothesis it was demonstrated that ATP facilitates the exchange of transferrin-bound carbonate without substituting it and without removing iron (9).

The aim of the present work is to obtain evidence for the direct involvement of ATP in the iron uptake by reticulocytes.

II. MATERIALS AND METHODS

Chemicals of analytical grade were used. Purified rabbit transferrin was purchased from Koch-Light Laboratories Ltd. Rabbit reticulocytes were prepared as previously described (1). Labelling of transferrin with 59Fe were carried out according to the method of Bates and Schlabach (10).

Reticulocytes mediated transferrin-bipyridine iron transfer and iron uptake by reticulocytes was measured as previously described (6). The ATP content of the cells was determined by the luciferin-luciferase method (11). Radioactivity was measured with a Beckman Biogramma spectometer.

Membrane ATPase activity of intact erythrocytes was measured according to Sarkadi et al. (12).

Reticulocytes were incubated at 37°C in the presence of KCN, fluoroacetate and NaF (5 mM; 16 mM; 4 mM, respectively) to inhibit all the ATP producing pathways of the cells. Samples were taken from the reaction mixture and the inhibitors were washed off from the cells with ice cold isotonic KCl. Then the intracellular ATP concentration was determined and the reticulocytes mediated transferrin-bipyridine iron transfer was measured. As it can be seen on Fig. 1 the intracellular ATP concentration rapidly decreases as a function of the incubation time. On the other hand, the iron detaching ability of the cells decreases slightly at the beginning (in some experiments no decrease in the first phase could be measured) then when the intracellular ATP concentration reaches a critical low value (50-100 μM), a rapid decline of iron detachment can be observed. The iron uptake by the cells changes in a similar fashion. The ATP depleted reticulocytes are not able to remove transferrin-bound carbonate, either.

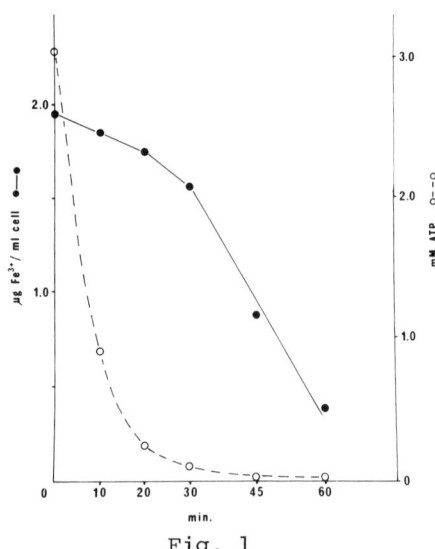

Fig. 1

Effect of ATP depletion on the reticulocyte-mediated transferrin-bipyridine iron transfer. Iron transfer: ●——● (20 min. 37°C); intracellular ATP concentration: o— —o. Abscissa shows the time of ATP depletion. For conditions see the text.

The above experiments were carried out in isotonic KCl medium to avoid the development of calcium dependent rapid K^+ efflux caused by ATP depletion (13). This precaution was very important in view of the possibility of calcium contamination in the solutions applied. The rapid K^+ efflux would result in the shrinkage of the cells due to the concomittant loss of water.

It has been demonstrated that DNP acts not only as an uncoupler of oxidative phosphorylation but also as an activator of latent membrane ATPases in mitochondria (14). In order to decide whether DNP activates the latent ATPases of the cell membrane, too, the membrane ATPase activity of intact mature erythrocytes was measured as a function of DNP concentration both in the absence and in the presence of oligomycin. As it is demonstrated on Fig. 2 DNP increases the ATPase activity of erythrocytes which do not contain mitochondria. Oligomycin inhibits completely the DNP-induced ATPases.

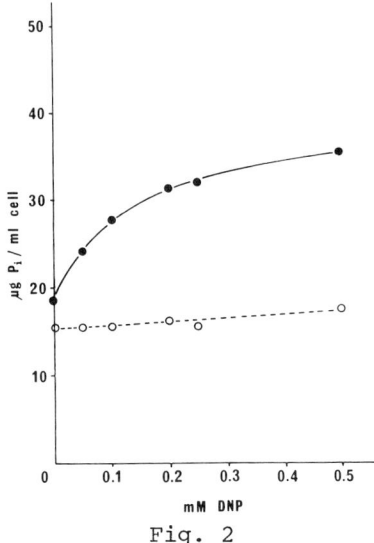

Fig. 2

Effect of DNP on the membrane ATPase activity of intact mature erythrocytes in the absence (●——●) and in the presence (o- -o) of 40 µg oligomycin/ml cell.

Further iron uptake by reticulocytes was measured in the presence of different concentrations of DNP (Fig. 3). The maximal inhibitory effect occurs at 0.15 mM.

It is known that oligomycin inhibits DNP induced ATPases as well as coupled respiration via inhibiting phosphorylation but does not alter uncoupled respiration.

In the following experiments the effect of DNP and oligomycin on the respiration of reticulocytes and their iron uptake process was studied. The concentrations of DNP and oligomycin

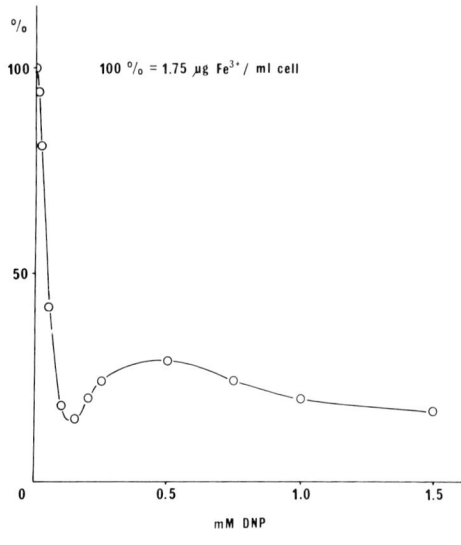

Fig. 3

Iron uptake by reticulocytes as a function of the concentration of DNP present in the reaction mixture. Incubation: 20 min. 37°C

were 0.15 mM and 40 µg/ml cell, respectively. In these concentrations DNP and oligomycin affected the respiration (measured by Warburg manometric technique) in the way as expected (Fig. 4A). Iron uptake was measured in identical conditions. As Fig. 4B shows oligomycin inhibits iron uptake by about 55 %, DNP inhibits this cell function by about 90 %.

However, when applying DNP and oligomycin simultaneously, the inhibition of iron uptake by reticulocytes was identical with the inhibition observed in the presence of oligomycin alone, in spite of the fact that the rate of respiration was identical with the one measured in the presence of DNP alone (Fig. 4A).

The effect of DNP on iron uptake is reversible. This enabled us to decide whether DNP acts on iron uptake via decreasing intracellular ATP cencentration or in some other way. The reticulocytes were incubated in the presence of 0.15 mM DNP and samples were taken at various time intervals. Then the cells were washed with ice cold saline and their ATP content and iron uptake ability were determined. As it can be seen on Fig. 5 the intracellular ATP concentration decreased rapidly but did not reach the critical low value (100 µM).

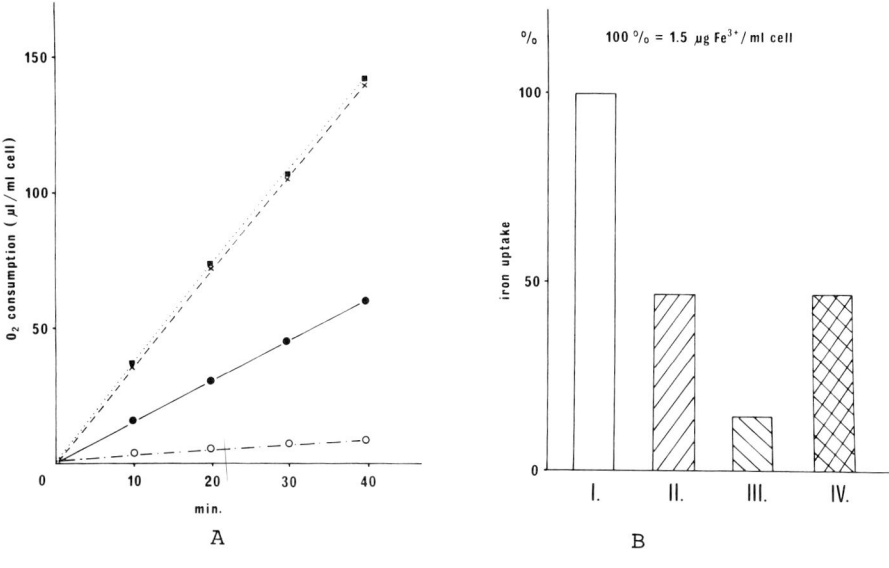

Fig. 4

Effect of oligomycin and DNP on respiration (A) and iron uptake (B). Concentrations applied: 0.15 mM DNP; 40 μg oligomycin cell.
A: Respiration of reticulocytes: ●——● control; x− −x + DNP; o−·−o + oligomycin; ■···■ + DNP + + oligomycin
B: Iron uptake: I. control; II; + oligomycin; III. + DNP; IV. + DNP + oligomycin

In spite of the great change in ATP content, the iron uptake by the treated reticulocytes decreased slightly. This slight decrease in iron uptake caused by longer DNP treatment is not comparable to the immediate and strong inhibition observed in the presence of the compound.

III. DISCUSSION

The involvement of ATP in iron uptake has been suggested by several researchers (7-9). As to the role of ATP in this process, views diverge. Mazur and Carleton (8) suggested that following the reduction of transferrin-bound iron ATP removes divalent iron from the protein and transfers it to ferritin. In our previous paper it was demonstrated, however, that ATP facilitates the exchange of transferrin-bound carbonate but without substituting it and without removing iron. Recently Kailis and Morgan suggested that ATP is involved in

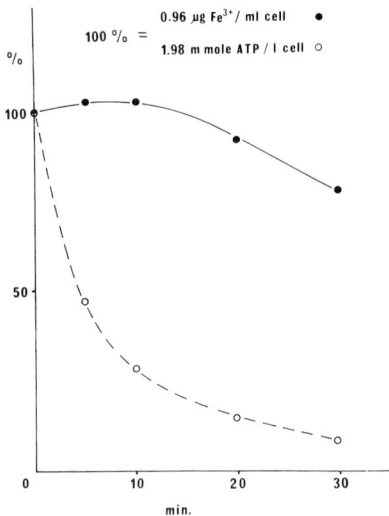

Fig. 5
Effect of pretreatment with 0.15 mM DNP on iron uptake by reticulocytes (●——●) and on their ATP content (o- -o). Iron uptake was measured for 20 min at 37°C after removal of DNP. Abscissa shows the time of pretreatment.

iron uptake as an energy source for endocytosis (15).

In the present work the relationship between the intracellular ATP concentration and iron uptake was studied.

If the ATP generating pathways of the reticulocyte were inhibited, the intracellular ATP concentration decreased rapidly during incubation at 37°C. However, the iron detaching ability of the cells changed slightly at the beginning and diminished rapidly only after the intracellular ATP concentration was decreased below a critical low value (100 µM; that is about 4 % of the control) (Fig. 1). These results indicate that the iron detachment by reticulocytes requires only a very small amount of ATP. This fact contradicts the hypothesis which suggests the endocytotic uptake of transferrin-bound iron, since it has been demonstrated that the endocytotic activity of erythrocytes is strongly dependent on intracellular ATP concentration (16-18). It has to be emphasized that in our experiments the iron uptake by reticulocytes of different ATP content was measured after the removal of inhibitors from the cells. It is important because in the presence of inhibitors only the joint effect of ATP depletion and inhibitors on iron uptake can be measured.

According to our present knowledge the iron detachment by cells is a membrane process. The results obtained by ATP depletion indicate that this cell function requires some kind of ATP membrane interaction. Due to the almost complete ATP depletion a drastic change in the whole membrane structure cannot be excluded.

In order to approach this question in another way the mode of action of DNP on iron uptake was studied in detail. As it is known DNP acts not only as an uncoupler of oxidative phosphorylation but induces latent membrane ATPases in mitochondria as well. As Fig. 2 shows, DNP activates latent membrane ATPases also in erythrocytes and oligomycin is able to inhibit these ATPases.

DNP exerts its maximal inhibitory effect on iron uptake at 0.15 mM concentration (Fig. 3). The question whether DNP inhibits iron uptake through uncoupling the oxidative phosphorylation or by inducing latent membrane ATPases can be answered with the aid of oligomycin. Fig. 4A and B show that the inhibitory effect of neither the oligomycin nor that of DNP can be explained by their effect on mitochondrial respiration. This is in accordance with our earlier finding that there is no direct connection between respiration and iron uptake (6). Recently Kailis and Morgan confirmed the latter conclusion (15).

Fig. 4B shows that oligomycin prevents the strong inhibitory effect of DNP on iron uptake by reticulocytes. This fact indicates that DNP inhibits iron uptake via activation of membrane ATPases. However, the effect of DNP on iron uptake cannot be explained with the decrease of intracellular ATP concentration. DNP inhibits iron uptake almost immediately after addition but this inhibition is reversible, that is after removing the inhibitor, the iron uptake by reticulocytes remains decreased slightly, if at all, in spite of the relatively great decrease in intracellular ATP concentration. It has to be mentioned that during the time of this experiment the intracellular ATP concentration did not fall below the critical value (100 μM). It is probable that DNP acts on iron uptake through the activation of membrane ATPases, which results in a local decrease of ATP concentration in the membrane or in its close environment.

These results justify the assumption that ATP is directly involved in iron uptake uptake by reticulocytes. One possible but not sole explanation for the role of ATP in iron uptake is that the nucleotide is a component of the membrane structure responsible for anion and iron detachment. The ATP-membrane interaction necessary for anion detachment loosens only at a very low ATP concentration which can be achieved by ATP depletion or locally, activating the membrane ATPases.

IV. REFERENCES

1. Egyed, A. Biochim. Biophys. Acta 304, 805 (1973)
2. Aisen, P. and Leibman, A., Biochim. Biophys. Acta 304, 794 (1973)
3. Martinez-Medellin, J. and Schulman, H. M., Biochem. Biophys. Res. Commun. 53, 32 (1973)
4. Williams, S. C. and Woodworth, R. C., J. Biol. Chem. 248, 5848 (1973)
5. Ohyama, I., J. Biochem. 61, 103 (1967)
6. Egyed, A., Acta Biochim. Biophys. Acad. Sci. Hung. 9, 43 (1974)
7. Mazur, A., Green, S., Carleton, A., J. Biol. Chem. 235, 595 (1960)
8. Mazur, A. and Carleton, A., J. Biol. Chem. 238, 1817 (1963)
9. Egyed, A., Biochim. Biophys. Acta 411, 349 (1975)
10. Bates, G. W. and Schlabach, M. R., J. Biol. Chem. 248, 3228 (1973)
11. Stanley, P. E. and Williams, S. G., Anal. Biochem. 29, 381 (1969)
12. Sarkadi, B., Szász, J., Gerlóczy, A., Gárdos, G., Biochim. Biophys. Acta 464, 93 (1977)
13. Gárdos, G., Biochim. Biophys. Acta 30, 653 (1958)
14. Hemker, H. C., Biochim. Biophys. Acta 73, 311 (1963)
15. Kailis, S. G., and Morgan, E. H., Biochim. Biophys. Acta 464, 389 (1977)
16. Ben-Bassat, I., Bensch. K. G., Schrier, S. L., J. Clin. Invest. 51, 1833 (1972)
17. Penniston, J. T., Arch. Biochem. Biophys. 153, 410 (1972)
18. Hayashi, H., Plishker, G. A., Vaughan, L., Penniston, J. T., Biochim. Biophys. Acta 382, 218 (1975)

ATP-INDUCED RELEASE OF Fe(III) FROM TRANSFERRIN

Franklin J. Carver
Earl Frieden

Florida State University

I. INTRODUCTION

A number of mechanisms have been postulated for the mobilization of iron from transferrin both on the plasma membrane and in the cytosol of erythroid precursor cells (1,2). The mobilization of iron directly from transferrin may require a modification of the iron binding site along with exposure of the transferrin-iron (III) complex to a biological chelator, e.g. ATP.

This study was undertaken to determine the role of nucleotides on iron release from transferrin by acid titration of the transferrin molecule. ATP was selected as a likely intracellular chelator since it is abundant in red blood cells (~2mM) (3) and a significant fraction of red blood cell ATP is present as Fe(ATP)$_2$ (4,5). It was found in the present study that pyrophosphate was the key structure in accelerating the release of iron from transferrin.

II. METHODS AND MATERIALS

Human transferrin (Behringwerke, A.G.) was dissolved in 10mM NaHCO$_3$/150mM NaCl pH 7.8 and iron as Fe(III)-NTA added to 115% of the total iron binding capacity of the apoprotein. The transferrin-iron(III) [Tf-Fe(III)] complex was allowed to equilibrate at room temperature for one hour then dialysed against 20mM Hepes pH 7.4 for three days. The concentration of the Tf-Fe(III) for iron release studies was based on an $E^{1\%} = .57$ at 465nm and a molecular weight of 76,000. For iron recovery experiments the formation of bipyridine-iron(II) [Bi-Fe(II)] was based on $\varepsilon = 8600$ at 520nm.

Reactions were initiated by Tf-Fe(III) (30nmoles) in a final volume of 1.0ml and optical density recorded 15 seconds later on a Cary model 15 spectrophotometer by measuring the loss in 465nm or gain in 520nm absorbing material. The rate of iron release, based on the loss of 465nm absorbing material, was calculated from the first minute of the reaction at 20° using a fast chart speed. The pH of each solution was determined after each run and found not to vary more than $\pm$ .01 for at least two hours after the initiation of the reaction with transferrin.

III. RESULTS

Fig. 1 represents the effect of HCl on the rate of release of Fe(III) from transferrin in the presence and absence of phosphate, pyrophosphate, and the three adenine ribonucleotides. From pH 7.4-6.7 the rate of loss of 465nm absorbing material was minimal. At 1mM the rates for three reactants were rapid from pH 6.6-6.2 and were in the following order: pyrophosphate > ATP > ADP. For those samples run with either pyrophosphate, ATP, or ADP at a pH < 6.0-6.2 the rate of reaction decreased. This apparent decrease was due to the rapid loss of 465nm absorbing material within the first non-recorded 15 seconds of the reaction. At pH 6.0 1mM phosphate and 1mM AMP had minimal releasing activity. The control curve represents transferrin titrated with HCl; no appreciable activity was observed until a pH < 5.4.

To determine if ATP was specific for the release of Fe(III) from transferrin 1mM ATP, CTP, TTP, and GTP were each titrated with HCl in the presence of Tf-Fe(III) (Fig. 2). This study indicates no difference in the rate of Fe(III) release for the four nucleotide triphosphates.

The role of ATP and ascorbate in the removal and reduction of Fe(III) from transferrin was studied at pH 6.1 using 25mM MES/160mM NaCl buffer

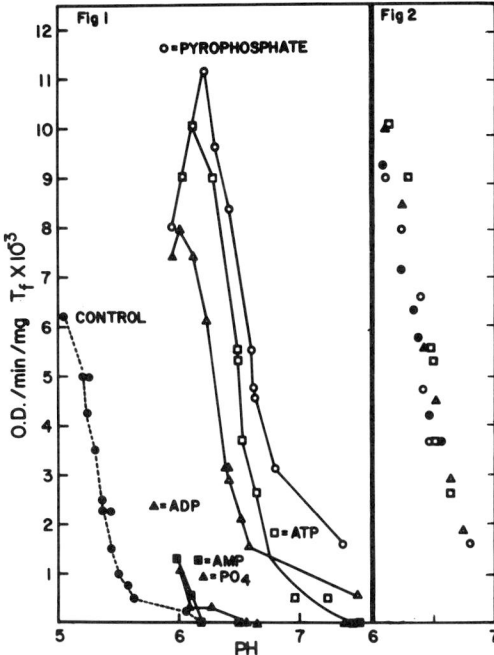

Fig. 1 The effect of pH on the rate of iron release from 30uM transferrin in the presence of 1mM adenine ribonucleotides, sodium pyrophosphate, and sodium monohydrogen phosphate. o=pyrophosphate, □=ATP, △=ADP, ■=AMP, ▲=phosphate, ●=Tf only.

Fig. 2 The equivalence of four nucleotide triphosphates in their effect on the rate of pH-induced release of iron from 30uM transferrin. O=CTP, ●=TTP, △=GTP, □=ATP.

Nucleotide, pyrophosphate, and phosphate solutions for experiments in Figs. 1&2 were prepared in 20mM Hepes pH 7.4. Final buffer concentration for each assay was 20mM Hepes/160mM NaCl. Each point represents the addition of 10ul HCl (in a .6-1.0 M series), 200ul reactant, 590ul 20mM Hepes/271mM NaCl, and 200ul Tf-Fe(III). Samples were incubated at 20°.

(Fig. 3). Based on absorbance at 465nm and assuming two iron atoms/mole transferrin, each sample contained 60nmoles of Fe(III). A control run without ATP showed a 3% loss of Fe(III) after 90 minutes. 0.1mM ascorbate removed a total of 17% of the iron, 90% of which was recovered (not shown) as Bi-Fe(II). 1mM ATP removed about 50% of the Fe(III) within the first five minutes; the loss of iron leveled off after 60 minutes. At the end of 90 minutes 77% or 46.2nmoles of iron was released by ATP. In a second experiment with ATP, bipyridine was added at 60 minutes at a final concentration of 0.2mM and did not effect a change in absorbance at 465 or 520nm. At the end of 90 minutes,

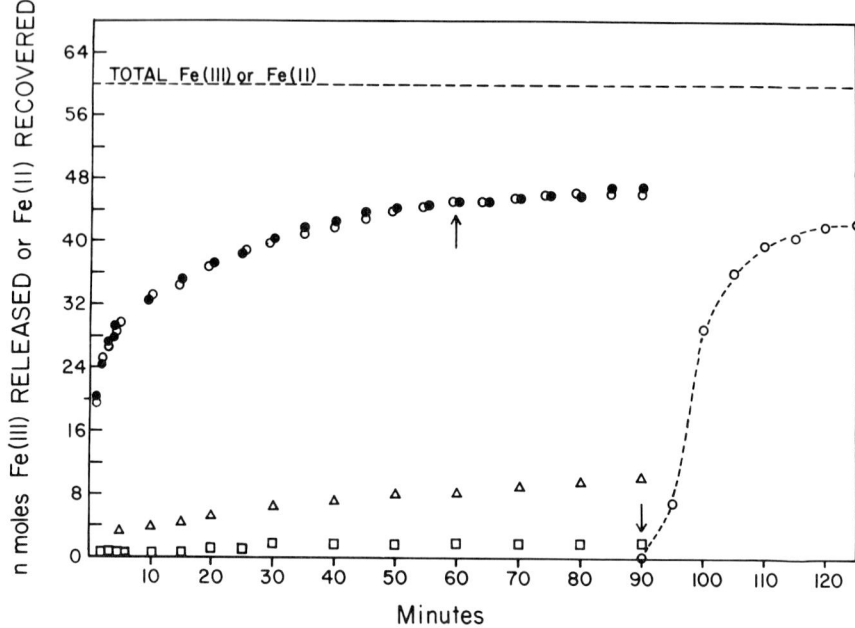

Fig. 3 The time dependence of Fe(III) release from 30uM transferrin and the recovery of reduced iron by bipyridine. ATP, ascorbate, and bipyridine solutions were prepared in glass distilled water and each adjusted to pH 6.2. Reactions were run at pH 6.1 ± .01 in 25mM MES/160mM NaCl pH 6.1 buffer in a total volume of 1.0ml and incubated at 20°. □ =Tf only, △=0.1mM ascorbate, ●=1mM ATP, ○=1mM ATP with 0.2mM bipyridine added at 60 minutes (indicated by ↑), and 0.1mM ascorbate added at 90 minutes (indicated by ↓).

ascorbate at 0.1mM final concentration was added to the sample containing ATP and bipyridine and monitored at 520nm over an additional 35 minutes. The concentration of the Fe(II) recovered by bipyridine was 42.3nmoles or 92% of the total released by ATP. In a separate study 1mM ascorbate and 1mM ATP showed the same rate of iron release as 1mM ATP. Bipyridine added at 60 minutes effected the recovery of 90% of the reduced iron. The results from a comparative study using 1mM glutathione were similar to 1mM ascorbate.

IV. DISCUSSION

Based on the results presented in this study we suggest the following chemical scheme for the release of iron from transferrin aided by ATP.

(1). Tf-Fe(III) $\xrightarrow{H^+}$ Tf----Fe(III)

(2). Tf----Fe(III) $\xrightarrow{ATP}$ Tf + [ATP-Fe(III)]

(3). [ATP-Fe(III)] $\xrightarrow{\text{ascorbate + Bi}}$ ATP + dehydroascorbate + Bi-Fe(II)

Data to support the above scheme indicates that di and triphosphates release iron from transferrin under mild acidic conditions, i.e. pH 6.6-6.2 (Fig. 1&2). The phenomenon of polyphosphate mediated iron release from transferrin is clearly not the same as that caused by acid alone.

The weakly acidic environment used in this study may favor the release of bicarbonate, which may be necessary for the release of Fe(III) from transferrin bound to reticulocytes (1,6). It has been suggested that Fe(III) binds weakly to transferrin in the absence of bicarbonate (7). The acid titration of Tf-Fe(III) with di and triphosphates may also represent the titration of the histidine group imidazole (Pk_a= 6.0) which has been implicated in transferrin-Fe(III) ligands (1,2). The dissociation of imidazole-Fe(III) ligands may reduce the binding affinity of Tf-Fe(III) and allow di and triphosphates to chelate the iron. Other possible effects of transferrin acidification include the complete protonation of the tyrosine phenolic groups associated with Tf-Fe(III) ligands (1,2) and/or a conformational change of transferrin.

ATP will release 50% of the transferrin iron rapidly at pH 6.1, but it releases the remaining iron more slowly over an extended period. This suggests that ATP may have a higher affinity for one of the Fe(III) binding sites of transferrin. Further experiments are required to explore this possibility.

At pH 6.1, no significant Bi-Fe(II) formation was observed when bipyridine was added to a Tf-Fe(III) and ATP mixture indicating that ATP will release Fe(III) (Fig. 3). Bipyridine has also been shown to chelate more than 90% of the Fe(III) released by ATP but only in the presence of a reducing agent, i.e. ascorbate or glutathione. Ascorbate had only a small effect on the release of iron from transferrin. Future experiments will examine the effect of pH on the properties of HCO_3-Tf-Fe (III) complex to determine if, after the acid removal of bicarbonate, the ATP chelation process can occur at a pH approaching physiological ranges.

In summary, we have shown that the rate of release of Fe(III) from transferrin was enhanced by polyphosphate and nucleoside di and triphosphates. The mechanism and significance of the reactions involved have been discussed.

VI. REFERENCES

1. Aisen, P., in "Inorganic Biochemistry" (Eichhorn, G.L. ed.), p. 280. Elsevier Scientific Publishing Company, N.Y., 1975.

2. Bezkorovainy, A., Zschocke, R.H., Arzneim-Forsch 24, 3 (1974).

3. Brown, P.R., Agarwal, R.P., Gell, J., Parks, R.E., Comp. Biochem. Physiol. 43B, 891 (1972).

4. Konopka, K., Leyko, W., Gondko, R., Sidorczyk, Z., Fabjanowska, Z., Swedowska, M., Clin. Chim. Acta 24, 359 (1969).

5. Konopka, K., Szotor, M., Acta Haemat. 47, 157 (1972).

6. Aisen, P., Brit. J. Haem. 26, 159 (1974).

7. Aisen, P., Aasa, R., Malmstrom, B.G., Vanngard, T., J. Biol. Chem. 242, 2484 (1967).

Supported by NIH HD-01236 and NIH fellowship AM05408

A STUDY OF PLASMA TRANSFERRIN
IN NORMAL AND IRON DEFICIENT RATS

H. Huebers, E. Huebers, S. Linck, and W. Rummel
University of the Saarland

I. INTRODUCTION

Transferrin is a β_1-globulin consisting of a single polypeptide chain of molecular weight of approximately 80,000. Each transferrin molecule contains two specific iron(III) binding sites which mediate the exchange of iron between various body tissues in accordance with their needs. It has been proposed that the two iron binding sites may behave biologically different (1-5). Evidence for a structural difference between the two binding sites has been obtained by electron paramagnetic resonance (EPR) spectroscopy (6, 7). It has also been suggested that the EPR spectrum of the physiological Fe(III)-transferrin-carbonate complex is a composite obtained from two distinctive sites (8). The aim of this contribution was to search for any difference between the two binding sites by chemical methods and to look for the biological significance of any possible chemical heterogeneity of the transferrin molecule.

II. MATERIALS AND METHODS

Animals: Normal and iron deficient Wistar rats, female, 180-200 g body weight were used. Iron deficiency was produced by repeated bleeding from the tongue vein in addition to a low iron diet (3.3 mg Fe/kg) within 12 to 15 days (9). Serum iron was 30-50 µg/100 ml and reticulocytosis ranged from 30 to 45% at a hematocrit from 23 to 32%.

In vivo labelling of plasma

a) *by intestinal absorption*--Blind segments of the upper small intestine (duodenum and jejunum, 25 cm) were filled with a ^{59}Fe containing saline (900 nmol ^{59}Fe-FeCl$_3$, pH 2.0). After an exposure period of 30 min the amount of ^{59}Fe absorbed was measured in a body-counter for small animals after removal of the tied-off segment. The ^{59}Fe tagged blood was collected by heart puncture and heparin was added as anticoagulant. The cooled blood was centrifuged and the ^{59}Fe plasma was mixed at 0° C with untreated blood sediment from fasted iron deficient or normal animals. The hematocrit was adjusted to 30%.

b) by intravenous injection--0.5 ml of ^{59}Fe-FeCl$_3$ or ^{59}Fe-FeSO$_4$ solutions (140 or 400 ng-atom Fe) were injected in the tail vein of rats. After 10 min exposure in vivo the animals were sacrificed and the blood was obtained and prepared for incubation studies as described above.

Incubation studies: ^{59}Fe tagged blood samples were kept under moderate shaking in a water bath at 37° C. At different intervals samples were cooled in an ice bath and centrifuged at 2,000 g for 10 min. The radioiron was measured in plasma samples and in the saline washed reticulocyte-rich sediment. This sediment will be referred to as "reticulocytes."

Isolation procedures: rat diferric plasma transferrin consisting of the two species transferrin fast (Tf$_f$) and transferrin slow (Tf$_s$) was isolated using a previously described method (10).

Monoferric transferrin species were isolated using the diferric transferrin species Tf$_f$ and Tf$_s$ as starting material. Samples of these preparations dissolved in 50 ml buffered saline (pH 7.4) were applied individually to Sephadex G50 chromatography (column 2.5 x 89.5 cm, gel equilibrated with 0.25 M acetate/acetic acid buffer, pH 5.1; void volume: 145 ml). Elution was performed with the same buffer as taken for equilibration at 1 ml/min. The UV-absorption of the eluate was monitored continuously using a Uvicord II detector. The ^{59}Fe content of the collected fractions was measured in a γ-spectrometer. The protein carrying ^{59}Fe fractions coming at the void volume were applied to a column (2.5 x 17 cm) filled with Whatman CM52 cation exchange cellulose (equilibrated with 0.25 M acetate/acetic acid buffer, pH 5.1). Linear gradient elution at 0.5 ml/min flow rate was performed adding salt (0.01→0.15 M NaCl) to the acetate buffer.

Electrophoretic separation: Electrophoresis on polyacrylamide-gel was carried out as previously described (10).

Kinetic studies: Iron depletion kinetics from the various transferrin samples were followed spectrophotometrically at 535 nm after addition of bathophenanthroline (50 μg/ml) and ascorbic acid (1 mg/ml) to the samples dissolved in 10 ml of 0.25 M acetate buffer, pH 5.6.

III. RESULTS

a) Incubation studies with ^{59}Fe blood: In iron deficient and normal rats the iron uptake by reticulocytes from in vivo via absorption loaded plasma is linear by time (Fig. 1).

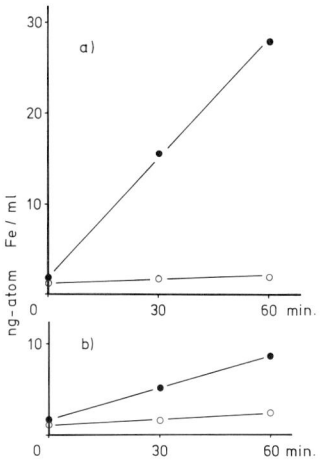

Fig. 1. Fe uptake by reticulocytes from normal (-o-) and iron deficient (-•-) rats. a) plasma from iron deficient rats (Fe-loaded in vivo via absorption); b) plasma from normal rats (Fe-loaded in vivo by intravenous injection of ferrous ammonium sulfate).

The slope is 0.4 ng-atom Fe/ml x min for the reticulocyte-rich suspensions and 0.02 mg/ml x min for the normal control. This makes a difference in the uptake rate by a factor of 20. In normal rats the in vivo via intravenous injection of ferrous ammonium sulfate loaded plasma is taken up by the reticulocyte-rich suspension at a rate of 0.14 ng-atom Fe/ml x min and 0.02 ng-atom Fe/ml x min for the normal control. The difference in the slopes is seven fold and thus considerably less than using Fe-loaded plasma from iron deficient animals. These results raise the possibility that there is a qualitative difference in the iron donor properties of the transferrin pool in normal and iron deficient states. The following studies were designed to obtain arguments for this possibility by chemical analysis of the transferrin molecule.

b) <u>Isolation of one iron transferrin species</u>: As previously reported (10, 12, 13) (^{59}Fe)-iron saturated rat plasma resulted in the isolation of two diferric-transferrin species (flow diagram, Table 1). In plasma from iron deficient rats the amount of Tf$_s$ is about three times increased whereas the increase in the Tf$_f$ species is not so pronounced (Fig. 2).

Table 1
Flow Diagram for the Isolation of Diferric Transferrin Species from Rat Plasma

^{59}Fe <u>in vivo</u> labelled rat plasma

saturation with Fe^{2+} at pH 8.0

Sepharose 6B chromatography
(0.125 M Tris/HCl buffer, pH 8.2)

DEAE Sephadex A50
(gradient elution: 0.05 to 0.16 M Tris/HCl, pH 8.4)

isoelectric focusing
(pH range 5-7)

Tf$_f$: pI 5.7; Tf$_s$: pI 5.6

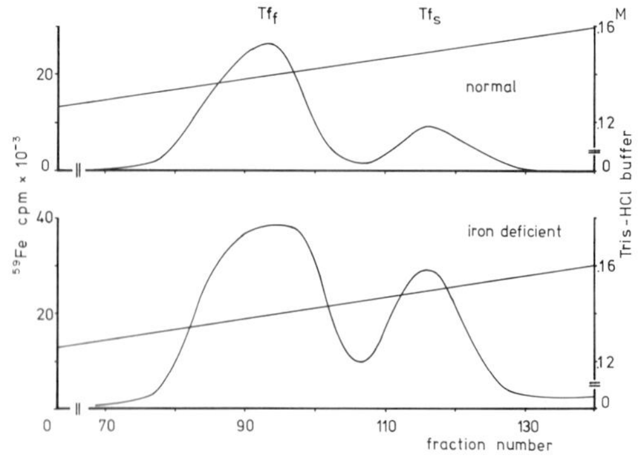

Fig. 2. Sephadex A50 chromatography of the (^{59}Fe)-iron saturated transferrin fractions from iron deficient and normal rats.

No appreciable difference was found in the amino acid analysis between Tf_f and Tf_s; further, these species were found to have identical behavior on SDS electrophoresis and in their antigenic properties (10). The absorption spectra showed maxima 280 nm and in a range from 460 to 465 nm (Fig. 3). For Tf_f the A465/A280 ratio was 0.0463 and for Tf_s 0.0474. In the following the isolation of one iron transferrin species is exemplified with Tf_f only. The behavior of the diferric Tf_f and Tf_s species in the isolation process for the one iron containing species is very similar.

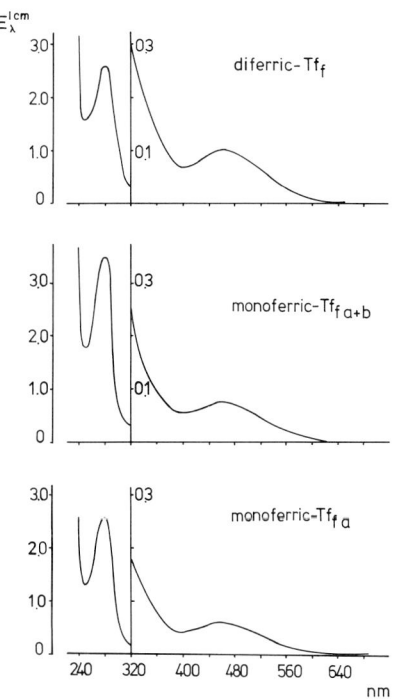

Fig. 3. Absorption spectra of some iron transferrin species isolated from rat plasma.

During gel chromatography on Sephadex G50 (pH 5.1) the diferric Tf_f loses just half of its total iron content (A465/A280 = 0.0229) and half of its radioactive iron moiety (Fig. 4), thus giving the impression that the fractions coming at the void volume contain a transferrin fraction with the overall composition being monoferric. The remaining part of ^{59}Fe radioactivity adheres to the gel and can be washed away by the addition of ascorbic acid to the elution buffer (arrow in Fig. 4). Electrophoresis of the fraction with the overall composition being monoferric reveals that there are two bands with equal protein staining and equal ^{59}Fe content (Fig. 5).

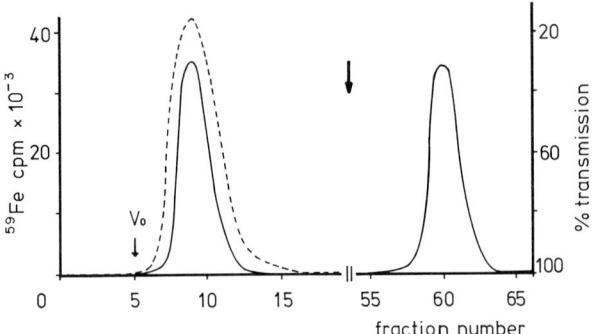

Fig. 4. Gel chromatography of rat plasma diferric Tff on Sephadex G50 at pH 5.1. arrow--addition of 20 mg ascorbic acid to the elution buffer.

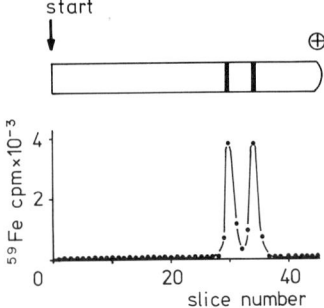

Fig. 5. Polyacrylamide-electrophoresis of the ^{59}Fe tagged transferrin fraction derived from diferric Tff after Sephadex G50 chromatography. 8.5% polyacrylamide gel; Tris/glycine buffer, pH 8.1; time 90 min at 80V, thereafter 300 min at 160V.

By cation exchange chromatography on CM-cellulose using a buffered salt gradient for elution there emerged two protein peaks with equal amounts of protein as judged by UV absorption and by protein determination with the Biuret method (Fig. 6). Only the first protein peak contained radioiron and the A465/A280 was 0.0230, thus indicating that this peak contained a monoferric transferrin (Fig. 3). The other protein fraction was found to be apotransferrin by immunological and chemical methods.

Proteins of Iron Metabolism

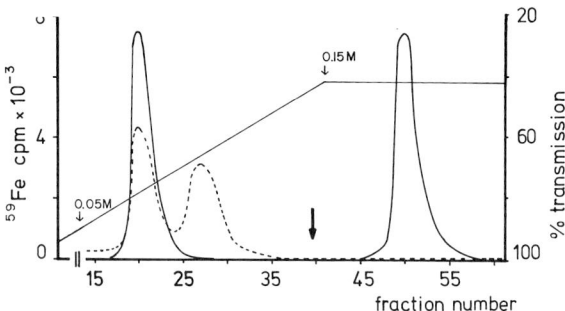

Fig. 6. Cation exchange chromatography (Whatman CM52) of the ^{59}Fe tagged transferrin fraction derived from diferric Tf$_f$ after Sephadex G50 chromatography.

The ^{59}Fe activity in the first peak (referred to as Tf$_{fa}$) amounted to 50% of the applied activity. The remaining activity adheres to CM-cellulose and can be eluted by adding some ascorbate to the elution buffer. The radioiron from the monoferric transferrin fraction can be removed completely by the addition of ascorbate and by coming to pH 4 at 4° C within 3 hrs. For the sake of clarity the results of these procedures leading to the isolation of one iron transferrin species are summarized in Fig. 7. To the explanation of the results the two plasma transferrins Tf$_f$ and Tf$_s$ must be considered to contain each two latent diferric transferrins, e.g. Tf$_{fa}$ and Tf$_{fb}$. These latent species become apparent by lowering the pH to 5.1, thereby allowing the removal of one iron from each of these species which results in the appearance of two distinct monoferric transferrin species on polyacrylamide electrophoresis. By CM-cellulose, the species Tf$_{fb}$ loses its remaining iron and can be isolated thereafter only as apo-Tf$_{fb}$. The same scheme as for Tf$_f$ is held true for Tf$_s$. The only difference was the fact that the color of the monoferric species Tf$_{sa+b}$ was less stable than that of the Tf$_{fa+b}$ control (after 2 weeks storage at 4° C at pH 5.1). Considerable differences between the various iron containing transferrins were found when looking to iron depletion kinetics, e.g. the diferric and monoferric transferrin iron depletion kinetics (pH 5.6) are quite different; the diferric transferrin species shows a much faster release of its iron at the beginning. After 4 hrs the amount of released iron is about the same (Fig. 8).

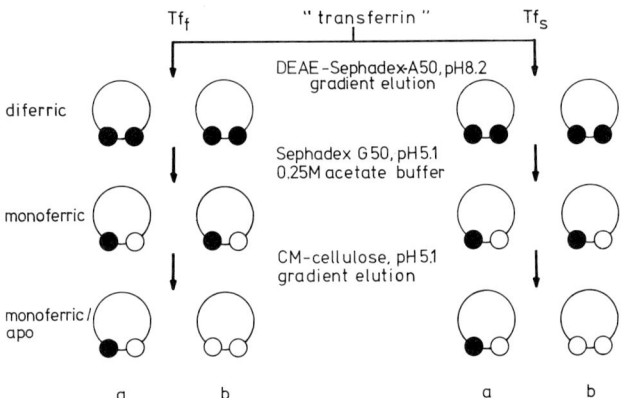

Fig. 7. Isolation scheme for the various monoferric transferrin species from rat plasma.

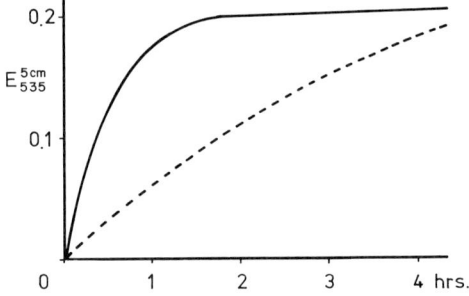

Fig. 8. Iron depletion kinetics of diferric and monoferric transferrin species. solid line--diferric transferrin Tf_{ff}; broken line--monoferric transferrin Tf_{fa}; (0.25 M acetate buffer, pH 5.6; ascorbic acid 1 mg/ml; bathophenanthroline 50 µg/ml; temperature $37°$ C).

IV. DISCUSSION

Besides the confirmation of the findings that rat plasma transferrin consists of two species (for literature see ref. 11) which can be separated by ion exchange chromatography (11) this work presents evidence that there are four well defined one-iron carrying transferrin species (Tf_{fa}, Tf_{fb}, Tf_{sa}, Tf_{sb}). In addition, it could be revealed that the behavior of the two iron binding sites on the transferrin molecule are different insofar confirming the spectroscopic studies (6-8). Simply by change of the pH just one iron atom is released from its specific binding site. A further heterogeneity of the four one iron containing transferrin species is observed under the conditions of CM-cellulose chromatography where two species are depleted of their iron. It has been established that the amount of Tf_s and Tf_f species and their related iron containing molecules is considerably altered in the iron deficient animal. No attention has been paid to this situation when assessing the Fletcher and Huehns hypothesis with the rat as an experimental animal. Indeed, this heterogeneity of the transferrin pool in rats is expected to complicate the assessment of this hypothesis by introducing several transferrin species which seem to have different iron exchange rates. Moreover, the situation is aggrevated by the fact that there is a cooperation between the various one iron transferrin species leading in general to a higher stability of the iron transferrin complex against reducing agents (13).

In conclusion, it is expected that the availability of the distinct diferric and monoferric transferrin species isolated from rat plasma will create more precise experimental conditions to obtain more insight into the heterogeneity of the iron binding protein of the plasma that is simply referred to as an entity, "transferrin."

V. ACKNOWLEDGEMENTS

Investigations in the author's laboratory were supported by a grant of the SFB38 "Membrane Research" of the University of the Saarland, West Germany.

VI. REFERENCES

1. Fletcher, J., and Huehns, E. R., Nature 215, 584 (1967).
2. Fletcher, J., and Huehns, E. R., Nature 218, 1211 (1968).
3. Brown, E. B., in "Proteins of Iron Storage and Transport in Biochemistry and Medicine," (R. R. Chrichton, ed.), p. 97. North-Holland, Elsevier, Amsterdam (1975).
4. Ganzoni, A. M., Hahn, D., and Spati, B., Blut 24, 269 (1972).
5. Hahn, D., Eur. J. Biochem. 34, 311 (1973).
6. Aisen, P., Aasa, R., and Redfield, A. G., J. Biol. Chem. 244, 4628 (1969).
7. Price, E. M., and Gibson, J. F., J. Biol. Chem. 247, 8031 (1972).
8. Aasa, R., Biochem. Biophys. Res. Comm. 49, 806 (1973).
9. Forth, W., et al., Arzneim.-Forsch. 19, 363 (1969).

10. Huebers, H., Huebers, E., Rummel, W., and Chrichton, R. R., Eur. J. Biochem. 66, 447 (1976).
11. Morgan, E. H., in "Iron in Biochemistry and Medicine," (A. Jacobs and M. Worwood, eds.), p. 29. Academic Press, New York (1974).
12. Huebers, H., in "Proteins of Iron Storage and Transport in Biochemistry and Medicine," (R. R. Chrichton, ed.), p. 381. North-Holland, Elsevier, Amsterdam (1975).
13. Huebers, H., and Huebers, E., 16th International Hematology Congress, Kyoto, Japan, in press.

VARIABLES IN THE RAT IRON-TRANSFERRIN RETICULOCYTE SYSTEM

Shigeru Okada
Elmer B. Brown

*Washington University School of Medicine
St. Louis, Missouri*

I. INTRODUCTION

Controversy persists about the Fletcher-Huehns concept of functional difference of the two iron binding sites of transferrin (1,2). Several reasons for this controversy can be identified: 1) There may be species variation and many studies have used transferrin and reticulocytes from mixed species; 2) The biologic test systems for preparing and assaying selectively-labeled transferrin are not precise; and 3) The intactness of transferrin and the specificity of binding of its iron are often poorly defined. Although we have presented extensive evidence supporting the Fletcher-Huehns theory in the rat (3-6), we now report studies of two elements that are important in the interaction of transferrin-bound iron and reticulocytes of the rat. The first study explores the effect of the iron status of the reticulocyte on its iron uptake; the second study examines the iron donating properties of two isotransferrins in the rat.

II. EXPERIMENTAL PROCEDURE

Reticulocytes were obtained by bleeding Sprague-Dawley rats. Iron-replete reticulocytes came from rats given intraperitoneal replacement of iron dextran (0.5 mg/ml of blood removed) after each bleeding; iron-deficient reticulocytes came from rats given only saline injections after bleeding. Reticulocyte uptake of ^{59}Fe and ^{125}I was measured at intervals after in vitro incubation at $37°$ as previously described (3). ^{59}Fe incorporation into heme, membrane, and non-heme fractions was also measured.

For purification and separation of isotransferrins, serum pooled from iron deficient or normal rats was used without prior saturation with iron. The fraction of serum precipitated between 34 and 60% saturation with ammonium sulfate was dissolved in water and dialyzed against five changes of 0.05 M Tris buffer, pH 8.0. After passage through a column of Sephadex G-100 equilibrated with the same buffer, the pink-colored eluate was applied to a column of DEAE-Sephadex A-50 and eluted with Tris buffer, pH 8.0, in a linear gradient ranging from 0.05 M to 0.35 M. Proteins were measured by the Lowry method (7) and iron content and iron binding capacity were measured as previously described (3). Protein solutions from each peak were pooled and dialyzed for 14 to 16 hours against citrate 0.1 M, pH 4.6, to remove endogenous iron. The apotransferrin was then dialyzed against 0.01 M HEPES-saline buffer, pH 7.5. The two isotransferrins "S" (slow) and "F" (fast) were diluted appropriately to produce the same protein concentration and were then labeled with ^{59}Fe and carrier Fe at 90% saturation for reticulocyte uptake studies. ^{125}I-labeling of transferrin was performed by the chloramine-T method (8) modified by terminating the reaction after 5 to 10 seconds to avoid denaturation. Specific iron binding to transferrin was achieved using Fe-NTA as previously described (3).

III. RESULTS AND DISCUSSION

Fig. 1 shows a comparison of the uptake of transferrin and iron by iron-replete and iron-deficient reticulocytes. Both the initial rate and total amount of transferrin taken

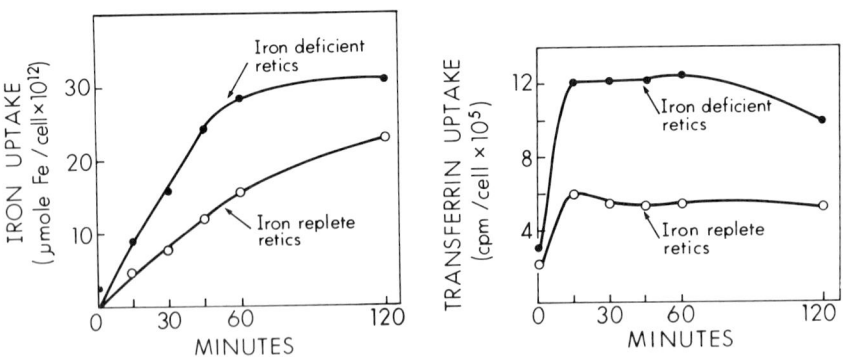

Fig. 1. Iron and transferrin uptake by reticulocytes at 37° C. Iron deficient retics from rat with SI 23, UIBC 626, retics 8.6%. Iron replete retics from rat with SI 103, UIBC 452, retics 14.1%.

up by the iron-deficient reticulocytes is greater than by the iron-replete reticulocytes. Furthermore, the rate of radio-iron uptake was faster and reached higher levels in the iron-deficient reticulocytes.

Previous studies have demonstrated that radioiron incorporation into heme is more efficient in the iron-replete reticulocytes, which suggested that nonspecific iron binding to the surface membrane or interference with heme synthesis was occurring in the iron-deficient reticulocytes (6). This was examined directly when both types of reticulocytes were incubated with ^{59}Fe-transferrin for 30 minutes, washed 3 times with cold saline, and resuspended in ^{56}Fe-transferrin to measure the remaining radioiron in the isolated ghost membranes, and in the heme and non-heme fractions of the hemolysate. As shown in Table 1, iron-replete reticulocytes

Table 1. *Effect on ^{59}Fe kinetics of incubation in isotope-free transferrin media at 37° C after 30 minutes incubation of reticulocytes with ^{59}Fe-transferrin (cpm/cell x 10⁶).*

	Reincubation time (min)				
	0	5	10	15	20
Iron Deficient Reticulocytes[a]					
Whole reticulocyte uptake	168	169	171	172	172
^{59}Fe in membrane	40	38	28	30	28
" " heme	111	115	123	125	131
" " non-heme	15	15	17	16	12
recovery	166	168	168	171	171
Iron Replete Reticulocytes[b]					
Whole reticulocyte uptake	141	144	146	149	147
^{59}Fe in membrane	30	30	25	25	22
" " heme	105	107	112	118	118
" " non-heme	8	8	7	8	5
recovery	143	145	144	151	145

Donor [a] SI 14, UIBC 664, Retics 28.4%
Donor [b] SI 92, UIBC 495, Retics 23.4%

achieved slightly higher percentages of heme synthesis with lower levels of non-heme radioiron but little difference in membrane-associated radioiron. These data point up the small but significant difference between iron-deficient and iron-replete reticulocytes. Added to the previously defined increase in the number of iron-receptor sites on younger reticulocytes (9), these observations call for caution in interpreting studies of transferrin function in which the mode of preparing reticulocytes is not defined. Workman and coworkers have also pointed out differences in reticulocytes obtained after bleeding and phenylhydrazine administration in rabbits (10).

Perhaps of even greater importance is the difference in iron delivery by the two isotransferrins isolated from rat serum. These isotransferrins were first described by Gordon and Louis in 1963 (11), and were recently re-examined by Huebers and Huebers in a report in Kyoto in 1976 (12). The isotransferrin electrophoretic pattern demonstrating the S and F component separated on DEAE-Sephadex is shown in Fig. 2. The iron content of each isotransferrin is also shown for a sample of serum that had not been stripped of iron by treatment with citrate.

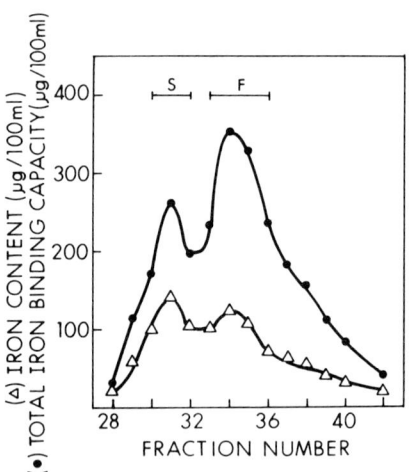

Fig. 2. Elution pattern of S and F transferrin with a 0.05 M to 0.35 M gradient of Tris buffer.

When the S peak protein labeled with ^{55}Fe and the F peak protein labeled with ^{59}Fe at the same protein concentration and iron saturation were incubated with iron-replete reticulocytes, there was a difference in the rate and total uptake of radioiron from the two isotransferrins (Fig. 3).

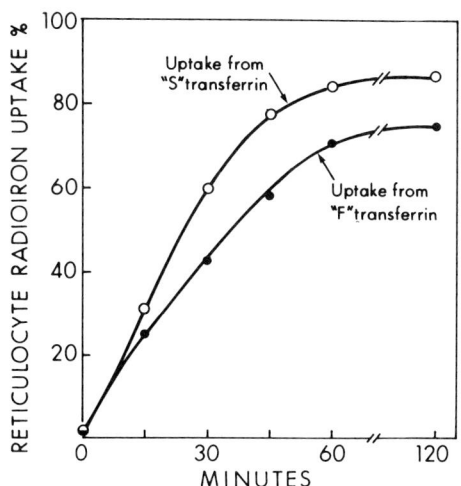

Fig. 3. *Uptake by reticulocytes of radioiron from "S" and "F" transferrins.*

S transferrin was a more efficient iron donor than the F transferrin. These results were consistent in three separate batches of purified isotransferrins with iron uptake from S significantly greater than from F ($p < 0.01$). When serum from iron-deficient rats was examined, the proportion of the S isotransferrin was approximately 2-fold greater than in the serum from normal rats, which would accentuate its efficiency of iron delivery to reticulocytes.

Many questions are raised by these observations that have not yet been answered by appropriate experiments. First, the exact relationship of these observations on isotransferrins and those of Huebers and Huebers (12) on the differences in stability of iron binding at lower pH is unclear, although the isotransferrins appear to be similar. Second, the more efficient release of radioiron to reticulocytes by S isotransferrin suggests an alternate explanation to that of Fletcher and Huehns for the differential release of iron by transferrin in the rat. Separate examination of the functional delivery of iron by sites A and B of each isotransferrin is required in the rat system as well as examination of other species for a similar phenomenon.

IV. REFERENCES

1. Aisen, P., Brown, E.B., in "Progress in Hematology" (Brown, E.B. ed.), Vol. IX, p. 25, Grune & Stratton, New York, 1975.

2. Brown, E.B., Ciba Foundation Symposium No. 51 on Iron Metabolism, 1977 (in press).

3. Awai, M., Chipman, B., Brown, E.B., J. Lab. Clin. Med. 85, 769 (1975).

4. Awai, M., Chipman, B., Brown, E.B., J. Lab. Clin. Med. 85, 785 (1975).

5. Brown, E.B., Okada, S., Awai, M., et al, J. Lab. Clin. Med. 86, 576 (1976).

6. Okada, S., Chipman, B., Brown, E.B., J. Lab. Clin. Med. 89, 51 (1977).

7. Lowry, O.H., Rosebrough, J.J., Farr, A.L., et al, J. Biol. Chem. 193, 265 (1951).

8. Tollefson, D.M., Feagler, J.R., Majerus, P.W., J. Biol. Chem. 249, 2646 (1974).

9. Kornfeld, S. Abstracts of the simultaneous sessions. XII Congress, International Society of Hematology, New York, 1968, p. 86.

10. Workman, E.F., Graham, G., Bates, G.W., Biochim. Biophys. Acta 399, 254 (1975).

11. Gordon, A.H., Louis, L.N., Biochem. J. 88, 409, (1963).

12. Huebers, H., Huebers, E., Abstracts of the XVI Congress, International Society of Hematology, Kyoto, 1976, p. 113.

PLASMA IRON KINETICS IN MAN

Karl Skarberg, Alan Christensen,
George Marsaglia, and Clement Finch
*University of Washington and
Karolinska Sjukhuset, Sweden*

Evidence has been published that iron exchange from transferrin to immature erythrocytes and to other body tissues is influenced by the plasma iron concentration (1). The purpose of this study is to examine the phenomenon in more detail with particular attention to its quantitative aspects.

Initial studies were carried out in vitro with rat plasma and rat reticulocytes (2). Radioiron labeled ferrous ammonium sulfate was added to plasma dialyzed free of iron to achieve a saturation in one instance of <10% and in the other of over 90%. The two plasmas were then adjusted to contain the same amount of iron and transferrin but with the first sample (originally low saturation) representing monoferric transferrin and the second sample (originally saturated) representing diferric transferrin. Radioactivity and iron uptake by reticulocytes were twice as great from the diferric transferrin plasma.

In vivo studies were then carried out employing two isotopes in which, again, iron was loaded on transferrin at low and high concentrations. The simultaneous tissue uptake of each isotope injected intravenously was followed in iron deficient, normal, iron injected, and hyper-transfused animals (2). Again, a more rapid tissue uptake of diferric transferrin iron was observed. Variations between tissue ratios appeared to relate also to the exposure of the individual tissue to plasma transferrin; thus, in the iron deficient animal an uptake ratio of 3.7:1 between di- and monoferric transferrin was found in circulating red cells, while in the bone marrow the ratio was 1.2:1.

These studies confirm the greater delivery of iron by the transferrin molecule containing two atoms of iron and suggest in addition that this difference was only moderately influenced by the iron status or level of erythropoiesis in the recipient animal (Table I).

TABLE I. Tissue uptake Ratios in Rats

$$\frac{\text{diferric transferrin iron}}{\text{monoferric transferrin iron}}$$

Recipient	Erythron	Liver
Iron deficient	2.1	2.6
Normal	1.9	2.5
Hypertransfused	1.6	2.4

The third study undertaken concerns plasma iron kinetics in man (3). In subjects of Group I tracer radioiron was added to plasma at the saturation of the subject. A second isotope was added and its plasma aliquot subsequently saturated so that virtually all of that radioiron was in the diferric transferrin form. Simultaneous clearance and red cell utilization of both isotopes was determined. With the assumption that loading of transferrin binding iron sites was random, calculations were made of the rate of clearance of pure monoferric and diferric transferrin iron (Table II). Diferric iron was cleared more rapidly but red cell utilization of the two isotopes did not differ. Additional studies were carried out in a second group of subjects where again the two separately labeled aliquots of plasma were injected, but in this instance the plasma iron pool of the normal subject was first increased by the intravenous injection of ferrous ammonium sulfate. Again, differences were observed between the clearance of mono- and diferric transferrin, but these differences were less than those found when the plasma iron pool was normal. A third group of patients having various hematologic disorders, some having chronic elevation of plasma iron, were studied. Ratios corresponded to those already obtained in normal subjects, depending on the plasma iron level.

TABLE II. Blood Clearance Ratios in Man

$$\frac{\text{diferric transferrin iron}}{\text{monoferric transferrin iron}}$$

	Mean Plasma Iron ($\mu g\%$)	Ratio
Normal subjects	114	2.2
Normal subjects	278	1.3

DISCUSSION

There has been evidence that iron turnover increases as the plasma iron concentration increases (1, 4, 5). This could result from the presence of more iron, or could relate to the amount of iron loading the individual transferrin molecule. The present studies, both _in vitro_ and _in vivo_, show that changes occur according to whether iron is in the diferric or monoferric form, and that these differences are found regardless of whether or not the plasma iron pool has been altered. It is tempting to suggest, based on _in vitro_ studies, that interaction of transferrin with the cell results in unloading of whatever iron the transferrin molecule holds. However, variable ratios have been observed in different groups of experimental animals, suggesting that additional factors influence the uptake process. Indeed, the ratio of reticulocyte uptake in the iron deficient animal of >2 clearly indicates a preference for the diferric transferrin molecule. In man, distribution between erythroid and non-erythroid tissues was unaffected by changes in transferrin saturation, even though the total plasma iron turnover was altered, so that the effect is a general one, independent of individual tissues.

These observations have certain implications concerning the transferrin mediated dispersal of iron within the body. It would appear essential in carrying out studies of internal iron kinetics that radioiron be distributed between transferrin molecules in the same pattern as that of the subject's plasma iron pool in general. Thus, if a patient has a saturated transferrin, it is not sufficient to label normal transferrin and inject this; rather, the labeled plasma should also be saturated. The second implication is that expressions of erythron iron turnover should be adjusted so as to permit comparison between individuals. Assuming that the purpose of the measurement is to evaluate erythropoiesis, the effect of a variable plasma iron needs to be "read out" if erythropoiesis is to be accurately evaluated. Past measurements in normal subjects have shown considerable variation. This may be considerably reduced if corrections are made for changes induced by plasma iron variation. The previous study by Cook _et al_. (6) which attempted to separate erythroid from non-erythroid iron turnover employed a correction which is quite similar to the one which may be derived from the present experiments.

REFERENCES

1. Aisen, P., Brown, E. B., Progress in Hematology IX, 25 (1975).

2. Christensen, A., Finch, C. A., unpublished data.

3. Skarberg, K., Marsaglia, G., and Finch, C. A., unpublished data.

4. Fletcher, J., Clin. Sci. 37, 273 (1969).

5. Brown, E. B., Okada, S., Awai, M., et al., J. Lab. Clin. Med. 86, 576 (1975).

6. Cook, J. D., Marsaglia, G., Eschbach, J. W., Funk, D. D., and Finch, C. A., J. Clin. Invest. 49, 197 (1970).

Section VI

CELLULAR AND SUBCELLULAR IRON METABOLISM

FACTORS INFLUENCING THE RATE OF IRON RELEASE
FROM Fe^{3+}-TRANSFERRIN-CO_3^{2-}

Gary A. Graham
George W. Bates
*Texas A&M University
College Station, Texas*

I. INTRODUCTION

In carrying out its biological role as an iron transport agent, transferrin must exhibit three important functions. First, it must be able to sequester iron assimilated from the diet or from iron storage sites; second, transferrin must carry iron through the circulatory system in a form that is nontoxic and unavailable to the nonspecific complexing agents of serum; and third, the protein must be able to recognize and donate iron to specific cellular receptors. It is becoming increasingly evident that the chemistry of transferrin is highly suited to these functions. Transferrin is able to rapidly sequester iron from a variety of ferric and ferrous complexes; the iron is bound tightly and in a fashion that is nonaccessible to nonspecific complexing agents; and three, mechanisms exist which allow the rapid and active release of iron from the specific metal binding site. In this paper we wish to report and review some of the salient chemical aspects involved in the release of iron from transferrin.

While a description of the exact molecular processes and mechanisms involved in the release of iron from transferrin to cellular receptor sites is no doubt several years away, we are certainly in a position now to examine the chemical fea-

---

The abbreviations used are: EDTA, ethylenediaminetetraacetic acid; DTPA, diethylenetriaminepentaacetic acid; NTA, nitrilotriacetic acid; BPS, bathophenanthroline sulfonate.

tures which will provide a broad outline for our understanding of these processes. We suggest five possible factors for promoting iron release from transferrin: (1) chelation, (2) reduction of Fe^{3+} to Fe^{2+}, (3) weakening of the anion-iron thermodynamic linkage, (4) disruption of the Fe^{3+}-amino acid residue bonds, (5) allosteric (comformational) alterations. It is quite likely that cellular receptor sites utilize a combination of these factors, perhaps synergistically, in promoting iron release.

II. METHODS AND MATERIALS

Human transferrin was obtained from Behring Diagnostics, Somerville, New Jersey. The protein was rendered chelate-free by dialysis against frequent changes of 0.155 $\underline{M}$ KCl (1). The protein was then dialyzed against the standard buffer, 50 m$\underline{M}$ Tris-HCl, 0.155 $\underline{M}$ NaCl, pH 7.5. The concentration was determined by spectrophotometric titration. Fully saturated Fe^{3+}-transferrin-CO_3^{2-} was prepared by the addition of FeNTA to a transferrin solution supplemented with 10 mM $NaHCO_3$. The resultant mixture was allowed to stand for 30 minutes and then dialyzed against frequent changes of 0.155 $\underline{M}$ KCl. After dialysis against the standard buffer, the concentration was determined using the absorptivity of 2.5 x 10^3 1. eq.$^{-1}$ cm.$^{-1}$ (2). The protein was then adjusted to the desired concentration by dilution with the standard buffer.

All chelates were of reagent grade and not further purified. EDTA and DTPA were obtained from Aldrich Chemical Company, Milwaukee, Wisconsin. NTA was purchased from Eastman Kodak Company, Rochester, New York.

Stock chelate solutions were made at a concentration of 0.2 $\underline{M}$ by dissolving them in dilute NaOH. The final pH was adjusted to 7.5 with 50% NaOH. Appropriate dilutions were made with the standard buffer when needed.

Absorbance measurements were carried out on a Cary Model 118C spectrophotometer. The general procedure used for following the iron removal reaction by chelates was to place 0.5 ml. of a Fe^{3+}-transferrin-CO_3^{2-} solution in a semimicro cuvette. Next, 0.5 ml. of a chelate solution was added and mixed well. The change in absorbance at 470 nm. was recorded.

Rapid kinetics were monitored on a Durrum Model 110 stopped flow spectrophotometer (2). CO_2-free techniques and preparation of Fe^{3+}-transferrin-anion complexes have been described (3, 4). The reduction-chelation solution used in the iron removal experiments consisted of 50 m$\underline{M}$ ascorbic acid and 14 m$\underline{M}$ BPS. The transferrin solutions were in 5 m$\underline{M}$ Tris at pH 7.5 at a concentration of 2.7 x 10-5 $\underline{N}$. Fe^{3+}-transferrin-CO_3^{2-} or an Fe^{3+}-transferrin-anion complex was placed in one drive syringe of the stopped flow while the desired reducing agents, chelate, anion mixture was placed in the other syringe. Spectra and pH measurement of the products were made by pooling material obtained from the collection syringe.

III. RESULTS AND DISCUSSION

A. Chelation

The role of competing chelating agents in the removal of iron from transferrin has been the best studied aspect of the iron release reaction. Studies in several laboratories (5, 6, 7) have shown a disappointingly low reaction between chelating agents such as EDTA, DTPA, and NTA in the removal of iron from Fe^{3+}-transferrin-CO_3^{2-}. Early work in Saltman's laboratory established that the iron release reaction was first order with regard to chelating agent and protein bound iron. The iron release reaction in the presence of 0.1 $\underline{M}$ EDTA, for example, requires over 12 hours to reach completion. This indicates that the metal is well shielded from attack by the chelate. It also indicates that factors other than competitive chelation alone must be operative in the physiological reaction.

In view of the importance of chelation therapy in the treatment of iron overload, we have investigated the iron release reaction mediated by DTPA. In Fig. 1 is shown the time course of this reaction at several DTPA concentrations. While the highest concentrations remove significant amounts of iron over a several hour period, theraputic levels of DTPA (approximately 0.1 m$\underline{M}$) can be seen to be quite ineffective in promoting the iron release reaction.

It has been suggested (6) that a combination of chelating agents can promote the iron release reaction. We investigated this using a combination of NTA and DTPA. Our results indicated that the rate and the extent of the reaction in the presence of two chelating agents is an additive function of the activity of the chelating agents.

B. Reduction of Fe^{3+} to Fe^{2+}

Since transferrin exhibits only a weak affinity for Fe^{2+} (8), an apparent mechanism for iron release from the protein is reduction of the metal. Johnston and Cottingham (9) exam-

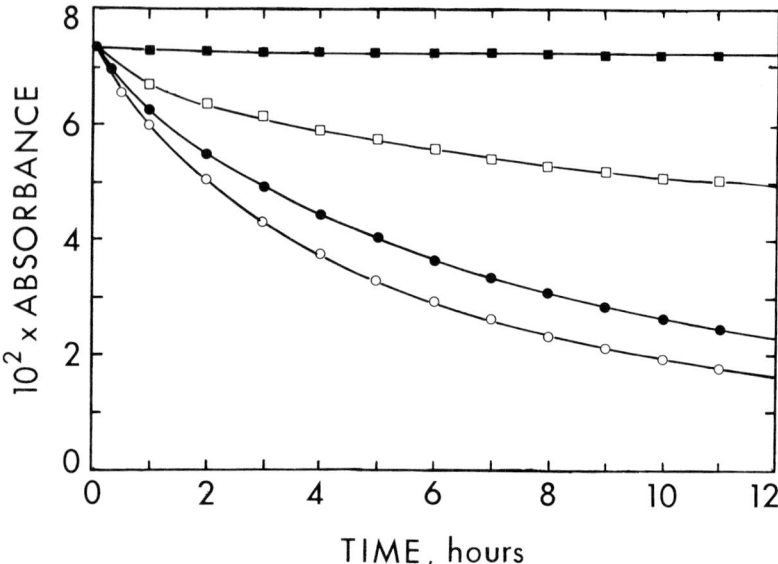

Fig. 1. The time course of the removal of iron from transferrin by DTPA. The final concentration of Fe^{3+}-transferrin-CO_3^{2-} was 3.0×10^{-5} $\underline{N}$. The final concentrations of DTPA were: 0.1 m$\underline{M}$ (■), 0.01 $\underline{M}$ (□), 0.05 $\underline{M}$ (●), and 0.1 $\underline{M}$ (○).

ined this effect in a system utilizing reduction by ascorbic acid and chelation by the chromogen 1,10-phenanthroline. We have adapted their system as a model for studying the effect of the reducing agents, however, we have modified it by using BPS as the ferrous ion acceptor. The reaction equation is given below:

$$Fe^{3+}\text{-}TRF\text{-}Anion \xrightarrow{BPS \,+\, ASC} Fe^{2+}\text{-}BPS + APOTRF + Anion$$

We have examined this reaction as a function of BPS concentration, ascorbate concentration, the presence or absence of citrate, pH, and $NaHCO_3$. The data suggest that the reaction is zero order in BPS and that there is little or no effect of added citrate. This information indicates that the chelation step is not rate limiting, but that reduction may be the controlling factor. Consistent with this is our observation that the rate is dependent on ascorbate concentration. The effect of carbonate substitutes in enhancing this reduction-chelation reaction will be treated in a separate section below.

Using the reaction system described above, we monitored the rate of iron release in a stopped flow spectrophotometer in which the transferrin was contained in one drive syringe at a pH of 7.5, while the reduction-chelation mixture was held in the other at various pH values. Upon completion of the reaction, the contents of the acceptor syringe were obtained and the pH of the reaction determined. At pH values below 7.0 a marked increase in the rate of iron release is observed. The slope of this log-log plot is 1.3 through the linear region, which suggests that protonation of more than one site may be involved. The half life of the reaction at pH 7.5 is 14 hours, much too long for physiological significance. At pH 4.1, however, the half life is only 2 seconds. It is apparent that protonation has a tremendous enhancement effect on iron release rates. Proton donation sites might include amino acid residues which are ligands to the iron, or protonation of the carbonate in the anion binding site, which could cause a reduced thermodynamic linkage between iron and anion binding.

C. Contribution of the Anion Binding Site

The elucidation of the absolute requirement of an anion for iron binding by transferrin, the confluence of the anion and iron binding sites, and description of the spectrum of carbonate substitutes are observations that are consistent with the anion playing a dominant role in the binding and release of iron by transferrin (3, 4). The twenty synergistic anions which can act as carbonate substitutes allow a wide range of stabilities of the Fe^{3+}-transferrin-anion complexes (4). The rate at which these ternary complexes release iron to citrate (a nonsynergistic anion) under CO_2 free conditions was examined. In Fig. 2 of reference 4 the very marked effect of the resident anion on the iron release reaction to citrate is exhibited. The Fe^{3+}-transferrin-oxalate complex which is more stable than the physiological carbonate complex is correspondingly slower in the iron release reaction. The relatively unstable Fe^{3+}-transferrin-maleate exhibits an enhancement of iron release by a factor of 3,600. Under physiological conditions in which transferrin is being acted upon by a cellular iron receptor complex, weakening of the CO_3^{2-}-Fe^{3+} thermodynamic linkage could enhance the rate of iron release. In the experiment described here, substitution of weaker anions enhances the rate of release of iron to a ferric chelate. We were also interested in the effect of weaker anions on the reduction-chelation reaction. In a series of experiments using the BPS plus ascorbate reaction mixture with various Fe^{3+}-transferrin-anion complexes, the relative rates of reac-

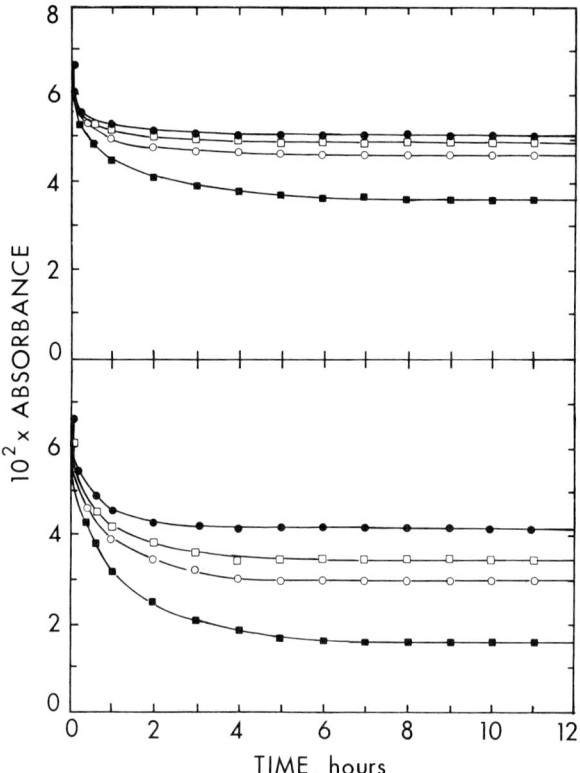

Fig. 2. The time course of the removal of iron from transferrin by NTA as a function of $NaHCO_3$ concentration. The final concentrations of Fe^{3+}-transferrin-CO_3^{2-} and NTA were 2.75×10^{-5} N and 30 mM, respectively. The final concentrations of $NaHCO_3$ were: none (■), 5 mM (○), 10 mM (□), and 25 mM (●). The reaction mixtures depicted in the top panel contained 1 mM Ca^{2+}. Those in the bottom panel had no Ca^{2+} present.

tions were enhanced in a manner that corresponded to the decrease in stability of the ternary complex. It is apparent that the anion binding site can dictate the reactivity of iron release via chelation and via reduction mechanisms. In experiments with Fe^{3+}-transferrin-NTA and the BPS-ascorbate system, we found that the addition of 50 mM $NaHCO_3$ simultaneous with the initiation of the reaction resulted in an enhancement factor of 3,000. Thr iron removal reaction under these conditions had a half life of less than 1 second. During attack by carbonate on the anion binding site an intermediate might be formed in which the anion binding site is not fully occupied. It has been suggested (3) that a vacant anion binding site may actually lead to expulsion of the metal from the protein and this could account for the very rapid kinetics that we have observed.

D. The Effect of Calcium-

It is important in relating in vitro iron release studies to chelation therapy that the work be done in an ionic environment that closely resembles that of plasma. We have examined the effect of physiological levels of bicarbonate ion and calcium ion on iron removal from transferrin by NTA. In Fig. 2 is shown the results of these studies. Two important effects are observed. First, bicarbonate concentration affects both the rate and the extent of the reaction. Physiological concentrations of bicarbonate ion decrease the amount of iron released by transferrin by a factor of 2. The stabilization of the Fe^{3+}-transferrin-CO_3^{2-} complex would be anticipated from mass action considerations. Certainly, in the evaluation of chelating agents for treatment of iron overload, physiological levels of bicarbonate ion should be included.

A rather unanticipated effect was the influence of calcium ion on iron removal. Since the NTA concentration in these experiments is 30 times that of the calcium level in the top panel, simple competition for the chelating agent is not a consideration. In the "no bicarbonate" experiment the fraction of iron removed from transferrin in the presence of calcium is 48%, while in the absence of calcium 75% of the iron is removed. Calcium apparently has a strong stabilizing effect on the Fe^{3+}-transferrin-CO_3^{2-}.

We investigated this phenomena further by carrying out a series of reactions with constant NTA and Fe^{3+}-transferrin-CO_3^{2-} and variable amounts of Ca^{2+}. A plot of the first order rate constant for the iron removal reaction vs the Ca^{2+} to Fe^{3+} ratio is shown in Fig. 3. An almost twofold decrease is noted in the reaction rate constant. Interestingly, the effect appears to plateau near 0.5 Ca^{2+} per Fe^{3+}. This suggests that perhaps 1 calcium ion is bound per molecule of Fe^{3+}-transferrin-CO_3^{2-}. The possibility that Ca^{2+} acts as an allosteric effector of the metal binding site of transferrin must be considered.

This paper has concentrated on the effects of chelating agents, reduction, carbonate substitutes, and Ca^{2+} in the reaction in which iron is released from Fe^{3+}-transferrin-CO_3^{2-}. Our future studies are aimed at an examination of iron release when a combination of several factors is operative.

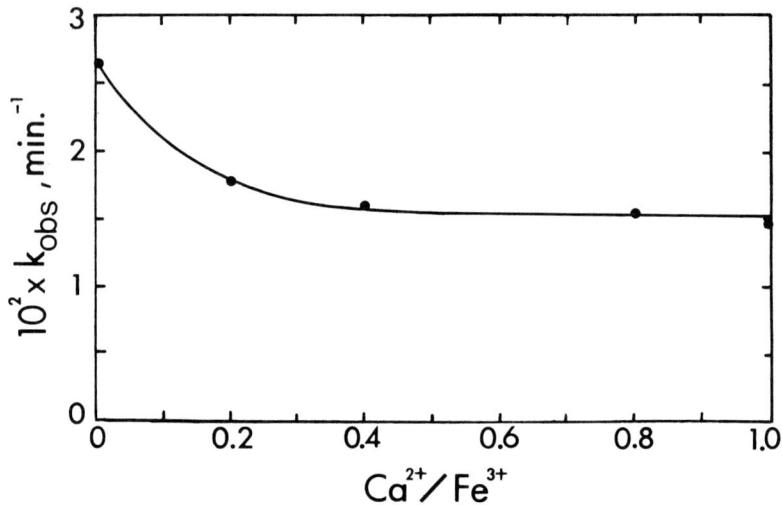

Fig. 3. *The observed first order rate constants obtained for the iron removal reaction. NTA was the competing chelate. Calcium was added to the transferrin solutions 10 minutes prior to the addition of the chelate. The rate constants were measured from plots of log fraction unreacted vs time. The final concentration of Fe^{3+}-transferrin-CO_3^{2-} was 2.0×10^{-4} $\underline{N}$. The final concentration of NTA was 10 m$\underline{M}$.*

V. REFERENCES

1. Price, E.M., and Gibson, J.F., Biochem. Biophys. Res. Comm. 46, 646 (1972).
2. Bates, G.W., and Wernicke, J., J. Biol. Chem. 246, 3679 (1971).
3. Bates, G.W., and Schlabach, M.R., J. Biol. Chem. 250, 2177 (1975).
4. Schlaback, M.R., and Bates, G.W., J. Biol. Chem. 250, 2182 (1975).
5. Bates, G.W., Billups, C., and Saltman, P., J. Biol. Chem. 242, 2816 (1967).
6. Pollack, S., Aisen, P., Lasky, F.D., and Vanderhoff, G., Brit. J. Haemat. 34, 231 (1976).
7. Harmuth-Hoene, A.E., Valdar, M., and Ohrtman, R., Chem. Biol. Int. 1, 271 (1969).
8. Bates, G.W., Workman, E.F., and Schlabach, M.R., Biochem. Biophys. Res. Comm. 50, 84 (1973).
9. Johnston, D.O., and Cottingham, A.B., Biochim. et Biophys. Acta 177, 113 (1969).

This work was supported by a Robert A. Welch Foundation research grant # A-430. G.G. is a TAMU Health Fellow.

STUDIES ON TRANSFERRIN RECEPTORS OF
ERYTHROID CELLS

Philip Aisen, Adela Leibman, Hsiang-Yun Yang Hu
and Arthur I. Skoultchi

*Departments of Biophysics, Medicine and Cell Biology
Albert Einstein College of Medicine
Bronx, New York   10461*

I. INTRODUCTION

The initial event in the delivery of iron from transferrin to erythroid cells is the binding of the protein to specific receptors on the cell surface (1-7). These receptors are susceptible to proteolytic attack, so that cells treated with Pronase or trypsin no longer bind transferrin or take up its iron. As immature red cells differentiate into mature cells they lose, along with their ability to synthesize hemoglobin, the capacity to take up transferrin and remove its iron. The fate of the transferrin receptors during cell maturation is not known.

Because of its accessibility in peripheral blood of animals made anemic by bleeding or treatment with hemolytic agents, the reticulocyte has been a convenient model for studying the interaction of transferrin with immature red cells requiring iron for the biosynthesis of hemoglobin. The reticulocyte, however, is a late stage in the development of the mature, circulating red blood cell. We have also turned, therefore, to the Friend erythroleukemic cell as a useful system for investigating the transferrin-to-cell cycle in iron metabolism. The Friend cell behaves in many ways like a transformed erythroid precursor cell arrested at an intermediate, hemoglobin-poor stage of development resembling the proerythroblast. When grown in media supplemented with dimethyl sulfoxide (DMSO), the Friend cell differentiates over the course of five days into a hemoglobinized form with characteristics of the orthochromatic erythroblast.

Thus, Friend erythroleukemic cells cultured in vitro provide a means of studying aspects of the transferrin-red cell interaction in a differentiating cell line.

II. EXPERIMENTAL PROCEDURE

Standard methods were used to prepare rabbit transferrin and to label transferrins with ^{59}Fe and ^{125}I (8). Reticulocytosis in rabbits was induced by repeated bleeding rather than by use of acetylphenyhydrazine, to avoid possible alterations in membrane structure induced by the hemolytic agent (9). Mouse transferrin was prepared by a combination of methods previously described (8). Friend erythroleukemic cells were grown at 37°C in Dulbecco-modified Eagle's medium supplemented with 10% fetal calf serum, 100 U/ml penicillin, 100 µg/ml of streptomycin, in the presence or absence of 280 nM DMSO. Incubations of mouse transferrin and Friend cells were carried out in a buffer of 0.15 M NaCl, 0.005 M KCl, 0.0074 M $MgCl_2$, and 0.01 M HEPES, enriched with 0.1% glucose and 0.2% bovine serum albumin, at pH 7.5. Incubations were terminated by addition of 16 volumes of ice-cold buffer, following which the cells were washed 3 times prior to counting or to preparation and fractionation of cell membranes.

A small amount of transferrin is adsorbed non-specifically to reticulocytes, even at 4°C, but upon incubation at 37° there is a progressive uptake of transferrin until a steady-state is reached at 10-20 minutes (10). This state is characterized by continued iron incorporation into the cell, but the amount of cell-bound transferrin remains relatively constant. The level of steady-state binding, therefore, has been taken as a measure of the number of specific cellular receptors available to complex transferrin.

Membranes were prepared by the procedures of Dodge et al (11) and solubilized and chromatographed on gel columns by methods largely derived from the work of Speyer and Fielding (3, 4). Electrophoretic analyses of membrane fractions followed the methods of Fairbanks et al (12).

III. RESULTS AND DISCUSSION

A. A Transferrin-Binding Component of Rabbit Reticulocyte Membranes.
Using LKB ultragel AcA 22 as a support medium, a macromolecular complex of transferrin and a membrane constituent was isolated from Triton X-100 solubilized membranes of

rabbit reticulocytes incubated with ^{125}I- and ^{59}Fe-labelled transferrin (Fig. 1).

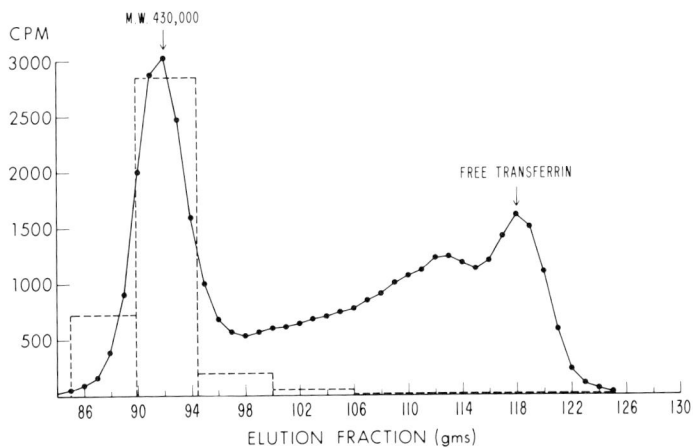

Fig. 1. Gel filtration chromatogram of Triton X-100 solubilized membranes from reticulocytes incubated with doubly-labelled transferrin. The 1.6 x 100 cm column was maintained at 18°, and eluted with 5 imosm phosphate, pH 7.4 (11), containing 1% Triton X-100. o—o, ^{125}I; o—o, ^{59}Fe. Also shown is the relative transferrin-binding activity of unincubated reticulocyte membrane fractions (---).

About 55% of the total ^{125}I activity applied to the column, representing 22% of the transferrin-binding activity of intact reticulocytes, eluted as a single, symmetrical peak. The remainder of the activity appeared in a second, broader peak at the elution position of free transferrin. The possibility that the high molecular weight transferrin-bearing complex is an artifact of transferrin aggregation was excluded by chromatographing pure transferrin, as well as a mixture of transferrin and a digest of mature circulating erythrocytes. In neither case was a high molecular weight fraction detected. Recovery of ^{59}Fe paralled that of ^{125}I, although the ratio of the isotopes differed in the two peaks. Taking as unity the ratio of the two isotopes in the transferrin used for incubation, a ratio of 0.77 is obtained in the high molecular weight peak, and of 0.54 in the lighter fraction.

Since the high molecular weight transferrin-containing fraction elutes very near the void volume of a Sephadex G-200 column, while transferrin itself emerged well within the included volume, it proved possible to use such a column to test for transferrin-binding activity of solubilized membranes of cells not incubated with transferrin. Membranes of washed, non-incubating reticulocytes were solubilized and chromatographed on a column of Ultragel AcA 22. Successive fractions were incubated with ^{125}I-labelled transferrin and applied to a second column of Sephadex G-200. Transferrin-binding activity, as measured by the appearance of ^{125}I near the void volume, was centered in a fraction with an apparent molecular weight near 400,000 (Fig. 1). This is in substantial agreement with a molecular weight of 445,000-75,000, or 370,000, for the transferrin-binding component observed in the membranes of reticulocytes incubated with transferrin. No transferrin-binding activity was demonstrable in membrane fractions from mature erythrocytes.

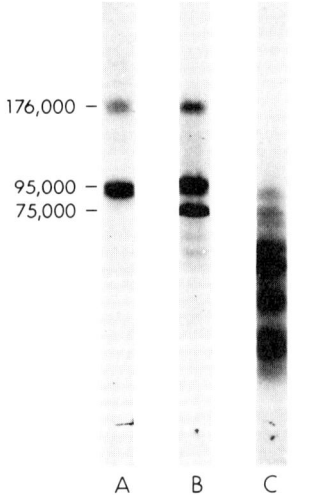

Fig. 2. SDS gel electrophoresis patterns of solubilized membranes. A. 400,000 MW fraction of mature erythrocytes. A strongly positive PAS reaction is present only in the 95,000 band. B. 445,000 MW peak of Fig. 1. Over 95% of the ^{125}I activity added to the gel is present in the 75,000 MW band. All three major bands gave strongly positive PAS reactions. C. 400,000 MW fraction of Pronase-treated reticulocyte membranes.

B. SDS-Gel Electrophoresis of Membrane Fraction.

Three distinct protein bands were present in the gel electrophoresis profile of the high molecular weight transferrin-bearing fraction in the chromatogram of Fig. 1, as shown in Fig. 2B. One of these, with a molecular weight near 75,000, contained over 95% of the radioactivity applied to the gel, and therefore represents transferrin itself. The other two fractions had weights of 95,000 and 176,000, respectively. Each of these fractions gave a strong PAS stain for carbohydrate.

A 400,000 molecular weight fraction from solubilized membranes of mature erythrocytes was also subjected to SDS gel electrophoresis, as shown in Fig. 2A. The two heavier protein bands observed in the transferrin-bearing fraction of transferrin-incubated reticulocytes are clearly identifiable. However, an important difference was obtained with the PAS stain: virtually no carbohydrate could be demonstrated in the 176,000 molecular weight band. Neither the 176,000 band nor the 95,000 band was present in membrane extracts of Pronase-treated reticulocytes.

On the basis of these experiments, we suggest that the primary transferrin receptor of the rabbit reticulocyte is a glycoprotein of molecular weight in the range 370,000-400,000, composed of a combination of subunits of molecular weights 176,000 and 95,000. The transferrin-binding function may reside in the carbohydrate moiety of the 176,000 MW subunit since this appears to be lost as the reticulocyte completes its differentiation into a mature circulating erythrocyte.

Using similar methods based on the original observations of Garrett et al (2), other workers have reported transferrin-binding activity in membrane fractions with molecular weights ranging from 35,000 to 200,000, isolated from immature red cells of a variety of species (3-7). The apparent discrepancies between this work and ours may reflect methodologic differences, species variations, or other factors, and remains a subject for continuing research.

If, indeed, the 445,000 MW fraction represents a complex of transferrin and its primary receptor, the reduced ratio of ^{59}Fe activity to ^{125}I activity in the complex as compared to that in the transferrin used for incubation suggests that iron is removed from transferrin while it is still complexed to its receptor. Whether this occurs at the cell surface (1), or within the interior of the cell (13), cannot be decided from our experiments. The finding that the $^{59}Fe:^{125}I$ ratio is even lower in the free transferrin peak of membranes from washed cells also corroborates the report that iron-depleted transferrin binds less strongly to receptors than iron-saturated transferrin, so that the iron-loaded protein would be expected to compete successfully

with apoprotein for transferrin-binding sites on the
reticulocyte surface.

C. Transferrin-Binding by the Friend Erythroleukemic Cell.
The course of transferrin uptake by Friend cells
(Fig. 3) is similar to that observed with reticulocytes,
although the time required to reach a steady state is
somewhat longer.

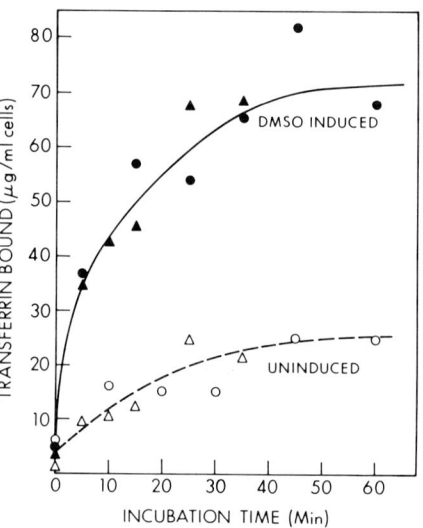

Fig. 3. Uptake of mouse
transferrin by three-day
DMSO-induced and uninduced
Friend erythroleukemic
cells from an incubation
medium containing 0.5 mg
transferrin per ml.

Cultures grown for three days in the presence of DMSO, containing about 50% benzidine-positive cells, bind nearly four times as much transferrin (per unit volume of cells) as cultures grown in the absence of DMSO, which contain less than 1% hemoglobinized cells. The capacity to take up transferrin-bound iron is also stimulated by exposure of Friend cells to DMSO: three-day cultures grown in the presence of DMSO take up 0.42 μg of iron per ml of cells per hour, while control cultures take up only 0.06 μg of iron per ml of cells. Thus, it appears that the capacity of Friend cells to interact with transferrin, a necessary event in the biosynthesis of hemoglobin, is also inducible by DMSO.

A mutant Friend cell line, which failed to synthesize hemoglobin when incubated with DMSO, likewise failed to enhance its transferrin-binding activity.

A comparison was then undertaken of the transferrin-binding components of Friend cell membranes with the putative transferrin receptor of the rabbit reticulocyte. Cells from three-day cultures, both induced and non-induced, were incubated with ^{125}I transferrin and their membranes isolated and solubilized as before. When chromatographed on a column of Ultragel AcA 22, patterns shown in Fig. 4 were obtained. In each case, two peaks with ^{125}I were observed, one eluting at approximately the same position as that of the heavier transferrin-bearing complex from reticulocyte membranes, and the other behaving like transferrin.

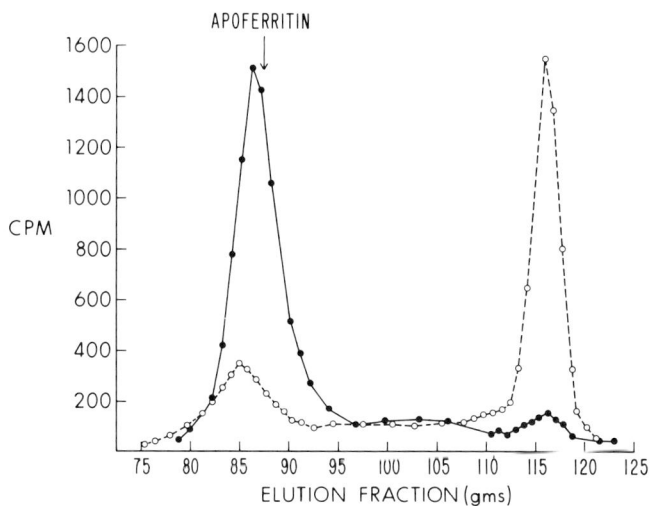

Fig. 4. Gel filtration chromatograms of Triton-X-100 solubilized membranes from DMSO-induced (o—o) and uninduced (o---o) Friend cells incubated with ^{125}I-labelled mouse transferrin.

An important difference in the two preparations was evident, however. The heavy peak was much larger than the free transferrin peak in membrane digests from DMSO-induced cells, while the reverse was true in untreated cells. If the assumption is made, therefore, that the 445,000 MW peak represents the transferrin receptor complex of the Friend cell, then the binding of transferrin to its receptor would appear to be substantially stronger in the induced than in the undifferentiated cell.

Attempts to analyze the structure of the 445,000 MW fraction were less satisfactory with the Friend cell than with the rabbit reticulocyte. In solubilized membranes from the DMSO treated cell, prominent bands were seen corresponding to molecular weights of 97,000, 62,000, 53,000 and 15,000, in addition to the transferrin band near 75,000 MW, while in extracts from untreated cells bands were present at 115,000, 95,000, 62,000, 21,000 and at the position of transferrin. We presume the complexity of the patterns reflects the complex nature of the cell membrane in a relatively primitive cell with a wide spectrum of metabolic functions, and we cannot say with confidence which components are integral to the receptor itself. The fact that the 95,000 band is seen in all preparations may suggest that this, at least, is a unit of the receptor. Despite this uncertainty, we feel that the Friend cell may offer a unique system for studying the transferrin-red cell interaction in a differentiating cell.

ACKNOWLEDGEMENT

Portions of these studies have been reported in detail elsewhere (14, 15).

These studies have supported, in part, by Grants CA 16368 and AM 15056 from the National Institutes of Health.

REFERENCES

1. Jandl, J.H., Inman, J.K., Simmons, R.L., Allen, D.W. J. Clin. Invest. 38, 161 (1959).

2. Garrett, N.E., Garrett, R.J., Archdeacon, J.W. Biochem. Biophys. Res. Commun. 52, 466 (1973).

3. Fielding, J., Speyer, B.E. Biochem. Biophys. Acta 363, 387 (1974).

4. Fielding, J., Speyer, B.E. in "Proteins of Iron Storage and Transport in Biology and Medicine" (R.R. Crichton, ed.), 121, North-Holland Publ. Co, Amsterdam, 1975.

5. Van Bocksxmeer, F., Hemmaplardh, D., Morgan, E.H. ibid., 111.

6. Witt, D.P., Woodworth, R.C. ibid., 133.

7. Sly, D.A., Grohlich, D., Bezkorovainy, A. ibid., 141.

8. Harris, D.C., Aisen, P. Biochemistry 14, 262 (1975).

9. Workman, E.F., Jr., Graham, G., Bates, G.W. Biochim. Biophys. Acta 399, 254 (1975).

10. Baker, E., Morgan, E.H. Biochemistry 8, 1133 (1969).

11. Dodge, J.T., Mitchell, C., Hanahan, D.J. Arch. Biochem. Biophys. 110, 119 (1963).

12. Fairbanks, G., Steck, T.L., Wallach, D.F.H. Biochemistry 10, 2606 (1971).

13. Morgan, E.H., Appleton, T.C. Nature, 223, 1371 (1969).

14. Leibman, A., Aisen, P. Biochemistry, in press.

15. Hu, H.-Y. Y., Gardner, J., Aisen, P., Skoultchi, A., submitted for publication.

THE ISOLATION OF TRANSFERRIN RECEPTORS FROM RETICULOCYTE MEMBRANES

B. Ecarot-Charrier, V. Grey, A. Wilczynska and
H.M. Schulman

*Lady Davis Institute for Medical Research
of the
Jewish General Hospital, Montreal, Canada*

Introduction:

The attachment of transferrin to specific membrane receptors on developing erythroid cells is thought to be the first step in a series of reactions which culminate in the incorporation of iron into hemoglobin (1). Recently, transferrin complexed with membrane components has been solubilized from reticulocyte ghosts and it was suggested that these complexes contained transferrin attached to specific membrane receptors (2,3,4). We wish to report here on the partial purification of the putative transferrin receptor from rabbit reticulocyte cell membranes and to present evidence that such material has the properties of a specific receptor -- namely, saturable and reversible binding of transferrin.

Materials and Methods:

Reticulocytes were obtained from anemic rabbits, purified on dextran gradients (5) and washed and incubated as described previously (6).

Iron-free human transferrin (Behringwerke) was saturated with iron (6) and labelled with ^{125}I using lactoperoxidase (Calbiochem) coupled to CNBr-activated Sepharose 4B (Pharmacia) (7).

Rabbit antisera to human transferrin were obtained as described elsewhere (8). Specific antibodies were purified by affinity chromatography of the immunoglobulin fractions from the immune sera on CNBr-activated Sepharose 4B coupled with human transferrin.

Following incubation of reticulocytes with human transferrin for 30 min. at 37°C, washed cells were hemolyzed in an ice bath with 7 volumes of 20 mosM sodium phosphate or Tris-HCl buffer, pH 7.4. Ghosts were collected by centrifugation at 17,000 x g for 20 min. at 4°C and were washed 3 times by resuspension in the hemolysing solution and centrifugation as above. Washed ghosts were solubilized by suspension in two volumes of the hemolysing solution containing 0.7% (v/v) Triton X-100 (Sigma Chem. Co.). After stirring for 30 min. in an ice bath the solutions were centrifuged for 3 hrs. at 80,000 x g at 4°C. The supernatant fluids, containing 75-85% of the initial transferrin binding activity, were stored at 4°C.

Gel filtration of solubilized material and transferrin on Sephadex G-200 was done at room temperature with 0.9 x 30 cm columns and a flow rate of 6.0 ml/hr. The elution buffer was either 20 mosM sodium phosphate, pH 7.4 containing 0.1% Triton X-100 (pH 7.4 buffer) or 0.1 M citrate, pH 5.0 containing 0.1% Triton X-100 (pH 5.0 buffer) as indicated in the text.

For electrophoresis in polyacrylamide gels containing Triton X-100 (9), samples of up to 25 μl were made 10% with sucrose and subjected to electrophoresis in tubes containing 1 cm long 2.5% acrylamide stacking gels made with 0.62 M Tris-HCl, 0.1% Triton X-100, pH 6.7 and 7 cm long 5% acrylamide running gels made with 0.38 M Tris-HCl, 0.1% Triton X-100, pH 8.9. The buffer chambers contained 0.005 M Tris, 0.038 M glycine, pH 8.4. Triton X-100 (0.1%) was present in the upper chamber only. Electrophoresis was at 2.5 mA per gel for 2 hrs. at room temperature. The distribution of radioactivity was determined in 1 mm slices of gels which had been fixed with 12.5% CCl_3COOH. Proteins were located by spectrophotometric scanning of gels at 530 nm which had been stained with Coomassie blue (10).

SDS polyacrylamide gel electrophoresis was performed by the method of Fairbanks et al (10) as modified by Steck and Yu (11).

Sedimentation profiles were obtained from 150 μl samples layered on 6 to 21% (w/v) linear sucrose gradients made up in 20 mosM sodium phosphate, pH 7.4, containing 1% Triton X-100 and centrifuged at 4°C in a Beckman SW-56 rotor at 30,000 rpm for 17 hrs. Fractions collected through the bottoms of the cellulose nitrate tubes were analyzed for radioactivity.

Polyethylene glycol (Carbowax 6000) was a product of Union Carbide.

Results and Discussion:

When Triton X-100 extracts of ghosts from reticulocytes which had been incubated with ^{125}I-transferrin were chromatographed on Sephadex G-200, two peaks of radioactivity eluted from the column (Fig. 1) in a pattern similar to that

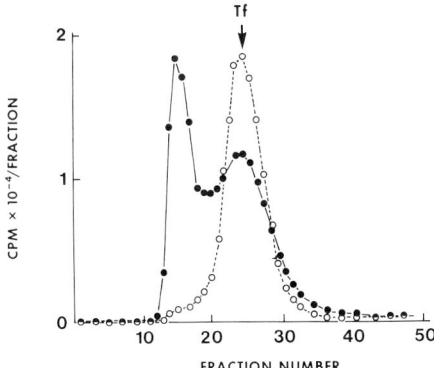

Fig. 1 *Sephadex G-200 gel filtration at pH 7.4 of ^{125}I-transferrin (○) and a Triton X-100 extract of membranes from cells incubated with ^{125}I-transferrin (●).*

reported previously (4). The lower molecular weight peak corresponded to free transferrin while the higher molecular weight peak eluted from the column just after the void volume. We could also demonstrate that these extracts contained radioactive material with a higher sedimentation coefficient (Fig. 2) and a lower electrophoretic mobility (Fig. 3) than

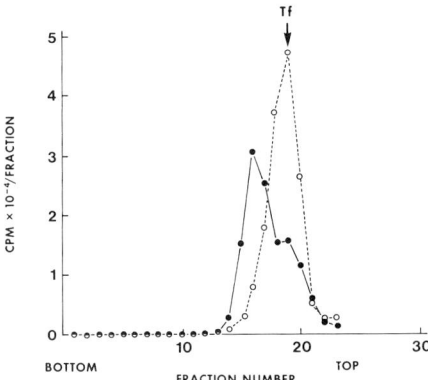

Fig. 2 *Sucrose density gradient centrifugation of ^{125}I-transferrin (○) and a Triton X-100 extract of membranes from cells incubated with ^{125}I-transferrin (●).*

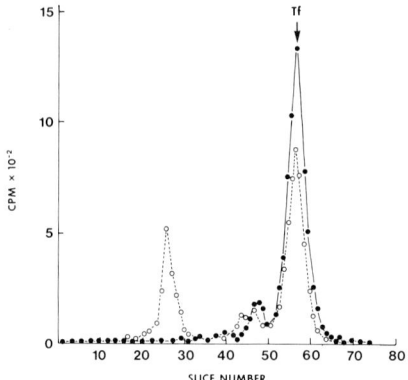

Fig. 3 Triton X-100 gel electrophoresis of ^{125}I-transferrin (●) and a Triton X-100 extract of membranes from cells incubated with ^{125}I-transferrin (○).

free transferrin. The presence of Triton X-100 (0.1 - 1.0%) was required to obtain two forms of soluble radioactivity by these techniques, for when it was removed from the extracts (12) or was absent during the analytical procedures, soluble radioactivity was associated only with free transferrin. Incubation of the extracts with excess free non-radioactive transferrin prior to analysis by gel filtration, ultracentrifugation or electrophoresis resulted in the appearance of radioactivity associated only with material corresponding to free transferrin. Conversely, if ^{125}I-transferrin was incubated with extracts prepared from cells which had been incubated with non-radioactive transferrin, radioactivity was present in the higher molecular weight fractions obtained by gel filtration or ultracentrifugation. The results of such exchange experiments suggest that the solubilized membrane material is capable of binding transferrin reversibly and that the free transferrin found after gel filtration, ultracentrifugation and electrophoresis of the solubilized membrane fraction might have resulted from the dissociation of transferrin from components in the membrane extract.

Since both insulin (13) and gonadotropin (14) are precipitated by polyethylene glycol when bound to their specific membrane receptors, the solubility of transferrin in polyethylene glycol in the presence and absence of membrane extracts was examined to see whether a convenient assay for transferrin-binding activity could be developed. The solubilized membrane fraction from reticulocytes which had been incubated with ^{125}I-transferrin was treated with various concentrations of polyethylene glycol in the presence of rabbit gamma-globulin as carrier and the amount of transferrin precipitated was measured. Fig. 4 shows that the presence of

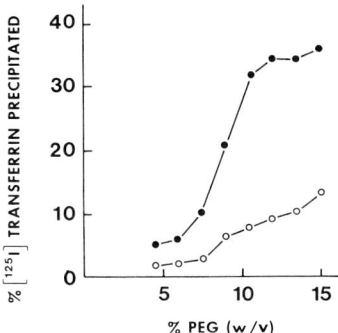

Fig. 4 The effect of polyethylene glycol (PEG) on the solubility of free ^{125}I-transferrin (○) and ^{125}I-transferrin present in a Triton X-100 extract of membranes from cells incubated with ^{125}I-transferrin (●).

the solubilized fraction from reticulocyte membranes increases the amount of transferrin precipitated by polyethylene glycol. The amount of transferrin precipitated by polyethylene glycol in the presence of a solubilized fraction from erythrocyte membranes was indistinguishable from the amount precipitated when transferrin alone was present. In the case of reticulocyte extracts the absence of transferrin bound to membrane components in the supernatant fluid following polyethylene glycol precipitation and its presence in the precipitate was confirmed by gel filtration on Sephadex G-200.

An assay for transferrin binding activity based on the differential solubility of free and bound transferrin in 12% (w/v) polyethylene glycol was utilized to study factors affecting the stability of the transferrin-receptor complex. It was found that while transferrin was readily exchangeable at pH 7.4 the complex was stabilized at pH 5.0 and that reversible binding of transferrin was restored when the complex which was at pH 5.0 was brought to pH 7.4. No difference between the binding of iron-saturated and iron-free transferrin could be detected. These effects of pH on the stability of the transferrin-receptor complex were used in our attempts to isolate the receptor from Triton X-100 extracts of reticulocyte membranes.

Our strategy for attempting to separate a transferrin binding component from other solubilized reticulocyte membrane constituents utilizes the observation that rabbit reticulocytes which had been incubated with human transferrin at

37°C and washed at 4°C could be specifically agglutinated by rabbit antibodies to human transferrin. Triton X-100 extracts of rabbit reticulocyte membranes containing an excess of human transferrin were brought to pH 5.0 and incubated for 30 minutes at 37°C and overnight at 4°C with purified antibodies to human transferrin at a concentration which precipitated all of the transferrin present in the mixture. The immunoprecipitates were washed exhaustively in 0.1 M citrate, pH 5.0 containing 0.1% Triton X-100 and then incubated in 0.1 M Tris-HCl, pH 8.0 containing 0.1% Triton X-100 at room temperature for 30 minutes. The supernatant fluids, in which transferrin was absent, contained about 50% of the transferrin binding activity which was present in the original extracts. Such preparations are capable of binding transferrin reversibly and as is shown in Fig. 5 their transferrin binding capacities are saturable.

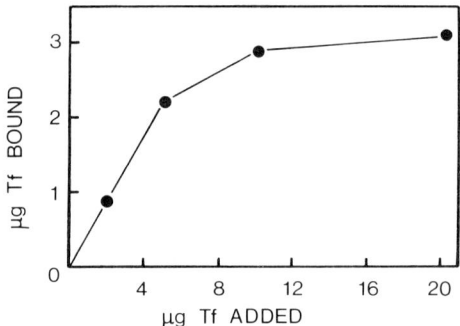

Fig. 5 The effect of transferrin concentration on the amount of transferrin found in the first peak eluted from Sephadex G-200 at pH 5.0 after incubation of receptor with transferrin at pH 7.4.

The material which was dissociated from antibody-transferrin-receptor complexes at pH 8.0 and which contained transferrin binding activity migrated as a single band during electrophoresis in polyacrylamide gels containing Triton X-100 (Fig. 6). When analyzed by SDS polyacrylamide gel electrophoresis it also migrated as a major band (upper half of Fig. 7) in contrast to the complex composition of the total Triton X-100 extract of reticulocyte membrane (lower half of Fig. 7).

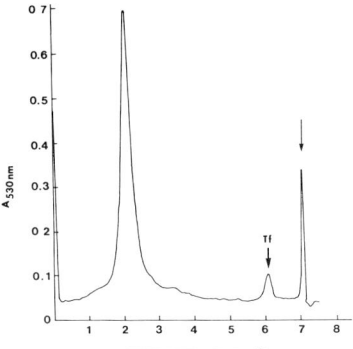

Fig. 6 Triton X-100 gel electrophoresis of receptor dissociated from antibody-transferrin-receptor complexes. The unlabelled arrow refers to the tracking dye.

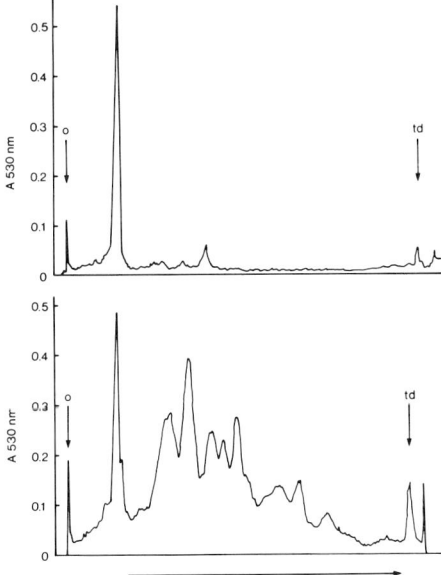

Fig. 7 SDS gel electrophoresis of receptor dissociated from antibody-transferrin-receptor complexes (upper curve) and the Triton X-100 extract of reticulocyte membranes (lower curve).

In summary, by utilizing specific antibodies to transferrin and the stabilizing effect of pH 5.0 on the association between transferrin and its solubilized reticulocyte membrane receptor, we have been able to isolate a fraction of the membrane which migrates as a single component on gel electrophoresis and has the properties of saturable and reversible binding of transferrin.

ACKNOWLEDGEMENTS

This work was supported by a grant from the Medical Research Council of Canada. Dr. Charrier was a Medical Research Council Fellow.

REFERENCES

1. Jandl., J.H., Inman, J.K., Simmons, R.L., Allen, D.W., J. Clin. Invest. 38, 161 (1959).
2. Speyer, B.E., Fielding, J., Biochim. Biophys. Acta 332, 192 (1974).
3. Fielding, J., Speyer, B.E., Biochim. Biophys. Acta 336, 387 (1974).
4. van Bockxmeer, F., Hemmaplardh, D., Morgan, E.H., in "Proteins of Iron Storage and Transport in Biochemistry and Medicine" (Crichton, R.R. ed.) p. 111. North-Holland Publishing Co., Amsterdam, 1975.
5. Schulman, H.M., Biochim. Biophys. Acta 148, 251 (1967).
6. Martinez-Medellin, J., Schulman, H.M., Biochim. Biophys. Acta 264, 272 (1972).
7. David, G.S., Biochem. Biophys. Res. Comm. 48, 464 (1972).
8. Galet, S., Schulman, H.M., Bard, H., Pediat. Res. 10, 118 (1976).
9. Dewald, B., Dulaney, J.T., Touster, O., in "Methods in Enzymology" (Fleischer, S., Packer, L. eds.) Vol. XXXII, p. 84. Academic Press, New York, 1974.
10. Fairbanks, G., Steck, T.L., Wallach, D.F.H., Biochemistry 10, 2606 (1971).
11. Steck, T.L., Yu, J., J. Supramol. Struct. 1, 220 (1973).
12. Holloway, P.W., Anal. Biochem. 53, 304 (1973).
13. Cuatrecasas, P., Proc. Natl. Acad. Sci. 69, 318 (1972).
14. Dufau, M.L., Charreau, E.H., Catt, K.J., J. Biol. Chem. 248, 6973 (1973).

ON THE BINDING SITES OF IRON-TRANSFERRIN ON RAT RETICULOCYTES

C. van der Heul, M.J. Kroos and H.G. van Eijk.

Departments of Internal Medicine I and Chemical Pathology,
Medical Faculty,
Erasmus University Rotterdam/The Netherlands.

I. INTRODUCTION.

The uptake of iron by reticulocytes is a problem which involves an adsorption of the iron-binding protein transferrin and an uptake of iron into the cell. A lot of work has been done to identify the membrane receptor for transferrin, however, untill now the real composition of this receptor is unknown. In 1974 Fielding and Speyer (1,2) described for human reticulocytes three membrane components which are involved in the iron transport into the cell. One of these components, classified as B_2, is probably the transferrin receptor complex, another component B_1 may function as an intermediate between B_2 and intracellular iron-transporting components. This work has partly been conformed for rabbit reticulocytes by Bezkorovainy (3) Being interested in the function and the composition of the membrane receptor, it is necessary to obtain in an easy way immature red cells for investigation. For practical reasons we focussed our attention on the rat reticulocytes and bone marrow cells, the more as it is known from literature (4,5,6) that there exists a striking resemblance between the incubation studies of transferrin with reticulocytes descended from man and several types of animals.

II. MATERIAL AND METHODS.

Reticulocytosis was produced in male Wistar rats by removal of 6 ml of blood, by means of orbitapunction each day on 5 days, with a 3 day-interval between the third and fourth specimens, which resulted in reticulocytosis of 25-65% (7). Rat transferrin and antibodies to rat transferrin were isolated as has been described (8). The rat transferrin was iodinated following Katz (9). Double labeled ^{59}Fe, ^{125}I-transferrin was prepared as has been described (10). For the incubation of reticulocytes, erythrocytes and reticulocyte ghosts with transferrin, the preparation of membrane complexes and the separation of the components over Sepharose we followed the methods of Fielding (1,2).

Rabbits were immunised with rat reticulocyte ghosts and the B_2-fraction. The γ-globulin fraction, free of the antitransferrin fraction by CNBr-activated Sepharose-transferrin column (8), was digested in Fab- and Fc-fragments following standard methods.
Immunofluorescence studies on reticulocytes and bone marrow cells were performed with pure rat transferrin, antitransferrin and fluorescent γ-globulin (TRITC). Glycoprotein fractions of the reticulocyte membranes were isolated following the methods of Hamaguchi and Cleve (11). Selective solubilisation of reticulocyte membranes was performed following the work of Rosenberg and Guidotti (12, 13).

III. <u>RESULTS</u>.

Bone marrow cells of normal rats incubated with transferrin and studied with indirect immunofluorescent techniques showed a distinct fluorescence in about 34% of the cells. Differentiation of the bone marrow cells resulted in 36% of erythroid cells. After inducing an anaemia the fluorescence increased to 50%, while the erythroid cell content increased to 52%, indicating that the fluorescence is specific for the immature red cells and that during the process of iron uptake transferrin is bound at the outside of the membranes.

Following the work of Fielding and Speyer we were able to isolate two radioactive membrane components from rat reticulocytes on Sepharose 2B (Fig. 1). Rechromatography of the B-peak on Sepharose 6B resulted in three radioactive peaks B_1, B_2 and B_3 (Fig. 2). Our fractions B_1 and B_2 are, based on the position in the elution diagram, similar with Fielding's fractions B_1 and B_2 for human reticulocytes. B_3 represents partly an unspecific iron- and transferrin-binding membrane component, because of the fact that this fraction reacts with antibodies against reticulocyte ghosts, giving use to a ^{125}I- and ^{59}Fe-labeled precipitate. Another part of fraction B_3 represents transferrin. Like Fielding we found in fraction B_1 nearly only ^{59}Fe-activity and no ^{125}I-activity. Kinetic studies showed that ^{59}Fe in fraction B_1 reaches its peak value after B_2. It was shown that the transport of ^{59}Fe from B_2 to B_1 was time- and temperature-dependent.

Incubation of rat erythrocytes and reticulocyte ghosts and ghost fragments after solubilisation with Triton X-100, with

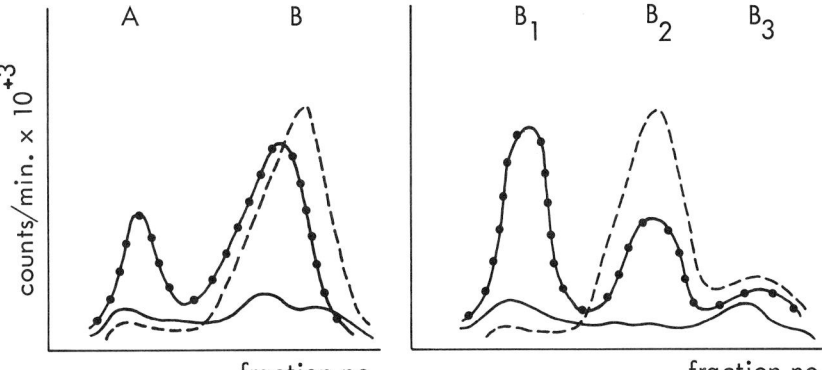

fig. 1. Solubilised membrane on Sepharose-2B

fig. 2. B-peak of fig. 1 on Sepharose-6B

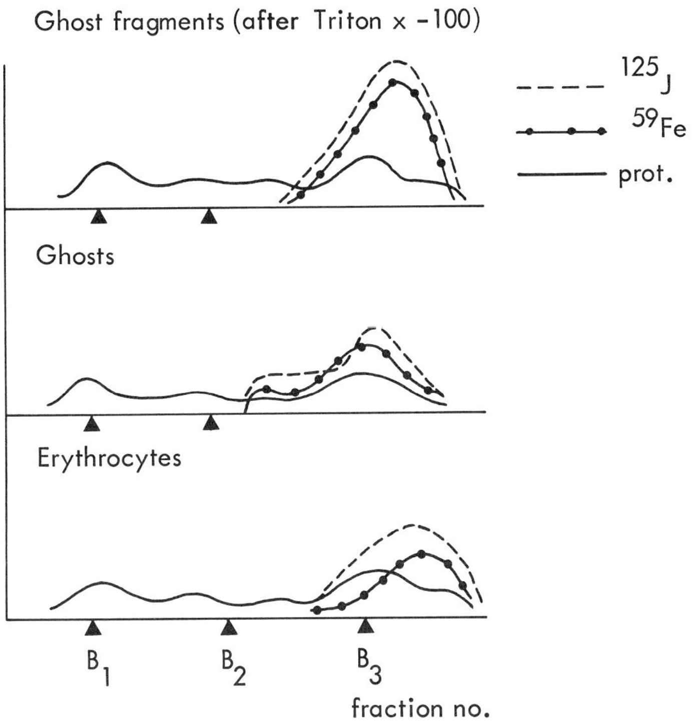

fig. 3. Chromatography patron on Sepharose-6B of incubated erythrocytes, ghosts and ghost fragments (after Triton x -100)

double labeled transferrin resulted in ^{59}Fe- and ^{125}I-activity in B_3 and no activity in A, B_1 and B_2, which indicates that for specific binding of transferrin to the receptor intact immature red cells are necessary (Fig. 3).

Sedimentation analysis of B_2, after partly removing Triton X-100, showed heterogeneity with, among other components, probably the receptor with a sedimentation coefficient of about 7 S.

To investigate the specific physiological role of the receptor during the process of transferrin binding and iron uptake by the cell we wanted to block the receptor with its antibodies. For that reason we produced antirat-B_2-γ-globulin in rabbits. The antirat-B_2-γ-globulin reacts among others with transferrin and the transferrin-receptor complex (Fig. 4).

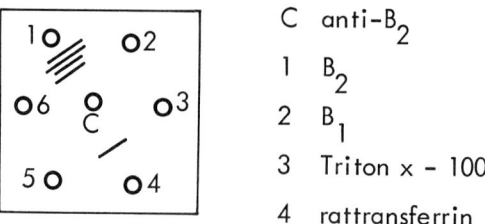

C anti-B_2
1 B_2
2 B_1
3 Triton x - 100
4 rattransferrin

fig. 4. Immunodiffusion plate of anti B_2 serum

The antitransferrin part was removed from this antiserum by affinity chromatography over a CNBr-activated Sepharose transferrin column. As expected the antirat-B_2 causes an agglutination with a reticulocyte-rich cell suspension. To avoid agglutination, in order to study the blocking effect, Fab-fragments of antirat-B_2 were prepared. An incubation of the cell suspension with these Fab-fragments before incubation with ^{125}I-^{59}Fe-transferrin causes a decrease in iron uptake.

In order to determine the position of our B_2-fraction on the membrane we tried to localise the B_2-fraction in one of the glycoproteins of the cell membrane, which possesses many antigenic determinants, or in the known and partly characterised membrane fractions, as described in the work of Rosenberg and Guidotti. The transferrin receptor on the cell

membrane is not present in the glycoprotein fraction, as all ^{125}I-activity remains in the interphase after extraction of the membranes with chloroform/methanol.
It was shown that the transferrin receptor probably is present in one of those membrane proteins which remains after extraction the reticulocytes with chelating agents, NaCl and organic solvents (Fig. 5).

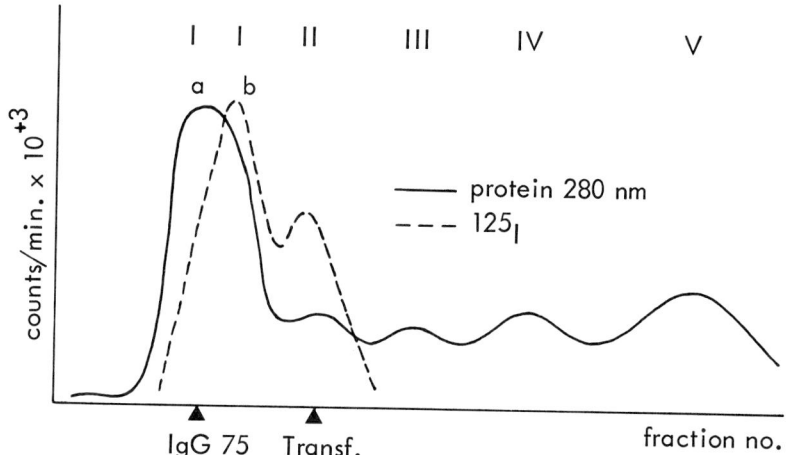

fig. 5. Lipid extracted rat reticulocyte membranes solubilised in 3% SDS and chromatographied on Sephadex G-150

All these proteins (I-V) react with antirat-B_2, indicating again that B_2 and antirat-B_2 are heterogeneous.
We expect the receptor component(s) to be present in fraction I_b (Fig. 5). The retention on Sepharose G-150 indicates a molecular weight of about 150,000 for the transfer-receptor complex, which means a molecular weight for the receptor of about 80,000 in agreement with a sedimentation analysis (6 S).

REFERENCES.

1. Speyer, B.E. and Fielding, J., Biochim. Biophys. Acta 332, 192-200 (1974).
2. Fielding, J. and Speyer, B.E., Biochim. Biophys. Acta 363, 387-396 (1974).
3. Sly, D.A., Grohlich, D. and Bezkorovainy, A. in "Proteins of Iron Storage and Transport in Biochemistry and

Medicine", ed. R.R. Crichton, p. 141-145, North-Holland Publishing Company, Amsterdam (1975).
4. Jandl, J.H., Inman, J.K., Simmons, R.L. and Allen, D.W., J. Clin. Inv. 38, 161-185 (1959).
5. Jandl, J.H. and Katz, J.H., J. Clin. Inv. 42, 314-326 (1963).
6. Verhoef, N.J., Kremers, J.H.W. and Leijnse, B., Biochim. Biophys. Acta 304, 114-122 (1973).
7. Verhoef, N.J. and Noordeloos, P.J., Cli. Sci. and Mol. Med. 52, 87-96 (1977).
8. v. Eijk, H.G. and v. Noort, W.L., J. Clin. Chem. Clin. Biochem. 14, 475-478 (1976).
9. Katz, J.H., J. Clin. Inv. 40, 2143-2152 (1961).
10. Verhoef, N.J. and v. Eijk, H.G., Cli. Sci. and Mol. Med. 48, 335-340 (1975).
11. Hamaguchi, H. and Cleve, H., Biochem. Biophys. Res. Commun. 47, 459-464 (1972).
12. Rosenberg, S.A. and Guidotti, G.J., Biol. Chem. 243, nr. 8, 1985-1992 (1968).
13. Rosenberg, S.A. and Guidotti, G., J. Biol. Chem. 244, nr. 19, 5118-5124 (1969).

NEW EVIDENCE FOR THE INTERNALIZATION OF

FUNCTIONAL TRANSFERRIN IN RABBIT RETICULOCYTES

J. Martinez-Medellin, H.M. Schulman*, E. De Miguel*
and L. Benavides

*Departamento de Biologia, Facultad de Ciencias
Universidad Nacional Autonoma de Mexico*

Introduction:

The mechanisms involved in the delivery of iron by transferrin to red cell precursors for the synthesis of heme are still not completely understood. There is some controversy as to whether (1-4) or not (5-8) transferrin is internalized by the cells prior to release of its iron.

The uptake of transferrin by reticulocytes can be divided into temperature-independent and temperature-dependent steps. The first has been ascribed to the binding of transferrin to specific cell membrane receptors and the second to a metabolically-dependent uptake of transferrin into the cells (9), an interpretation which has been disputed (8). Two of the most direct demonstrations of the internalization of transferrin by red cell precursors were electron microscope studies which showed transferrin entering the cells by what appeared to be a pinocytosis-like mechanism (10-11). However, in these experiments ferritin-conjugated transferrin was used and although it was shown that unconjugated ferritin was not bound to the cell membranes or taken up by pinocytosis, the cell-mediated release of iron from ferritin-conjugated transferrin was not demonstrated.

The experiments to be reported here were undertaken to determine whether or not transferrin molecules are internalized by reticulocytes under conditions in which the cell-mediated release of iron from the transferrin is known to occur (2). Our approach has been to compare the accessibility

---

*Lady Davis Institute for Medical Research

of reticulocyte-bound transferrin to iodination by lactoperoxidase (LPO) with cells which had been incubated with transferrin at 4°C and 37°C.

Materials and Methods:

Rabbit apotransferrin, purified and saturated with iron (2), was iodinated with ^{131}I (New England Nuclear) using lactoperoxidase (Calbiochem) coupled to CNBr-activated Sepharose 4B (Pharmacia) (12). Approximately one iodine per transferrin molecule was incorporated.

Goat antiserum to rabbit transferrin was prepared as described elsewhere (2). The gamma globulin fraction, isolated by ammonium sulfate precipitation, was coupled to CNBr-activated Sepharose 4B (13).

Red blood cells, obtained from anemic rabbits (25-30% reticulocytes) were washed twice in phosphate buffered saline, pH 7.2 (PBS), before incubation. For the incorporation of transferrin into reticulocytes, 0.5 ml of packed cells were incubated for 20 min. at either 4°C or 37°C in 1.5 ml of serum-free medium (2) containing 7.4 µM ^{131}I-transferrin, after which the cells were washed three times in 5 ml of PBS at 4°C.

For the iodination of cells with LPO, the washed cells were resuspended at 4°C in 2.2 ml of PBS containing 0.3 mCi of Na ^{125}I (New England Nuclear), 5 x 10^{-9} M KI and 91 µg/ml of LPO. (At 4°C, LPO concentration is rate limiting. The concentration used resulted in an efficiency of iodination of approximately 5%). Iodination was started by the addition of 10 µl aliquots of 10^{-2} M H_2O_2 at 15 min. intervals four times, after which the cells were washed three times at 4°C with 5 ml of PBS. The cells were then incubated for 30 min. at 37°C in 2 ml of PBS. After removal of the cells the supernatant fluids were dialyzed at room temperature against 1 mM Tris-HCl, pH 8, containing 10^{-6} M KI and then against PBS.

Transferrin was isolated from the dialyzed supernatant fluids by specific adsorption to CNBr-Sepharose coupled with antibodies to rabbit transferrin. The supernatant fluids, containing 5 mg of bovine serum albumin were incubated with 200 µl of the Sepharose for 24 hrs. at room temperature. Controls containing 5 mg of unlabelled transferrin in addition to the supernatant fluids were prepared in order to account for non-specific binding to the Sepharose beads. The Sepharose beads were then washed twice in PBS and once each in 0.1 M borate (pH 8.5), containing 1 M NaCl, and 0.1 M acetate (pH 4), containing 1 M NaCl and then assayed for ^{125}I and ^{131}I in a two channel automatic gamma counting system.

About 1% of the total cell bound ^{125}I was recovered as transferrin by these methods.

Results and Discussion:

Briefly, our experiments involved incubating reticulocytes with ^{131}I-transferrin at either 4°C or 37°C, washing away the excess transferrin at 4°C, and reacting the cells with LPO and ^{125}I at 4°C -- a procedure which iodinates only cell surface components (14). After removal of LPO and unbound ^{125}I, transferrin was released from the cells by incubating them at 37°C and was purified from the incubation media by adsorption to specific solid phase antibodies.

As shown in Table I almost 3.7 times more transferrin was taken up by the cells at 37°C than at 4°C, confirming what had been reported previously (2). It can also be seen that very little transferrin was lost from the cells during the LPO reaction and washing and that a large proportion of the transferrin was released from the cells during the subsequent incubation at 37°C. Seventy-seven percent and 61% respectively of the transferrin present after LPO treatment and washing was recovered from the cells which had been incubated at 37°C and 4°C. The absence of transferrin in the medium probably accounted for the substantial amount of transferrin which remained bound to the cells.

TABLE I Amount of Transferrin Taken Up and Released by Cells Incubated at 37°C and 4°C.

Temperature at which cells were incubated with Tf*	37°C	4°C
Tf bound to cells after initial incubation	41.7 µg (100%)	11.3 µg (100%)
Tf bound to cells after iodination with LPO and washings at 4°C	38.1 µg (91%)	8.3 µg (73%)
Tf present in supernatant fluids after incubation at 37°C	29.4 µg (71%)	5.1 µg (45%)

*Specific activity was 14,000 cpm ^{131}I/µg Tf.

Table II shows that the total LPO-catalyzed ^{125}I incorporation into cell-bound transferrin for the cells incubated at 4°C is about double the amount obtained with the cells which had been incubated at 37°C, in spite of the fact that the latter contained almost six times more transferrin than the former. This suggests that twice as many transferrin molecules were attached to the cell surfaces at 4°C than at 37°C.

TABLE II Specific Activity of Transferrin Present in the Supernatant Fluids After Incubation of the Cells at 37°C

Temperature at which cells were incubated with Tf	μg Tf released at 37°C	LPO catalyzed ^{125}I in Tf released (cpm x 10^{-3})	Specific Activity (cpm ^{125}I/μg Tf)
37°C	29.4	45.1	1540
4°C	5.1	98.7	19500

$$\frac{\text{Specific Activity 37°C}}{\text{Specific Activity 4°C}} = 0.08$$

The ratio of 0.08 (Table II) of the specific activities, in terms of LPO iodination, of the transferrin from the cells incubated at 37°C and 4°C suggests that only about 8% of the transferrin taken up at 37°C was on the external surface of the cells (assuming that 100% of the transferrin taken up at 4°C was so situated). This ratio should have been one if all of the transferrin molecules were bound to the cell surface at both temperatures. This conclusion, that most of the transferrin taken up at 37°C is either inside the cells or very deep within the cell membranes, and thus inaccessible to the action of LPO, is consistent with the results of Hemmaplardh and Morgan (15) who used proteases to liberate transferrin bound to the cell surface.

The fact that only about half as much transferrin was bound to the cell surface at 37°C compared to 4°C, in spite of the presence of saturating concentrations of transferrin in the incubation media, may indicate that each transferrin molecule binds to two cell surface receptor sites at 37°C and only one site at 4°C. This interpretation of the data leads to the prediction that cells saturated with transferrin at

4°C would lose about half of their transferrin molecules immediately upon incubation at 37°C. Such an unexplained effect of temperature has been observed by Hemmaplardh and Morgan (15).

Based on these data and other available evidence we would like to postulate that the following series of steps, which precede the release of iron from transferrin, are involved in transferrin uptake by reticulocytes. (i) Transferrin binds to a specific membrane receptor by an energy independent process. (ii) The transferrin-receptor complex undergoes an energy-dependent change resulting in the appearance of a second binding site on transferrin with a high affinity for a second cell receptor, forming a receptor-transferrin-receptor complex. (iii) The receptor-transferrin-receptor complex is internalized into the cell by a second energy-dependent process.

Acknowledgements:

J.M. is grateful to Dr. George Bates for helpful discussions. This work was supported by grants from OAS, Proyecto Multinacional de Bioquimica, Metabolismo del Hierro and the Medical Research Council of Canada.

References:

1. Morgan, E.H., and Appleton, T.C., Nature 223, 1371 (1969).

2. Martinez-Medellin, J., and Schulman, H.M., Biochim. Biophys. Acta 264, 272 (1972).

3. Borova, J., Ponka, P., and Neuwirt, J., Biochim. Biophys. Acta 320, 143 (1973).

4. Neuwirt, J., Borova, J., and Ponka, P., in "Proteins of Iron Storage and Transport in Biochemistry and Medicine" (Crichton, R.R. ed.), p. 161, North-Holland Publishing Company, Amsterdam, 1975.

5. Fielding, J., and Speyer, B.E., Biochim. Biophys. Acta 363, 387 (1974).

6. Speyer, B.E., and Fielding, J., Biochim. Biophys. Acta 332, 192 (1974).

7. Workman, E.F., and Bates, G.W., Biochem. Biophys. Res. Comm., 58, 787 (1974).

8. Verhoef, N.J., and Noordeloos, P.J., Clinical Science and Molecular Medicine, 52, 87 (1977).

9. Baker, E., and Morgan, E.H., Biochemistry 8, 1133 (1969).

10. Appleton, T.C., Morgan, E.H., and Baker, E. in "The Regulation of Erythropoiesis and Hemoglobin Synthesis" (Travinicek, T., Neuwirt, J. eds.), p. 310. Universita Karlova, Praha, 1971.

11. Sullivan, A.L., Grasso, J.A., and Weintraub, L.R., Blood 47, 133 (1976).

12. David, G.S., Biochem. Biophys. Res. Comm. 48, 464 (1972).

13. Pharmacia Fine Chemicals brochure on CNBr-activated Sepharose 4B.

14. Hubbard, A.L., and Cohn, Z.A., J. Cell. Biol. 55, 390 (1972).

15. Hemmaplardh, D., and Morgan, E.H., Biochim. Biophys. Acta 426, 385 (1976).

FERRITIN AS A CYTOSOL TRANSPORT PROTEIN: DOES TRANSFERRIN ENTER RETICULOCYTES?

J. Fielding and Barbara E. Speyer

St. Mary's Hospital, London, United Kingdom

Ferritin is considered essentially as an iron storage protein which may be mobilised when metabolic need exceeds that available from other sources; conversely iron is deposited as ferritin when supply exceeds metabolic need.

In previous work (1, 2) we showed the presence of labelled cytosol components in reticulocytes after incubation with doubly labelled iron-transferrin. One of these was haemoglobin; the other was referred to as component C, which appeared similar to that observed by others (3, 4). This component behaved as a transport intermediate for iron in the pathway to haemoglobin; it also contained labelled ferritin. In this previous work cytosol was fractionated on Sephadex, and component C was eluted in the void volume. This provided no evidence as to whether it was complex in its composition. We therefore studied the cytosol preparation in greater detail using 150 cm. columns of Sepharose 6B from which all component C eluted within the column volume, clearly separated from haemoglobin. The cytosol was prepared by centrifugation of haemolysate at 20,000g: thus it contained no membrane or mitochondrial elements. Figure 1 shows a typical fractionation. Component C is a single iron labelled peak on Sepharose 6B as it is on Sephadex G-200, and is thus presumably a single substance. It eluted from the column at the same place as pure marker ferritin; it was precipitated completely by anti-ferritin serum but not by anti-transferrin serum. This component may thus be referred to as cytosol ferritin.

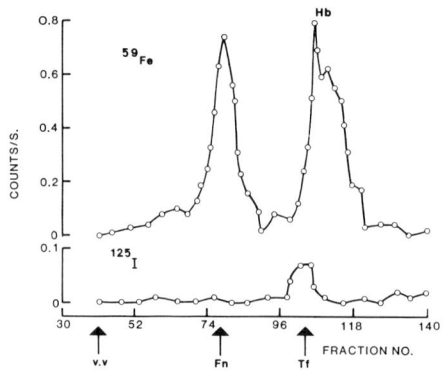

Fig. 1. Cytosol fractionation on Sepharose 6B.
v.v. = void volume;
Markers:
Fn = ferritin,
Tf = transferrin.

Since component C was shown to be an iron transport intermediate (1, 2) and is now identified as ferritin alone, this gives ferritin the role of the principal cytosol iron transport intermediate. It was therefore important to confirm this view.

Reticulocytes were exposed for 10 minutes with ^{59}Fe-^{125}I-transferrin and followed by a chase with "cold" iron transferrin for varying times up to 90 minutes. The components previously described were isolated and in particular the cytosol fractionated on Sepharose 6B (Fig. 2).

These chaser experiments have fully confirmed

that cytosol ferritin is indeed an active intermediate. Figure 2 shows that the whole of this component elutes in the position of pure ferritin. It decreases with increasing time of chase, after an initial increase, while haemoglobin continues to increase linearly. These results are more clearly seen in figure 3 in which the total counts under each curve have been plotted against time of chase.

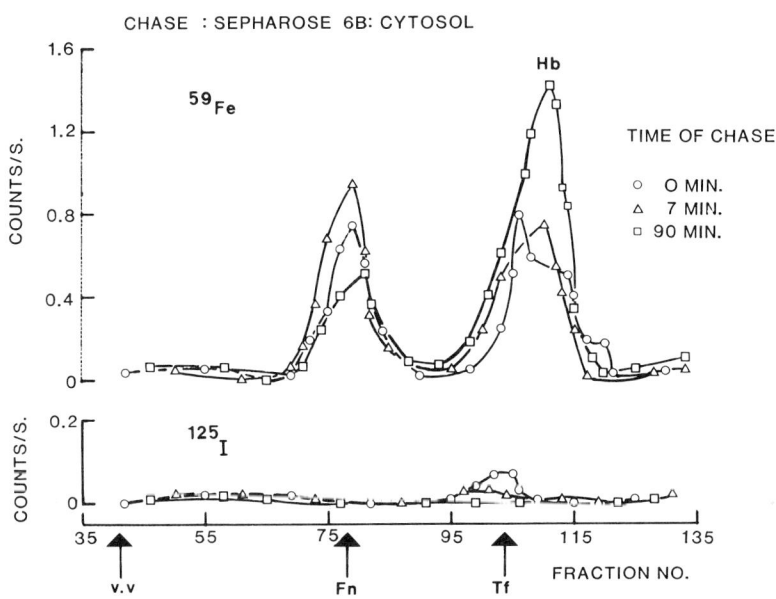

Fig. 2. Markers: Fn = ferritin, Tf = transferrin.

In additional experiments iron supply was reduced by blocking uptake from transferrin with sulphydryl inhibitors. Figure 4a shows that, as already demonstrated, cytosol ferritin and haemoglobin have a reciprocal quantitative relationship during a chase. Figure 4b shows, in the same cells, that when haemoglobin production was considerably

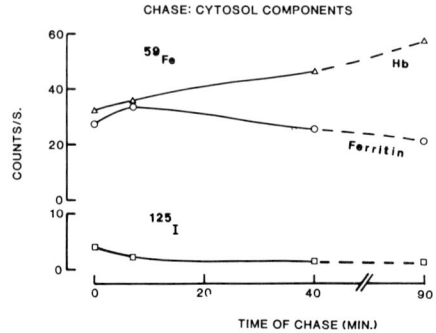

Fig. 3. Total counts in cytosol components during chase.

below the cells' synthesizing capacity, the available iron was still provided via cytosol ferritin which maintained its function as an intermediate under these conditions. Cytosol ferritin was formed first and its iron then transported to haem. The quantitative relationships of this experiment are shown in Table 1. In these experiments recovery of ^{59}Fe from the cytosol columns was complete and no other intermediates were identified. If such intermediates, for example small molecular weight iron carriers exist, they must have an extremely short half life. The results show that cytosol ferritin is the major intra-cellular intermediate for iron transport in reticulocytes.

This finding has significance for current theories on the interaction of iron transferrin with reticulocytes. There are two main hypotheses: the classical theory of Jandl and Katz (5) which visualises a membrane receptor binding iron transferrin, the subsequent release of iron into the cell and of apo transferrin or, as some would now have it, Fe_1-transferrin, from the membrane surface. Since 1969 this concept has been questioned by Morgan and Baker (6) who suggested that iron transferrin may enter into the reticulocyte by diffusion and reach the mitochondria. This was later modified to include specific receptor binding and entry by endocytosis (7).

The definition of what is inside or outside a cell or its membrane may not be of much help in elucidating the problem. The essential question

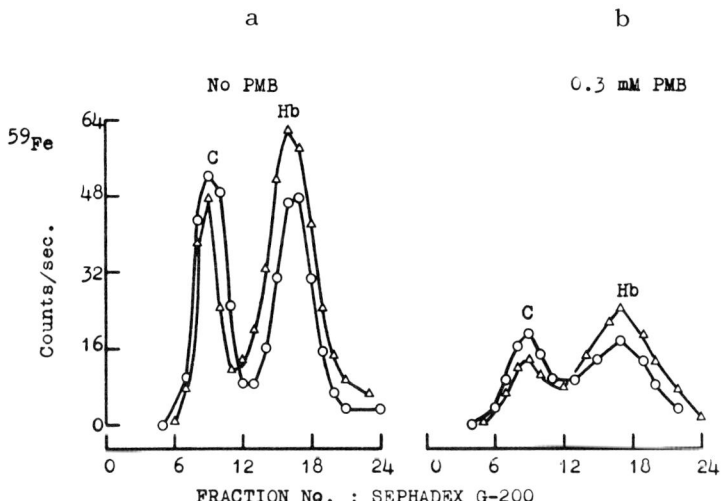

Fig. 4. Effect of sulphydryl inhibition by PMB on cytosol ferritin and Hb during chase: O, zero time; Δ, after 40 min. chase.

to be answered is: does iron transferrin reach the mitochondrial surface or does dissociation of iron from transferrin occur before this location?

The demonstration that ferritin is a constant cytosol transport intermediate makes it unlikely that iron transferrin reaches the mitochondria.

We have therefore examined Morgan's hypothesis more closely. The theory requires that, together with transferrin, the specific transferrin receptor should be taken into the cell interior and thus implies constant regeneration of specific receptors. The hypothesis appears to have originated at the same time as observations were made on iron transferrin uptake by reticulocytes in the presence

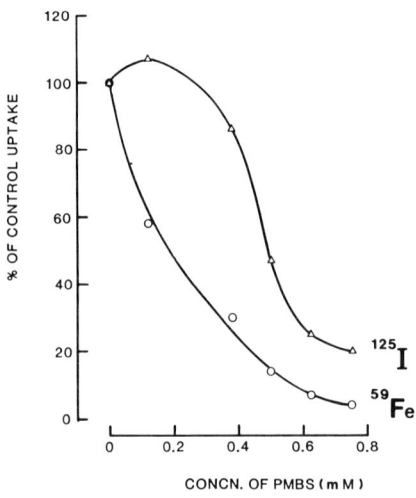

Fig. 5. Reticulocyte uptake from ^{59}Fe-^{125}I-transferrin: effect of PMBS concentration.

of inhibitors (6). Morgan reported that both iron and transferrin protein uptake were reduced by sulphydryl inhibitors, and concluded that these inhibitors acted by reducing the permeability of the membrane to iron transferrin.

Our results (8) using sulphydryl inhibitors at concentrations of 0.3 mM or more, agree with these observations of Morgan and Baker. However at lower concentrations these inhibitors have a quite different effect. Binding of transferrin protein to the reticulocyte is not reduced, indeed it

Table 1: Distribution of components after chase with and without sulphydryl inhibition (PMB): cytosol fractionation on G-200; counts/sec.

		No PMB		PMB	
		0	40 min.	0	40 min.
^{59}Fe	Ferritin	312	216	121	86
	Hb	374	601	238	294
	Total	686	817	359	380
	Recovery	93%	115%	103%	111%
^{125}I	Cytosol	52	10	54	10

usually increases to some extent.

At the same low concentrations, iron uptake into the cell is markedly reduced (Fig. 5). At low concentration of inhibitor the movement of transferrin protein takes a separate pathway from that of its contained iron. The essence of the question to be asked is at what location does this separation take place. On this question the effect of PMBS is particularly relevant. This sulphonated inhibitor is markedly hydrophilic and has been shown to act only at the membrane surface; penetration of the membrane is minimal (9). It seems, therefore, that the separation of iron and apotransferrin in PMBS inhibited cells takes place at the membrane surface and thus defines the location at which iron is released from transferrin.

Our conclusions are therefore as follows:
1. That the significant cytosol iron transport intermediate in reticulocytes, and therefore

presumably in other red cell precursors, is ferritin.
2. That the presence of such an intermediate in the cytosol excludes the hypothesis that iron transferrin enters deeply into the cell as far as mitochondria, but lends support to the classical hypothesis of Jandl and Katz.

REFERENCES

1. Speyer, B. E. and Fielding J., 1974, Biochim. et Biophys. Acta, 332, 192.

2. Fielding, J. and Speyer, B.E., 1974, Biochim. et Biophys. Acta, 363, 387.

3. Allen, D. W. and Jandl, J.H., 1960, Blood, 15, 71.

4. Greenough, W. B., Peters, T. and Thomas, E. D., 1962, J. Clin. Invest., 41, 1116.

5. Jandl, J. H. and Katz, J. H., 1963, J. Clin. Invest. 42, 314.

6. Morgan, E. H. and Baker, E., 1969, Biochim. et Biophys. Acta, 184, 442.

7. Hemmaplardh, D. and Morgan, E. H., 1976, Biochim. et Biophys. Acta, 426, 385.

8. Edwards, S. A. and Fielding, J., 1971, Brit. J. Haemat., 20, 405.

9. Vansteveninck, J., Weed, R. I. and Rothstein, A., 1965, J. Gen. Physiol., 48, 617.

THE ROLE OF MITOCHONDRIA IN THE CONTROL OF IRON DELIVERY TO HEMOGLOBIN MOLECULES

PŘEMYSL POŇKA
JAN NEUWIRT
JITKA BOROVÁ
OTA FUCHS

Department of Pathological Physiology
Faculty of General Medicine
Charles University
Prague, Czechoslovakia

I. INTRODUCTION

Immature erythroid cells obtain iron directly from transferrin. The transfer of iron into erythroid cells involves reversible uptake of transferrin, release of iron and its incorporation into heme in the mitochondria. The iron-transferrin complex initially attaches to a surface membrane receptor but further mechanisms by which the iron is delivered to the mitochondria has not yet been completely elucidated /1,2/.

There is certain evidence that transferrin transports iron directly to the mitochondria without the need of another intracellular carrier /2-5/. The mechanism of iron release from transferrin is largely unknown. Recent experiments suggest that ATP /6/ may play a role in iron release from transferrin or intracellular transport of iron. The rate at which the iron is released from transferrin is limited by the level of uncommitted heme in mitochondria. This represents a feedback mechanism by which heme controls the cellular supply of iron for hemoglobin synthesis /for details see ref. 2/.

In this contribution we report experiments indicating that mitochondria play a key role in intracellular iron metabolism. Evidence is presented to indicate that reticulocyte "stroma" or "ghosts" contain mitochondria that are able to synthesize heme. The complete omission of this fact /7,8/ may lead to misinterpretation of intracellular iron kinetics. The basic experimental approach involved the preincubation of rabbit reticulocytes with ^{59}Fe-transferrin /human tranferrin, Behringwerke, added in physiologic concentrations/ for 30 min and isolation of either mitochondria or insoluble particles containing stroma and mitochondria /PCSM/. Isonicotinic acid hydrazide /INH/ was added during the preincubation period /for 45 min/ to induce mitochondrial non-heme iron accumulation /9/. ^{59}Fe-labelled mitochondria or PCSM were then reincubated and various factors affecting heme synthesis or iron release from mitochondria were studied. The details of experimental procedure are described elsewhere /2,9/.

II. SOME ASPECTS OF IRON METABOLISM IN MITOCHONDRIA

Mitochondria /from INH and ^{59}Fe-transferrin preincubated reticulocytes/ were layered on the discontinual sucrose gradient and centrifuged for 15 hours at 40 000 rpm /2°C/. In separate tubes, containing similar sucrose gradients, the membranes isolated /10/ from reticulocyte-free erythrocytes were centrifuged under the same conditions. Fig. 1 shows that the ^{59}Fe-radioactivity corresponds to the activity of succinate-cytochrome c oxidoreductase which is a mitochondrial marker enzyme. It is apparent that the migration of erythrocyte membranes /shown as absorbance at 280nm in Fig. 1/ differs from that of ^{59}Fe-labelled reticulocyte mitochondria.

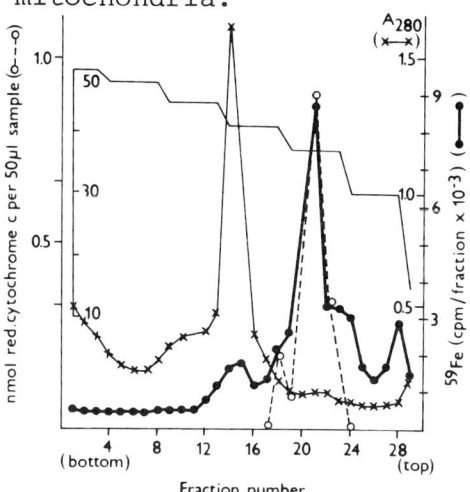

Fig. 1 *Sucrose gradient fractionation of reticulocyte mitochondria and reticulocyte-free erythrocyte "ghosts".* ^{59}*Fe-labelled reticulocyte mitochondria were prepared according to Guggenheim et al. /11/. For details see Poňka et al. /2/. Reproduced by permission of ASP Biological and Medical Press B.V., Amsterdam, The Netherlands.*

A similar sample of ^{59}Fe-labelled rabbit reticulocyte mitochondria was incubated /50 min, 37°C/ in unlabelled stroma-free reticulocyte hemolysate. After incubation mitochondria-free hemolysate was fractionated on Sephadex G-200 /Fig.2/. During a 50 minute incubation about 35-38% of ^{59}Fe was released from mitochondria and the addition of protoporphyrin IX did not enhance the rate of iron release. However, the addition of protoporphyrin IX considerably stimulated the amount of ^{59}Fe in the fraction that corresponds to rabbit hemoglobin. It was demonstrated that in this fraction ^{59}Fe is in the form of heme.

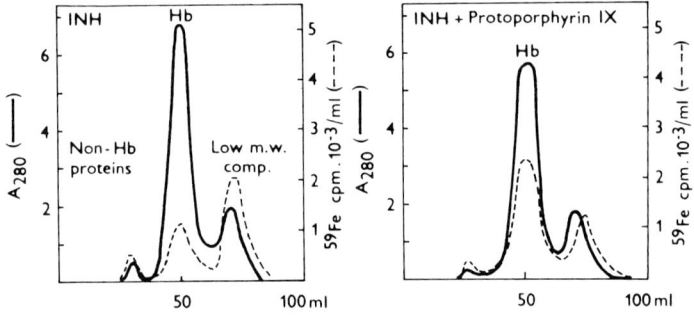

Fig. 2 *Sephadex G-200 elution profile of the supernatant from a mobilization experiment containing ^{59}Fe-mitochondria and unlabelled reticulocyte lysate. Additions: INH, 5mM, protoporphyrin IX, 0.1mM. Reproduced from Poňka et al. /2/ by permission of ASP Biological and Medical Press B.V., Amsterdam, The Netherlands.*

The following experiment presents evidence that significantly more iron is released from mitochondria incubated at 37°C /Fig. 3 B, C/ than from mitochondria kept at 4°C /Fig. 3 A/. When solubilized ^{59}Fe-mitochondria - that had previously been incubated with protoporphyrin IX - were fractionated by gel filtration on Sephadex G-200, a new peak of heme radioiron appeared /Fig. 3 C/. This ^{59}Fe-heme was eluted in the same position as rabbit hemoglobin /Fig. 3 C/. This result suggests that heme formed *de novo* in mitochondria is perhaps bound to some carrier or even to globin chains. Here it should be stressed that all the factors affecting export of heme from mitochondria can essentially influence the level of mitochondrial heme and thus the uptake of iron by mitochondria /2/.

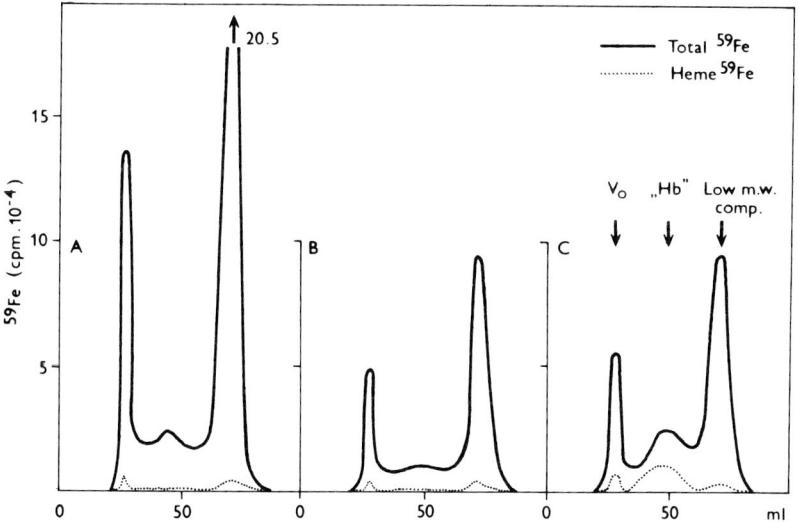

Fig. 3 *Chromatography of solubilized rabbit reticulocyte mitochondria on Sephadex G-200.* ^{59}*Fe-mitochondria were prepared /2/ and incubated /40 min/ at 37°C /B,C/ or 4°C /A/ without /A,B/ or with /C/ protoporphyrin IX /0.2mM/ and dithiothreitol /2mM/. Conditions of incubation, solubilization in Triton X-100 /with the exception that phosphate buffer was replaced by 2mM citrate-Na$_3$/ and chromatography were similar as described previously /2/. In each fraction total ^{59}Fe and heme ^{59}Fe were determined /9/.*

III. THE RELEASE OF IRON FROM PARTICLES CONTAINING STROMA AND MITOCHONDRIA /PCSM/

There seems to be sufficient evidence at hand that reticulocyte stroma contain mitochondria /2,5,9/. This conclusion is further supported by the following experiment. Reticulocytes were incubated with INH /10mM/ and ^{59}Fe-transferrin and the stroma fraction was isolated as described elsewhere /2/. When ^{59}Fe--labelled reticulocyte particulate fraction was reincubated with protoporphyrin IX, newly-formed ^{59}Fe-heme appeared both in stroma and stroma-free lysate /Table I/. In erythroid cells with inhibited heme synthesis non-heme iron apparently accumulates in mitochondria but not in membranes /9,12/. The utilization for heme synthesis of non-heme ^{59}Fe present in PCSM after addition of protoporphyrin IX, is certainly due to mitochondrial heme synthetase. It may be of interest to mention that addition of bovine serum albumin /1g/100ml/ to ^{59}Fe-PCSM, incubated with protoporphyrin IX, increased the export of newly-formed heme from PCSM without affecting the overall rate of ^{59}Fe-heme formation.

During a 50 minute period essentially more radioiron is mobilized from ^{59}Fe-labelled PCSM into reticulocyte lysate than into the buffer /Table I/. However, addition of iron chelating agents increases the rate of iron mobilization into lysate-free salt solution. For example, nitrilotriacetic acid added with a reducing agent /R.a./ mobilizes almost all the iron from PCSM during a 50 minute incubation /Table I/.

TABLE I Effect of various additions on the release of ^{59}Fe from PCSM and on heme synthesis

Exp.	Additions	Iron release^1	Heme$_2$ sup.	^{59}Fe in PCSM^1	Total heme ^{59}Fe^1
1	O	30.5	19.2	4.6	10.5
	INH	38.2	17.7	2.4	9.1
	INH + PROTO	41.6	46.9	10.0	29.5
2	O	9.7	6.0	6.4	8.0
	NTA	66.8	2.2	5.0	6.5
	NTA + R.a.	90.8	0.5	0.7	0.9
	PROTO	8.5	33.9	16.8	19.7
	PROTO + R.a.	48.9	32.0	15.2	30.8
	PROTO+R.a.+NTA	78.2	10.4	9.0	17.2

^{1}expressed as per cent of original ^{59}Fe in PCSM
^{2}expressed as per cent of total ^{59}Fe in supernatant

PCSM were incubated without /Exp.2/ or with /Exp.1/ reticulocyte hemolysate under similar conditions as described previously /2/. Additions: INH /5mM/, PROTO /protoporphyrin IX, 0.1mM/; NTA /nitrilotriacetic acid, 5mM/; R.a. /$Na_2S_2O_4$, 10mM/. Total heme ^{59}Fe represents the sum of heme ^{59}Fe found in supernatant and PCSM.

IV. CONCLUSIONS

There seems to be certain experimental evidence that transferrin binds to mitochondria /2,5/ and the iron is perhaps released from transferrin near or on the mitochondrial membrane. Both previous /2/ and present experiments suggest that mitochondria can, under certain conditions, release iron into reticulocyte lysate or salt solution. It is therefore possible that mitochondria take up iron directly from transferrin and provide it to various heme and non-heme iron compounds in both mitochondria and cytosol.

We wish to suggest that both mitochondria and the level of intramitochondrial uncommited heme play a central role in the control of iron movement in the erythroid cell.

REFERENCES

1. Morgan, E.H., in "Iron in Biochemistry and Medicine" /Jacobs, A., Worwood, M.eds./, p.29. Academic Press, London and New York, 1974.
2. Poňka, P., Neuwirt, J., Borová, J., Fuchs, O., in "Iron Metabolism", Ciba Foundation Symposium No. 51 /in press/. Elsevier /Excerpta Medica/ North-Holland, Amsterdam and New York, 1977.
3. Morgan, E.H., Baker, E., Biochim. Biophys. Acta 194, 442 /1969/.
4. Martinez-Medellin, J., Schulman, H.M., Biochim. Biophys. Acta 264, 272 /1972/.
5. Neuwirt, J., Borová, J., Poňka, P., in "Proteins of Iron Storage and Transport in Biochemistry and Medicine" /Crichton, R.R. ed./, p. 161. North-Holland Publ. Co., Amsterdam, 1975.
6. Egyed, A., Biochim. Biophys. Acta 411, 349 /1975/.
7. Fielding, J., Speyer, B.E., Biochim. Biophys. Acta 363, 387 /1974/.
8. Workman, E.F., Bates, G.W., Biochem.Biophys. Res.Commun. 58, 787 /1974/.
9. Borová, J., Poňka, P., Neuwirt, J., Biochim. Biophys. Acta 320, 143 /1973/.
10. Dodge, J.T., Mitchell, C.,Hanahan, D.J., Arch. Biochem. Biophys. 100, 119 /1963/.
11. Guggenheim, S.J., Bonkowsky, H.L., Harris, J.W., Webster, L.T., J. Lab. Clin. Med. 69, 357 /1967/.
12. Bessis, M.C., Jensen, W.N., Brit. J. Haemat. 11, 49 /1965/.

IRON MOBILISATION FROM ISOLATED RAT HEPATOCYTES

E. Baker, F.R. Vicary, E.R. Huehns
University College Hospital Medical School.

The mechanisms regulating the exchange of iron between plasma and storage sites are still uncertain. The naturally occurring iron chelators fructose and citrate, ceruloplasmin, xanthine oxidase, ascorbic acid and the plasma iron and transferrin levels have all been implicated in liver iron mobilisation. The physiological importance of these compounds is not certain, and is complicated by both species differences in their activity and differences in their effect on parenchymal and reticuloendothelial cells. Recent studies have shown that liver reticuloendothelial cells and hepatocytes can be selectively labelled with different radioiron markers (1). These markers have been used to gain more information on liver iron exchange in vivo (2,3) and in vitro in the isolated perfused liver (4,5). Isolated hepatocytes are a more convenient experimental system than the perfused liver, and this paper reports our initial studies on iron mobilisation from rat hepatocyte suspensions prepared by enzyme treatment after specific labelling in vivo with transferrin-^{59}Fe.

METHODS

Rat livers were labelled in vivo with plasma transferrin ^{59}Fe as described previously (4). Hepatocytes were isolated by the method of Berry and Friend (6), but incorporating modifications suggested by Cornell et al (7) and Jeejeebhoy et al. (8). The liver was perfused in situ with calcium-free Krebs-Henselheit saline containing collagenase and hyaluronidase and the resulting liver cell suspension washed in Krebs-Henselheit saline containing albumin (20 g.l^{-1}). Incubation was performed in MEM (minimum essential medium) containing Hepes (0.020M), bicarbonate (0.01M) and albumin (20 g.l^{-1}) and the reaction stopped by centrifuging at 4°. The release of hepatocyte ^{59}Fe was calculated from the proportion of radioactivity in the supernatant. All glassware was siliconised.

Cell viability was monitored during incubations by measuring trypan blue exclusion or release of cytoplasmic glutamic-pyruvic transaminase (GTP, Sigma Technical Bulletin No 55UV). Hepatocyte suspensions were discarded if initial viability measurements indicated cell damage during preparation. Data was rejected in individual experiments if trypan blue staining and/or GTP release increased when compared with the control value in the same experiment.

RESULTS

Iron Release from Hepatocytes

The release of iron from the hepatocyte suspension (Figure 1) was markedly dependent on incubation temperature and composition of the medium. There was always some radioactivity in the supernatant at the start of each incubation but this did not increase during incubation at $4°$; in contrast, there was a steady release of ^{59}Fe at $37°$ (Figure 1). The rate of ^{59}Fe release was calculated from the slope of the line obtained over the first hour of incubation at $37°$.

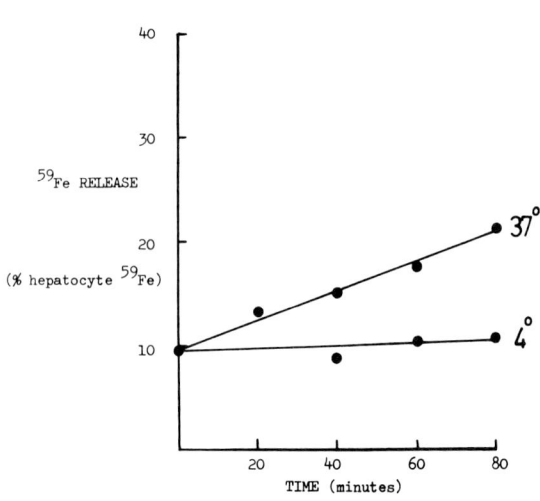

FIGURE 1.

The release of ^{59}Fe from rat hepatocytes during incubation at $4°$ or $37°$.

Effect of Serum and Apotransferrin

Serum produced a consistent increase in the rate of ^{59}Fe release from hepatocytes. As shown in Table I an increase in the proportion of rat serum in the incubation medium produced a corresponding increase in the rate of ^{59}Fe release.

TABLE I

The Effect of Serum Concentration on the Rate of ^{59}Fe Release.

COMPOSITION OF INCUBATION MEDIUM		^{59}Fe RELEASE
MEM (ml)	RAT SERUM (ml)	(% OF RATE IN MEM ALONE)
4	0	100
3	1	129
2	2	146
1	3	188

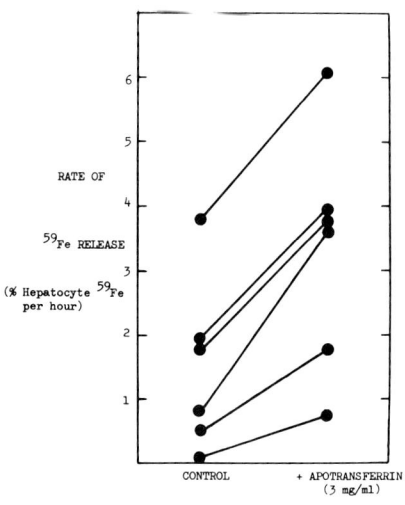

FIGURE 2.

The effect of apotransferrin on the rate of ^{59}Fe release.

Apotransferrin produced a marked increase in the rate of ^{59}Fe release. As shown in Figure 2, in each of 6 experiments where human apotransferrin (Behringwerke, A.G.) was added there was a corresponding increase in ^{59}Fe efflux. The average rate of release of ^{59}Fe increased significantly from 1.5 to 3.7% hepatocyte ^{59}Fe per hr. ($p < 0.01$).

Effect of Iron Chelators

Glycine and glucose had no effect on the rate of ^{59}Fe release however citrate produced a marked increase in ^{59}Fe release. The siderophores desferrioxamine and 2,3-dihydroxybenzoic acid also increase the rate of ^{59}Fe release (Table II). None of the chelators tested affected ^{59}Fe release at $4°$.

Several of the lipid soluble ionophores have the capacity to transfer iron across synthetic membranes and the erythrocyte membrane (9). A23187 was the only iónophore tested which increased ^{59}Fe release from hepatocytes without decreasing cell viability (Table II).

TABLE II

The Effect of Iron Chelators on the Rate of Release of ^{59}Fe.

CHELATE	CONCENTRATION ($M \times 10^3$)	^{59}Fe RELEASE (% OF RATE IN MEM ALONE)
Glucose	2	108 (3)*
Glycine	5	96 (2)
Citrate	5	142 (3)
	25	328 (3)
Desferrioxamine	0.08	141 (7)
Ionophore A23187**	0.001	125 (2)
	0.01	180 (2)
2,3-dihydroxybenzoic acid	5	138 (2)

* No of experiments.   ** Eli-Lilly Research Labs. Ind.

Effect of Gas Composition

Hypoxia has been reported to increase liver iron mobilisation in vivo (10). To determine the effect of hypoxia, hyperoxia and increase in CO_2 tension the release of hepatocyte ^{59}Fe was measured under gas phases of air, N_2, O_2, CO_2, air + 5% CO_2 and 95% O_2 + 5% CO_2. Flasks were gassed for 1 minute with the appropriate gas mixture, stoppered and incubated in triplicate for 30 minutes at $4°$ or $37°$ in the presence of 5 mM citrate. Under these incubation conditions no change was observed in cell viability or in ^{59}Fe release at $4°$, however marked changes occurred in the release of hepatocyte ^{59}Fe at $37°$ (Table III). Hyperoxia induced with a gas phase of 100% O_2 depressed ^{59}Fe release, while anoxia induced by a gas phase of 100% nitrogen increased the rate of ^{59}Fe release to double that observed in air. In contrast, anoxia induced by a gas phase of 100% CO_2 produced a marked depression of ^{59}Fe release to 22% of the value in air. There was also a reduction in ^{59}Fe release in air containing 5% CO_2, however CO_2 had no further effect on the depressed efflux observed in oxygen.

TABLE III

The effect of change in gas composition on the release of ^{59}Fe from hepatocytes. Net ^{59}Fe release at $37°$ was calculated from the mean value obtained at $37°$ minus the mean value obtained at $4°$.

GAS COMPOSITION	^{59}Fe RELEASE (% OF RATE IN AIR)
Air	100
Air + 5% CO_2	53
O_2	56
O_2 + 5% CO_2	58
N_2	201
CO_2	22

DISCUSSION

Liver perfusion with a solution of collagenase and hyaluronidase produced populations of liver cells useful in investigating liver iron metabolism. Iron release could be enhanced or depressed by manipulating the experimental conditions indicating that the hepatocyte suspension is a viable system. Iron release at $4°$ was completely inhibited irrespective of medium composition, while iron release at $37°$ was depressed by high levels of O_2 and CO_2 and increased by hypoxia and iron chelates. This suggests that the observed ^{59}Fe release from hepatocytes is part of a physiological process. However, there was a significant release of ^{59}Fe at $37°$ in most experiments, in the absence of added iron chelates. This may be due partly to natural iron chelates present in MEM or to transferrin or ceruloplasmin contaminants in the albumin preparation used and partly to iron release from deteriorating cells.

Serum and apotransferrin were effective mobilisers of ^{59}Fe from isolated hepatocytes, confirming observations made on the perfused liver (4) and in vivo (11). However, the iron chelators citrate, DFO and 2,3-DHB were also effective, indicating that transferrin may function simply by providing a receptor for iron in the plasma rather than by a more specific interaction with specialised transferrin receptors. The efficacy of the physiological iron chelator citrate in mobilising liver iron indicates its potential role in vivo possibly acting both as a ferroxidase (12) and iron chelator. Results obtained in the present study and in the perfused liver (5) confirm in vivo evidence for specific mobilisation of hepatocyte iron stores by DFO (1). The chelate 2,3-DHB also mobilised hepatocyte ^{59}Fe (without preincubation), in agreement with studies in Chang liver cells (13) and in vivo (14). The ionophore A23187 was also effective, however its high toxicity in vivo indicates the need for development of a safer analogue.

Hypoxia increased liver iron mobilisation without affecting cell viability. This confirms the studies of Mazur et al (10) suggesting that tissue hypoxia may cause the increase in serum iron observed after haemorrhage. Hyperoxia depressed iron mobilisation and hypoxia enhanced it, indicating that a ferric-ferrous reduction is important in iron mobilisation, whether from ferritin or from the chelatable iron pool. It seems likely that factors such as anoxic stress increase iron mobilisation by acting directly on NADH-linked ferric reduction implicated in iron removal from ferritin (15), while chelators such as citrate act indirectly by depleting the "chelatable iron pool", thus maintaining a concentration gradient for iron release.

REFERENCES

1. Hershko, C., Cook, J.D., Finch, C.A., J. Lab. Clin. Med. 81, 876 (1973).

2. Cook, J.D., Hershko, C., Finch, C.A., Brit. J. Haemat. 25, 695 (1973).

3. Fillet, G., Cook, J.D., Finch, C.A., J. Clin. Invest. 53, 1527 (1974).

4. Baker, E., Morton, A.G., Tavill, A.S., in "Proteins of Iron Storage and Transport in Biochemistry and Medicine" (Crichton, R.R. Ed.) p 173. North Holland Pub. Co., Amsterdam, 1975.

5. Baker, E., Morton, A.G., Tavill, A.S. Submitted to Brit. J. Haemat. (1977).

6. Berry, M.N., Friend, D.S., J. Cell Biol. 43, 506 (1969).

7. Cornell, N.W., Lund, P., Hems, R., Krebs, H.A. Biochem. J. 134, 671 (1973).

8. Jeejeebhoy, K.N., Ho, J., Greenberg, G.R., Phillips, M.J., Bruce-Robertson, A., Sodtke, U., Biochem. J. 146, 141 (1975).

9. Young, S., Baker, E., Gomperts, B.D., Huehns, E.R. in "Proteins of Iron Storage and Transport in Biochemistry and Medicine." (Crichton, R.R. Ed.) p 417. North Holland Pub. Co., Amsterdam, 1975.

10. Mazur, A., Baez, S., Shorr, E. J. Biol. Chem. 213, 147 (1955).

11. Hallberg, L., Solvell, L., Acta med. Scand. Vol. 168, Supp. 358, 89 (1960).

12. Lee, G.R., Nacht, S., Christensen, D., Hansen, S.P., Cartwright, G.E., Proc. Soc. Exp. Biol. Med. 131, 918 (1969).

13. White, G.P., Bailey-Wood, R., Jacobs, A. Clin. Sci. and Mol. Med. 50, 145 (1976).

14. Graziano, J.H., Grady, R.W., Cerami, A. J. Pharm. Exp. Therap. 190, 570 (1974).

15. Osaki, S., Sirivich, S., Fed. Proc. 30, Abstr. 1394.

THE INCORPORATION OF IRON INTO ISOLATED RAT HEPATOCYTES

*DIETMAR GROHLICH, COLIN G. D. MORLEY, ROBIN J. MILLER, AND ANATOLY BEZKOROVAINY**

BIOCHEMISTRY DEPARTMENT, RUSH-PRESBYTERIAN-ST. LUKE'S MEDICAL CENTER, CHICAGO, ILLINOIS 60612

Introduction

Liver and bone marrow are the major target tissues of exogenously administered iron in most animals. The interaction of iron with the various types of hemopoietic cells has been studied extensively since the discovery by Jandl et al. (1) of the role of transferrin in this process. Thus, in the rabbit reticulocyte system, as well as in the other bone marrow cells, transferrin carrying the iron first combines with a specific receptor on the cell surface, then through an as yet poorly understood mechanism, releases the iron to the cell for the biosynthesis of heme. The specific transferrin receptor has been purified and partially characterized (2, 3).

In addition to its transport into the erythropoietic cells, iron must also enter and exit into and from its principal storage sites in the liver cells. Several workers have dealt with the mechanisms whereby this may be accomplished, and it was proposed that a non-enzymatic process may be involved (4, 5, 6). Work in intact animals and in perfused livers has shown that iron-containing transferrin can be taken up by the liver cells (7), that liver can take up both ferric and ferrous iron (8), and that ferrous iron was taken up by perfused liver cells faster than was ferric iron in combination with transferrin. Finally, liver cells maintained in tissue culture were also able to take up iron from the iron-transferrin complex (10).

---

* To whom inquiries should be directed.

The work described in this paper deals with the mechanisms of iron incorporation into isolated rat hepatocytes prepared following the perfusion of liver with bacterial collagenase preparation.

Materials and Methods

Animals. Male white Sprague-Dawley rats (200-300 g) were obtained from local suppliers, and were housed in light-cycled rooms with rat chow and water ad libitum.

Isolation of hepatocytes. Hepatocytes were isolated following the perfusion of rat livers with the collagenase solution in situ essentially as described by Bissel et al. (11). The perfusion solution used was that described by Seglen (12). The final cell preparation, suspended in 8-9 ml of "washing buffer" (12) contained 3 to 5 x 10^6 cells per ml, with a viability of 70-90% as determined by the dye exclusion technique (13). In certain experiments, the perfusion medium also contained 0.05% soybean trypsin inhibitor (Sigma Corp., St. Louis, MO).

Iron solutions. Three types of iron solutions labelled with ^{59}Fe were used in these studies: ferric citrate, ferrous citrate, and ferric-trans-ferrin complexes (the latter in the form of rat serum containing 220 ug of iron/ml). Stock solutions containing from 12 to 65 ug iron/ml were prepared in case of the ferric and ferrous citrate experiments.

Incubation of hepatocytes with iron solutions. Hepatocytes were incubated in triplicate with solutions containing varying amounts of iron. A typical incubation mixture using ferric or ferrous citrate consisted of 2.5 ml of the L-15 Leibovitz tissue culture medium (14), 0.3 ml of the cell suspension, and 0.2 ml of the iron stock solution in a 25 ml Erlenmeyer flask. Where serum was used, there were 1.7-2.6 ml of the Leibovitz medium, 0.3 ml of the cell suspension, and 1 to 0.1 ml of the serum. The incubation was carried out at 37° in a Dubnoff shaker, and the reaction was terminated by quick cooling in an ice bath. After separating

and washing the cells, the latter were counted in
a Beckman Biogamma counter. The final concentration of iron in the incubation mixtures was between
0.37 and 3.7 ug/ml when the citrate complexes were
used, and 0.066 to 0.66 ug/ml when serum was used.
Normal human serum iron levels are 2.5-4.4 ug/ml,
whereas those for rats were 1.3-3.0 ug/ml (15).
In our experiments, therefore, iron levels offered
to the cells were at or slightly below those of
rat and human sera.

Results

Time-incorporation studies showed that iron
uptake by the hepatocytes occurred in 3 stages:
an instantaneous uptake, whose rate could not be
easily measured, followed by a linear rate lasting
for some 25 min, and finally a slower rate that
was followed for an additional 35-45 min. We
focused our attention on the second stage of iron
uptake, and utilized the Michaelis-Menten concept
to describe the kinetics observed. Throughout
these studies, a 20-min incubation period was
utilized.

Ferrous iron was absorbed faster than was
ferric iron, but it did not exhibit saturation
kinetics when examined by the Lineweaver-Burk or
the Eadie-Hofstee plots. Ferric iron, on the
other hand, was absorbed slower, but did exhibit
saturation kinetics with definite K_m- and V_{max}-
values. When iron was offered to the cells in the
form of transferrin complex, a low degree of incorporation was observed that also showed saturation kinetics.

When hepatocytes prepared in the presence of
the soybean trypsin inhibitor were used in these
experiments, the uptake of iron from serum increased dramatically, and approached the rate
observed with ferrous iron. A slight increase of
iron uptake was also observed with ferric iron.
The results obtained are summarized in table 1.

Discussion

Our work confirms qualitatively the results of

Table 1. Michaelis-Menten parameters characterizing the incorporation of iron into rat hepatocytes.[1]

Form of iron	$K_m \times 10^5$	$V_{max} \times 10^9$	$dv/dS \times 10^4$
Ferrous	-	-	2.4
Ferric	5.9	5.9	0.94
In serum	1.2	0.80	0.66
Ferric[2]	1.0	3.2	1.2
In serum[2]	1.3	2.5	1.7

[1] dv/dS is the rate of iron incorporation into hepatocytes as a function of iron (S) concentration in the reaction mixture.

[2] Hepatocytes prepared in the presence of soybean trypsin inhibitor.

Zimelman et al. (9) and those of Hoy and Harrison (8). our results further indicate that the ferrous iron is taken up well by a simple diffusion process requiring no enzymatic mechanism for its entry into the hepatocyte. On the other hand, ferric iron and transferrin-bound iron apparently utilize a system that may involve receptors on the hepatocyte membrane surface, and possibly other enzymatic steps.

The fact that the cells prepared in the presence of the trypsin inhibitor incorporated iron at a rate almost 3 times that observed in the absence of such an inhibitor indicates that there is a specific receptor for the iron-transferrin complex on the hepatocyte membrane surface, which is susceptible to a trypsin-like enzyme digestion. Significantly, the K_m-values observed with both types of cells remained the same with respect to transferrin-bound iron. This indicates that both types of cells had an identical receptor, though the cells prepared without the trypsin inhibitor had fewer such receptors on their membrane surfaces.

It may also be proposed that a receptor exists on the hepatocyte membrane surface that can combine with ferric iron. Judging from the K_m-values, observed in the two types of cells (1.0 and 5.0×10^{-5} M), it may be surmised that this receptor was in some way altered by the trypsin-like enzyme(s) present in the collagenase preparation, rather than being altogether destroyed as was the case with the iron-transferrin receptor. Whether or not the receptors combining with ferric iron and the iron-transferrin complex are one and the same remains yet to be elucidated.

In conclusion, it may be proposed that under physiological conditions, the incorporation of iron into rat hepatocytes takes place via a mechanism not dissimilar to that observed with rabbit reticulocytes and other hemopoietic cells.

Acknowledgement

This work was supported by American and Chicago Heart Associations, Grant No. 76-715.

References

1. Jandl, J. H., Inman, J. K., Simmons, R. L, Allen, D. W., J. Clin. Invest. 38, 161 (1959).

2. Speyer, B. E., Fielding, J., Biochim. Biophys. Acta 332, 192 (1974).

3. Sly, D. A., Grohlich, D., Bezkorovainy, A., in "Proteins of Iron Storage and Transport in Biochemistry and Medicine" (Crichton, R. R. ed) p. 141, North Holland/American Elsevier, Amsterdam, 1975.

4. Mazur, A., Green, S., Carleton, A., J. Biol. Chem. 235, 595 (1960).

5. Pape, L., Multani, J. S., Stitt, C., Saltman, P., Biochemistry 7, 613 (1968).

6. Miller, J. P. G., Perkins, D. J., Europ. J. Biochem. 10, 146 (1969).

7. Gardiner, M. E., Morgan, E. H., Austr. J. Exptl. Biol. Med. Sci. (Pt. 5) 52, 723 (1974).

8. Hoy, T. G., Harrison, P. M., Brit. J. Haematol. 33, 497 (1976).

9. Zimelman, A. P., Zimmerman, H. J., McLean, R., Weintraub,L. R., Gastroenterology 72, 129 (1977).

10. Beamish, M. R., Keay, L., Okigaki, T., Brown, E. B., Brit. J. Haematol. 31, 479 (1975).

11. Bissel, D. M., Hammaker, L. E., Meyer, U., J. Cell Biol. 59, 722 (1973).

12. Seglen, P. O., Methods in Cell Biol. 13, 30 (1976).

13. Girardi, A. J., McMichael, H., Henle, W., Virology 2, 532 (1956).

14. Leibovitz, A., Amer. J. Hyg. 78, 173 (1963).

15. Itzhaki, R. F., Belcher, E. H., Arch. Biochem. Biophys. 92, 74 (1961).

MITOCHONDRIAL IRON UPTAKE AND HEME SYNTHESIS IN
COPPER DEFICIENCY

D.M. Williams, A.J. Barbuto, C.L. Atkin and G.R. Lee
*Departments of Medicine, Biochemistry and Pathology
University of Utah College of Medicine*

I. INTRODUCTION

Copper deficiency in experimental animals results in severe anemia that appears to be due to restricted iron flow in at least 4 specific sites (1), one of which is located within the maturing erythrocyte. This intracellular defect leads to increased numbers of bone marrow sideroblasts, severe microcytic hypochromic anemia, reticulocytopenia and hyperferremia (2), and contributes to the development of severe hypochromic, microcytic anemia. Further, the abnormality appears to be due to reduced iron uptake and heme synthesis by the maturing erythrocyte (3). However, no defects in the porphyrin biosynthetic pathway have been identified in copper deficiency, and heme synthetase activity of reticulocyte sonicates, measured by incorporation of radioactive ferrous iron has been found to be normal (4).

In contrast, there is evidence to suggest that movement of iron into the mitochondrion is impaired in copper deficiency (3,5,6). The purpose of the studies reported herein is to define more clearly the role of copper in the intracellular movement of iron.

II. MATERIALS AND METHODS

Copper-deficient, iron replete swine were raised as described previously (7). Copper and iron levels in plasma, tissues, and cell fractions were determined by atomic absorption spectrometry following either trichloroacetic acid precipitation or wet digestion (3). Mitochondria were prepared in 0.25M sucrose from the livers of exsanguinated animals by the method of Schneider (8) and incubated (3-5 mg protein/ml) in a total volume of 3.0 ml containing

^{59}Fe (0.033mM), protoporphyrin (0.167mM), KH_2PO_4 (8mM) and Tris-HCl (8mM), all at pH 8.0. Reaction vessels were flushed with nitrogen, capped and incubated for 2 hours in the dark at 37°C. The reaction was stopped by adding the contents to 150 ml of acetone: acetic acid containing 0.5% $SnCl_2$. Heme was crystallized by modification of the method of Shemin et al (9) and heme synthesis was estimated by the incorporation of ^{59}Fe. Protein was measured by the method of Lowry et al (10), cytochrome oxidase activity was determined by the method of Wharton and Tzagaloff (11), and heme a was extracted and measured by the method of Rieske (12). Mitochondrial respiration was measured by standard manometric techniques (13).

III. RESULTS

A. Influence of Ionic State of Iron on Heme Synthesis

Heme synthesis by isolated hepatic mitochondria was studied to compare the effectiveness of Fe (III) with that of Fe (II) as substrates (Table 1). Ferrous iron was maintained in the reduced state by the addition of excess cysteine. When normal and copper deficient mitochondria were compared in the absence of mitochondrial substrates, there was little heme synthesized in either system when Fe (III) was used as the iron source. When Fe (II) was used as the iron source, more heme was synthesized by both the normal and copper deficient mitochondria. However, there was no significant difference in the amounts synthesized.

Table 1

Influence of Ionic State of Iron on Heme Synthesis by Isolated Hepatic Mitochondria

Iron State	Heme Synthesis (nmoles/mg protein/hr)	
	Normal	Copper Deficient
	N=9	N=8
Fe (II)	1.1+0.10	0.8+0.08
Fe (III)	0.3+0.06	0.3+0.04

Values refer to mean +1SE

B. Influence of Electron Transport Substrates on Heme Synthesis

Heme synthesis was studied in the presence of mitochondrial substrates to determine the effect on the utilization of Fe (III) for heme synthesis (Table 2). In both the normal and copper deficient mitochondria, heme synthesis in the absence of these substrates was virtually nil, but when succinate was added (4mM), heme synthesis was increased. However, there was significantly less heme synthesized by copper-deficient mitochondria than by control mitochondria ($p<0.05$). The addition of isocitrate to normal mitochondria also enhanced heme synthesis.

Table 2

Influence of Electron Transport Substrates on Heme Synthesis from Fe (III) and Protoporphyrin by Isolated Intact Hepatic Mitochondria

Additive	Heme Synthesis (nmoles/mg protein/hr)	
	Normal	Copper Deficient
	N=27	N=13
None	0.3±0.04	0.2±0.04
Isocitrate (4mM)	1.6±0.64	-
Succinate (4mM)	2.3±0.15	1.6±0.20

Values refer to mean ±1SE

C. Influence of Adenine Nucleotides on Heme Synthesis

Heme synthesis from Fe (III) and protoporphyrin was studied in normal mitochondria in the presence of high-energy nucleotides (Table 3). The addition of neither ATP nor ADP resulted in an increase of heme synthesis when compared with mitochondria alone. Addition of succinate resulted in stimulation of heme synthesis as before, but this stimulatory effect was diminished in the presence of either ATP or ADP.

Table 3

Influence of Adenine Nucleotides on Heme Synthesis from Fe (III) and Protoporphyrin by Isolated Intact Normal Hepatic Mitochondria

Additive	Heme Synthesis (nmoles/mg protein/hr)
None	0.3±0.13
ATP (100 µM)	0.3±0.45
ADP (100 µM)	0.1±0.07
ATP + Succinate	1.6±0.55
ADP + Succinate	1.4±0.15
Succinate	2.8±0.86

Values refer to mean ±1SE

D. Influence of Electron Transport Inhibitors and Uncoupling Agents on Heme Synthesis

Heme synthesis was studied in normal intact mitochondria in the presence of succinate (Table 4). As before, heme

Table 4

Influence of Electron Transport Inhibitors and Uncoupling Agents on Succinate - Stimulated Heme Synthesis from Fe (III) and Protoporphyrin by Isolated Intact Normal Hepatic Mitochondria

Inhibitor	Heme Synthesis (nmoles/mg protein/hr) Normal
None	2.2±0.11
Malonate (12mM)	0.8±0.31
Antimycin A (1nmol/mg protein)	1.0∓0.18
Sodium azide	1.1
Sodium cyanide (10µ moles/mg)	0
Carbon Monoxide	2.3±0.35
2,4 Dinitrophenol (1mM)	1.7±0.09
Oligomycin	2.6

Values refer to mean ±1SE

synthesis using Fe (III) as substrate was enhanced over heme synthesis in the absence of succinate. Addition of malonate, antimycin A, azide and cyanide resulted in significant inhibition of heme synthetic activity. However, carbon monoxide treatment of mitochondria did not result in inhibition of heme synthesis. Furthermore, although the uncoupling agent, 2,4 DNP reduced heme synthesis, oligomycin was without effect.

E. Characteristics of Intact, Isolated Normal and Copper-Deficient Hepatic Mitochondria

The rate of heme synthesis from Fe (III) and protoporphyrin, in the presence of succinate, was decreased in isolated copper-deficient hepatic mitochondria (N=20, (P<0.01) (Table 5). Mitochondria from copper deficient animals were depleted of copper, cytochrome oxidase activity and heme a. Oxygen consumption was less, and the rate of heme synthesis correlated with the cytochrome oxidase activity (N=12, r= + 0.78, p < 0.001).

Table 5

Characteristics of Isolated, Intact Normal and Copper-deficient Mitochondria

Determination	Normal	Copper-Deficient
Heme synthesis (nmoles/mg protein/hr)	2.6+0.19	1.3+0.10
Cytochrome oxidase (nmoles/mg protein)	244+14.5	72+13.8
Heme a (nmoles/mg protein)	0.28+0.026	0.07+010
O_2 consumption (μl O_2/mg protein/hr)	2.4+0.18	1.8+0.20
Copper (ng/mg protein)	61+30.1	4+0.9
Iron (μg/mg protein)	1.2+0.28	4.9+0.92

Values represent mean +1SE
By permission, Williams, DM, Loukopoulos, D, Lee, GR and Cartwright, GE: Role of copper in mitochondrial iron metabolism, Blood 48:77-85, 1976 (3)

IV. DISCUSSION

Recently, evidence has been presented that ferric iron is taken up by mitochondria in a process which employs both energy independent and energy dependent mechanisms (14).

Energy dependent iron uptake apparently results in the reduction of iron to the ferrous state since the process may be inhibited by respiratory inhibitors and is unaffected and even inhibited by the addition of ATP or ADP. This reduction of iron presumably prepares it for insertion into the protoporphyrin ring by the enzyme, heme synthetase, that is found in the inner mitochondrial membrane (16). There is as yet no direct evidence which permits the conclusion that the iron taken up either by energy-dependent or by energy-independent mechanisms is available for heme synthesis. However, our studies support the hypothesis that a functional electron transport system is required for the reduction of iron prior to its insertion into heme. Thus, heme synthesis is enhanced by respiratory substrates reduced by respiratory inhibition and unaffected or impaired with the addition of ATP or ADP. Similar observations by Koller et al also support the hypothesis (17). These studies further suggest that the physiologic iron precursor for normal mitochondrial iron metabolism may be in the ferric state.

The means by which the electron transport system effects the reduction of iron remains unsettled. Barnes et al have suggested that reduction may occur at the succinate dehydrogenase or NADH dehydrogenase sites (18). Romslo and Flatmark have proposed that Fe (III) ligands establish an oxidation-reduction equilibrium with the respiratory chain at the level of cytochrome $\underline{c}$ (19). However, these same authors have also presented evidence that calcium competes with mitochondrial iron uptake (20). Furthermore, uncoupling agents have been observed to diminish both mitochondrial iron uptake (15) and heme synthesis while treatment with carbon monoxide has no effect on heme synthesis. It is thus tempting to suggest that the copper enzyme, cytochrome oxidase, plays a direct role in the reduction of iron at the mitochondrial membrane. Alternatively, deficiency of cytochrome oxidase may simply result in a decrease of the respiratory rate of these mitochondria. In either case, it is clear that an intact electron transport system is essential for the normal mitochondrial processing of iron. Thus, it is proposed that the previously observed intracellular defect of iron processing in copper deficiency is due to a deficiency of cytochrome oxidase. This defect results in impaired synthesis of heme from Fe (III) and protoporphyrin.

V. REFERENCES

1. Lee, G.R., Williams, D.M., and Cartwright, G.E., in "Trace Elements in Human Health and Disease" (A.S. Prasad, Ed.), vol. 1, p. 373, Academic Press, New York, 1976.
2. Lee, G.R., Nacht, S., Lukens, J., and Cartwright, G.E., J. Clin. Invest. 47, 2058 (1968).
3. Williams, D.M., Loukopolos, D., Lee, G.R., and Cartwright, G.E., Blood 48, 77 (1976).
4. Lee, G.R., Cartwright, G.E., and Wintrobe, M.M., Proc. Soc. Exp. Biol. Med. 127, 977 (1968).
5. Hammond, E., Deiss, A., Carnes, W.H., and Cartwright, G.E., Lab. Invest. 21, 292 (1969).
6. Goodman, J.R. and Dallman, P.R., Blood 34, 747 (1969).
7. Roeser, H.P., Lee, G.R., Nacht, S., and Cartwright, G.E., J. Clin. Invest. 49, 2408 (1970).
8. Schneider, W.C., in "Manometric Techniques" ed. 4, (W.W. Umbreit, R.H. Burris, J.F. Stauffer, Eds.) p. 177, Burgess, Minneapolis, 1964.
9. Shermin, D., London, I.M., and Rittenberg, D., J. Biol. Chem. 183, 757 (1950).
10. Lowry, O.H., Rosebrough, N.J., Farr, A.L., and Randall, R.J., J. Biol. Chem. 193, 265 (1951).
11. Wharton, D.C., and Tzagaloff, A., in "Methods in Enzymology", (R.W. Estabrook and M.E. Pullman, Eds.) vol. 10, p. 245, Academic Press, New York, 1967.
12. Rieske, J.S., in "Methods of Enzymology," (R.W. Estabrook and M.E. Pullman, Eds.), vol. 10, p. 488, Academic Press, New York, 1967.
13. Potter, V.R., in "Manometric Techniques" ed. 4, (W.W. Umbreit, R.H. Burris, J.F. Stauffer, Eds.) p. 159, Burgess, Minneapolis, 1964.
14. Romslo, I. and Flatmark, T., Biochim. Biophys. Acta 305, 29 (1973).
15. Flatmark, T. and Romslo, I., J. Biol. Chem. 250, 6433 (1975).
16. Jones, M.S. and Jones, O.T.G., Biochem J. 113, 507 (1969).
17. Koller, M.E., Romslo, I., and Flatmark, T., Biochim. Biophys. Acta 449, 480 (1976).
18. Barnes, R., Connelly, J.L., and Jones, O.T.G., Biochem J. 128, 1043 (1972).
19. Flatmark, T., Romslo, I., J. Biol. Chem. 250, 6433 (1975).
20. Romslo, I. and Flatmark, T., Biochim. Biophys. Acta 325, 38 (1973).

MITOCHONDRIAL "NON-HEME NON-FeS IRON" AND ITS

SIGNIFICANCE IN THE CELLULAR METABOLISM OF IRON[1]

TORGEIR FLATMARK
ARILD TANGERÅS

Department of Biochemistry, University of Bergen

I. INTRODUCTION

Mitochondria hold a key position in the biosynthesis of heme and its regulation, and one of the basic prerequisits for normal heme synthesis, notably in erythroid cells, is a regulated supply of iron to mitochondria (1). Thus, ferrochelatase (EC 4.99.1.1), catalyzing the insertion of Fe(II) into porphyrin, is a mitochondrial enzyme (2-6) confined to the matrix side of the inner membrane (7-9). The importance of mitochondria in intracellular iron metabolism is particularly evident in disorders with deranged heme biosynthesis, in which the mitochondria are heavily loaded with iron (10).

Within the past few years, there has been considerable progress in understanding the mitochondrial uptake and utilization of iron (for review, see 11). Thus, mammalian mitochondria have evolved a transport system to accumulate iron from the environment (cytosol) and are able to utilize Fe(III) of certain synthetic low molecular weight complexes (11). Kinetic

---

[1]"Non-heme non-FeS iron" is tentatively used as a term to indicate the total amount of mitochondrial non-heme iron related to transport and storage; BPS, bathophenanthroline sulfonate; CCCP, carbonyl cyanide-$\underline{m}$-chlorophenylhydrazone; PIPES, piperazine-$\underline{N},\underline{N}$'-bis(2-ethanesulfinic acid); SDS, sodium dodecyl sulphate; TMPD,$\underline{N},\underline{N},\underline{N}',\underline{N}$'-tetramethyl-$\underline{p}$-phenylenediamine.

studies of the transport process have revealed that it requires energy, and when the iron donor is a Fe(III) complex, reducing equivalents are also required (12). Thus, iron appears to be transported across the inner membrane only as Fe(II) (12,13), but in contrast to the uptake of calcium mitochondria do not require energy to retain iron once it has been accumulated (12). However, the magnitude of this mitochondrial iron pool in vivo as well as its compartmentation, chemical nature, redox state and function is yet poorly understood.

II. MATERIALS AND METHODS

Rat liver mitochondria were prepared as described (14,15) except that they were washed 4-5 times. The matrix fraction was recovered from mitoplasts by sonication (16) in a medium containing 5 mM PIPES buffer, pH 6.5 and centrifugation (100 000$\underline{g}$, r_{av}=5.9 cm, 1 h at 4 °C); the yield of malate dehydrogenase (EC 1.1.1.37) in the supernatant was 85 %.

Total iron and copper was determined chemically (17). The reaction of bathophenanthroline sulfonate (BPS) with mitochondrial iron was followed by a dual-wavelength spectrophotometer at 37 °C in a medium (1ml volume) containing: 5 mM PIPES buffer (pH 6.5), 3.3 mM CN^-, 5 μM CCCP, 45 μM X-537A, 5 mM ascorbate and 0.1 mM TMPD. The absorbancy increments at 540 - 575 nm were converted to concentration of Fe(II) by titration of known amounts of Fe(II). The possible contribution of Cu(II) to the total spectral change was estimated from chemical analysis of the BPS-complex formed, to be less than 10 %.

The changes in light scattering were measured in a dual-beam instrument (Shimadzu MPS-50L recording spectrophotometer) at 575 nm. The content of heme proteins (cytochromes) was determined from the reduced minus oxidized difference spectra (18) using a dual-wavelength spectrophotometer.

All solutions containing interfering iron was passed through a column of Chelex 100 (BioRad).

III. RESULTS

At physiological levels of iron in the hepatocytes, the mitochondrial preparations revealed a relatively constant content of iron, i.e. 4.3 ± 0.6 nmol Fe·mg protein^{-1} (Table I). Since repeated washings and digitonin treatment at concentrations which disrupt lysosomes, did not significantly change this number, all the iron could be accounted for as being of mitochondrial origin. According to Table I approx. 2/3 of the iron is non-heme Fe.

TABLE I The Iron Content and Most Common Forms of Iron in Isolated Rat Liver Mitochondria[a]

Total Fe	Heme Fe	Non-heme Fe	"Non-heme non-FeS Fe"
4.3 ± 0.6[b]	1.24 ± 0.08[c]	3.0 ± 0.5[d]	1.4 ± 0.3[f]
		3.1 ± 0.6[e]	

[a] nmol·mg mitochondrial protein^{-1}; $\underline{n}$ = 4.
[b] Chemical analysis (17).
[c] Determined spectrophotometrically (18).
[d] Determined as a difference between total iron and heme iron.
[e] Determined as the total amount of iron which reacts with BPS in the presence of SDS and dithionite.
[f] For details, see text.

In order to differentiate between FeS iron and the non-heme iron which is related to transport and storage, experimental conditions in which the FeS-centers are stable have been selected. Thus, based on the reaction between the hydrophilic chelate BPS and iron in hypo-osmotically swelled mitochondria we have found[2] that rat liver mitochondria contain a substantial amount of non-heme iron, which is not related to the FeS-centers (Table I); on an average it accounts for approx. 1/3 of the mitochondrial iron. In the presence of SDS, i.e. under conditions where the FeS-centers are destroied (19), an additional 1/3 of the mitochondrial iron reacted with BPS. Whereas all the non-heme iron is complexed within seconds in the presence of SDS (Table I and Fig. 2), the time course is biphasic in its absence, consisting of a rapid phase (a few seconds) in which only 2-5 % of the "non-heme non-FeS iron" reacts and a very slow phase (completed in about 3 h) in which the remaining part of this iron pool is mobilized (Fig. 1).

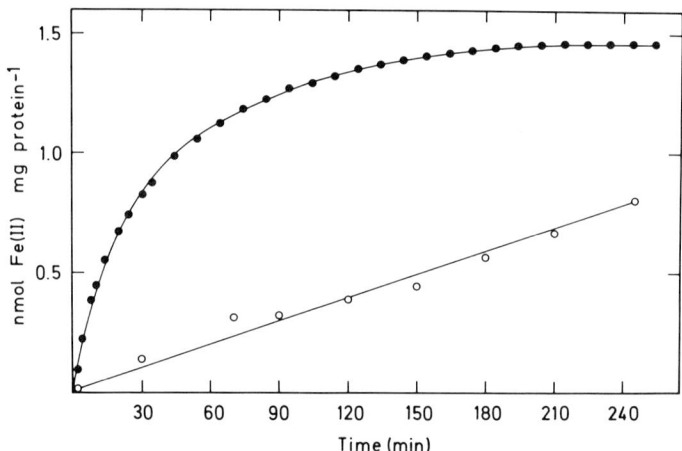

Fig. 1 The time course of the reaction of BPS with Fe(II) in hypo-osmotically swelled rat liver mitochondria (●) and the effect of preincubation time at 37 °C on the contribution of the rapid intial (30 s) phase of the reaction (o). Standard incubation conditions; 1.05 mg mitochondrial protein·ml^{-1}.

---

[2] Hypo-osmotic conditions were selected in order to eliminate spectral contributions from a slow swelling of mitochondria in 0.25 M sucrose induced by BPS.

The time course in the absence of SDS is little affected either by the presence of reducing agents (e.g. ascorbate/ TMPD), oxidizable substrates or the cation ionophore X-537 A in the incubation medium. On the other hand, the initial (30 s) phase increased to approx. 30 % of the total when dithionite was used as the reducing agent (Fig. 2).

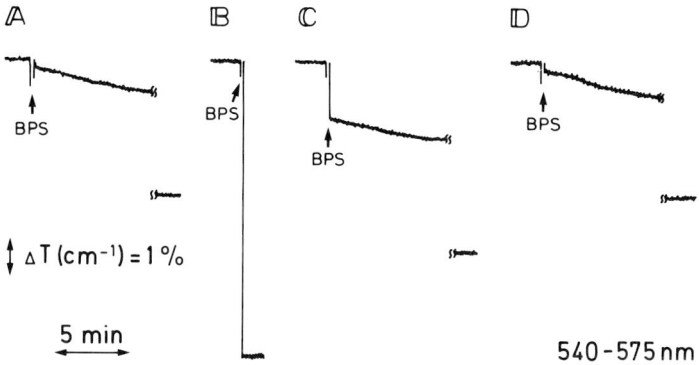

Fig. 2 The time course of the reaction of BPS with Fe(II) in hypo-osmotically swelled rat liver mitochondria in the absence (A) and presence (B) of SDS and dithionite; standard assay conditions. C and D: The time course in the presence of dithionite with mitochondria (C) and mitoplasts (D); 1.05 mg mitochondrial protein·ml^{-1}; 50 μM BPS.

The submitochondrial localization of the "non-heme non-FeS iron" was studied by two different approaches. First, preincubation of the mitochondria at 37 °C was found to increase the magnitude of the initial rapid (30 s) phase of the reaction, which indicates that this iron pool represents an efflux of relatively loosely bound Fe(II) and only to a small extent surface binding to the membranes (Fig. 1). Secondly, by fractionation of mitoplasts approx. 60 % of the "non-heme non-FeS iron" was recovered in the matrix phase. The total amount of iron in the matrix, which was mobilized by BPS, increased in the presence of SDS and dithionite.

IV. DISCUSSION

A. The Non-heme Iron Pool of Rat Liver Mitochondria

1. Non-FeS versus FeS Iron

The iron content of highly purified rat liver mitochondria was in this study found to be 4.3 ± 0.6 nmol·mg protein^{-1} and the iron is almost equally distributed between heme proteins, FeS-proteins and "non-heme non-FeS iron" (Table I). The latter fraction has been measured and to a certain extent characterized by its reaction with the hydrophilic chelate BPS in hypo-osmotically swelled mitochondria. Our data indicate that the FeS-centers are not affected at our incubation conditions: (a) the major pool of labile iron, which reacts slowly with BPS, revealed rather uniform kinetics; (b) the pH dependence of the reaction (with maximal formation of complex at pH$\gtrsim$ 6.5) is different from that generally observed for membrane-bound acid-labile FeS-centers; (c) the total amount of iron which reacts with the chelate increases two-fold in the presence of SDS which destroies the FeS-centers (19) and all the iron reacts within seconds; (d) a large proportion (approx. 60 %) of the measured "non-heme non-FeS iron" pool is confined to the matrix which contains only negligible amounts (< 0.05 nmol·mg protein^{-1}) of FeS protein of ferredoxin type (Pedersen, J.I.,personal communication). The possible contribution from the FeS protein of HIPIP type (20) and aconitase (21,22) is yet unknown; (e) our preliminary studies using EPR spectrocopy do not give any indication of destruction of the mitochondrial FeS centers (Flatmark,T. et al.,unpublished data) which is in good agreement with the recent demonstration that the FeS-centers of the respiratory chain is hydrophobically buried in the membrane and thus inacessible to hydrophilic molecules like BPS (23).

2. Compartmentation and Chemical Nature of the "Non-heme non-FeS Iron"

The "non-heme non-FeS iron" does not represent a single homogenous pool. One may roughly differentiate between a major (approx. 70 %) internal and a minor (approx. 30 %) external pool based on the criterion of accessibility to reaction with the hydrophilic iron chelate BPS. In addition, almost all of the "non-heme non-FeS iron", which reacts with BPS under standard incubation conditions, also reacts in the absence of exogenous reducing agents which indicates that this iron pool is present as Fe(II).

Whatever the ligands are, the iron confined to the inner compartment appears to be held in a rather specific manner which gives the metal a relatively high redox potential and favours its reduced form even under aerobic conditions. This conclusion compares well with the finding that iron is translocated across the mitochondrial inner membrane as Fe(II) (12), and that the $\underline{g}$ = 4.3 EPR signal, characteristic of Fe(III) in the high-spin state, is very low in respiratory inhibited or anaerobic mitochondria (Flatmark,T. et al.,unpublished data).

B. The Function of Mitochondrial "Non-heme Non-FeS Iron"

In the present study experimental evidence is given for a storage form of iron in mitochondria of functionally normal hepatocytes. So far only one metabolic function is known for the "non-heme non-FeS iron", i.e. to provide a source of iron which can be drawn on when required for biosynthesis of heme (11), and possibly also of FeS proteins (24). Based on the present experimental evidence (chemical reactivity and EPR spectroscopy) this iron pool mainly represents Fe(II) which is the form of iron utilized in heme biosynthesis (11).

It has recently been claimed by Ponka et al. (25) that mitochondria take up iron directly from transferrin and provide iron both for mitochondrial heme synthesis and for non-heme iron compounds in the cytosol,e.g. ferritin (25). In contrast to these authors, we have found no evidence of an "export" of iron from mitochondria. On the contrary, once iron has been accumulated by mitochondria in an energy-dependent process (even in saturating amounts),it is firmly retained and can not be released either by de-energetization (12) as is the case for calcium (26) or by hydrophilic and hydrophobic chelating agents (12). Secondly, the "non-heme non-FeS iron" pool can be mobilized only very slowly by the hydrophilic chelate BPS in hypo-osmotically swelled mitochondria (Figs. 1 and 2).

ACKNOWLEDGEMENTS

This study was supported by the Norwegian Research Council for Science and the Humanities and the Norwegian Cancer Society. We wich to thank Hoffman-La Roche for a gift of the ionophore X-537 A.

REFERENCES

1. Jacobs,A., in "Iron in Biochemistry and Medicine"(Jacobs, A. and Worwood,M. eds.), p. 405. Academic Press, New York, 1974.

2. Barnes,R., Jones,M.S., Jones,O.T.G. and Porra,R.J., Biochem.J. 124, 633 (1971).

3. Labbe,R.F. and Hubbard,N., Biochim. Biophys. Acta 41, 185 (1960).

4. Lochhead,A.C. and Goldberg,A., Biochem. J. 78, 146 (1961).

5. Minakami,S., J. Biochem. (Tokyo) 45, 833 (1958).

6. Nishida,G. and Labbe,R.F., Biochim. Biophys. Acta 31, 519 (1959).

7. Barnes,R., Connelly,J.L. and Jones,O.T.G., Biochem. J. 128, 1043 (1972).

8. Jones,M.S. and Jones,O.T.G., Biochem.J. 113, 507 (1969).

9. McKay,R.,Druyan,R., Getz,G.S. and Rabinowitz,M., Biochem. J 114, 455 (1969).

10. Cartwright,G.E. and Deiss,A., N.Engl.J.Med. 292,185 (1975).

11. Flatmark,T. and Romslo,I., Adv. Chem. Ser. in press (1977).

12. Flatmark,T. and Romslo,I., J.Biol.Chem. 250,6433 (1975).

13. Koller,M.E., Romslo,I. and Flatmark,T., Biochim. Biophys. Acta 449, 480 (1976).

14. Romslo,I. and Flatmark,T., Biochim. Biophys. Acta 305, 29 (1973).

15. Slinde,E. and Flatmark,T., Anal. Biochem. 56,324 (1973).

16. Romslo,I. and Flatmark,T., Biochim. Biophys. Acta 347, 160 (1974).

17. Van de Bogart,M. and Beinert,H., Anal. Biochem. 20, 325 (1967).

18. Vanneste,W.H., Biochim. Biophys. Acta 113, 175 (1966).

19. Davis,K.A. and Hatefi,Y., Biochemistry 10,2509 (1971).

20. Beinert,H., in "Iron-Sulfur Proteins" (Lovenberg,W. ed.), Vol. III, p. 61. Academic Press, New York, 1977.

21. Kennedy,S.C., Rauner,R. and Gawron,O. , Biochem. Biophys. Res. Commun. 47, 740 (1972).

22. Suzuki,T., Akiyama,S., Fujimoto,S., Ishikawa,M., Nakao,Y. and Fukuda,H., J. Biochem. (Tokyo) 80, 799 (1976).

23. Case, G.D., Ohnishi,T. and Leigh,J.S.,Jr., Biochem. J. 160, 785 (1976).

24. Brodrick,J.W. and Rabinowitz,J.C., in "Iron-Sulfur Proteins" (Lovenberg,W. ed.),Vol. III,p.   . Academic Press, New York, 1977.

25. Ponka,P., Neuwirt,J., Borova,J. and Fuchs,O., in "Iron Metabolism" (Ciba Foundation Symposium No. 51), in press (1977).

26. Drahota,Z., Carafoli,E., Rossi,C.S., Gamble,R.L. and Lehninger,A.L., J. Biol. Chem. 240, 2712 (1965).

A GENERAL MODEL OF INTRACELLULAR IRON METABOLISM

Benjamin F. Trump
Irene K. Berezesky

Department of Pathology
University of Maryland School of Medicine

The pathways of intracellular iron metabolism are fundamentally similar in widely diverse cell and species types (1-5). Thus the mechanisms of Fe uptake, storage and release in a macrophage in an area of old hemorrhage are remarkably similar to those in an hepatocyte, splenic sinusoid, normoblast, kidney tubule or intestinal epithelium. The purpose of this communication is to review the current "state of the art" and put forward a general hypothesis. Iron enters cells in vivo from specific carrier molecules such as transferrin (TF), as part of Fe-containing macromolecules, e.g., hemoglobin (Hb) or, in selected cases, as colloidal iron complexes such as Fe dextran. TF-Fe complexes appear to enter endocytic vacuoles prior to Fe release; the TF is subsequently secreted. Proteins such as Hb and colloidal Fe complexes appear to enter the lysosomes wherein the Fe is released back to cell sap. In both cases, Fe in cell sap is probably bound to a carrier from which it gains access to various pathways. Ferritin synthesis is stimulated principally on free polysomes and ferritin appears in the cell sap (2). Fe is also actively transported into mitochondria. Ferritin molecules as well as other Fe-containing macromolecules (e.g., cytochrome in membranes) then reenter the lysosomes via autophagocytosis (3) with release of iron, and gradually become converted to hemosiderin (4). Hemosiderin apparently represents a non-specific Fe macromolecular or colloidal complex which varies with lysosomal content (Fig. 1).

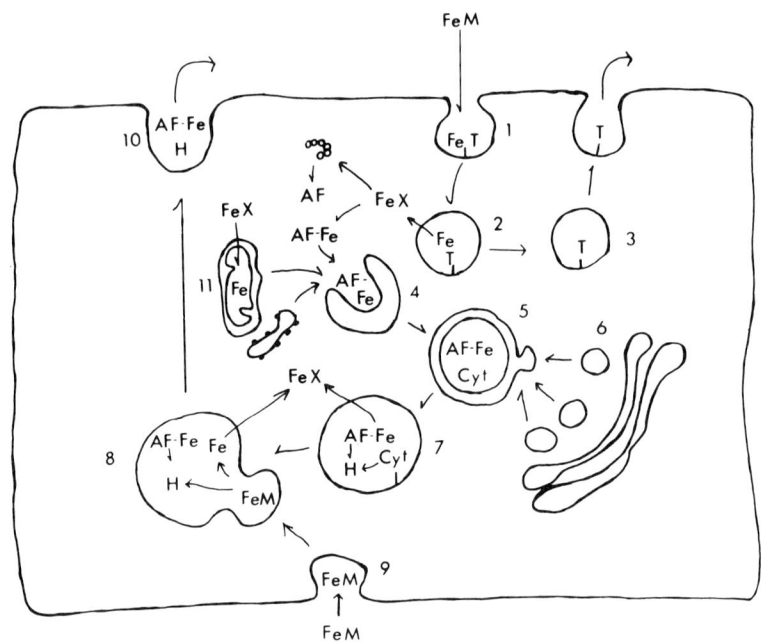

Fig. 1. A diagram showing the basic mechanisms involved in iron uptake distribution, turnover and excretion from a typical cell. (1) is an iron-containing macromolecule (FeM) being taken up by pinocytosis. In the case of transferrin (FeT), binding apparently occurs to specific membrane sites. As the incoming phagocytic or pinocytic vacuole is internalized (2) the iron is apparently released from the transferrin leaving a pinosome or phagosome containing the transferrin (3) which returns to the surface, fuses and releases the transferrin. The iron is believed to be released from the phagosome and carried in a soluble complex of unknown characteristics (FeX) in the cytoplasm. In the cell sap, the iron stimulates the formation of apoferritin (AF) and combines with it to form ferritin (AFFe) which is found predominantly in the cell sap. Iron-containing ferritin is then turned over by sequestration in autophagic vacuoles by budding into cisternae of ER (4) to form a body with sequestered pieces of cytoplasm containing ferritin and cytosol (5). These then fuse with Golgi vesicles or primary lysosomes (6) and convert to secondary lysosomes (7) containing ferritin. Other iron-containing proteins as well as cytochromes (Cyt) gradually become denatured, digested and form non-specific iron macromolecule complexes collectively called hemosiderin (8). Iron is also released to return to the cytoplasm forming more FeX. Iron macromolecules (step

9) can also be taken directly into the secondary lysosomes by fusion and direct conversion to hemosiderin. Iron release can occur as well. During subsequent phagocytic events, ferritin and presumably hemosiderin can pass by direct excretion to the extracellular space (step 10) as a possible leak during phagocytosis. The iron carrier (FeX) in the cytoplasm can also contribute iron for heme synthesis in the mitochondria (step 11). Mitochondria are also turned over by autophagocytosis and the cytochrome and other iron-containing proteins contribute iron to the intralysosomal iron pool which participate as mentioned above.

In iron-overload, which may damage cells through peroxidation, more and more hemosiderin is present. In iron deficiency, very little hemosiderin is seen. It is confined to the lysosomes except in sideroblastic anemia where similar appearing material accumulates in the mitochondrial matrix of red cell precursors and reutilization of stored lysosomal iron can apparently occur. Iron can be released from cells to some extent by defecation from lysosomes. This occurs by fusion of the lysosomes with the cell surface and then apparent extrusion of the molecules. The sequestration of ferritin may be a protective mechanism in the cell to prevent membrane lipid peroxidation and subsequent cell damage which presumably occurs in iron-overload. Normally, iron-overload does not occur in mitochondria. However, mitochondrial iron accumulations do occur in sideroblastic anemia, a disease of multiple causations (1).

Several steps lead to the formation of iron-loaded mitochondria in the so-called ring sideroblasts. It seems as if the mechanism for active transport of iron from the cytoplasm to the mitochondrial compartment is not sensitive to feedback inhibition and, therefore, iron continues to accumulate even though it cannot be utilized. Possible inhibition steps include inhibition of ALA synthetase or inhibition of ferrochelatase. Non-specific inhibition of mitochondrial enzyme synthesis by chloramphenicol can also result in deficiencies of several of these mitochondrial enzymes. Once loaded with iron, damage to inner mitochondrial membrane and mitochondrial function can occur, inducing further damage to the heme synthesis enzyme. Turnover of iron-loaded mitochondria by autophagocytosis presumably occurs. This leads to intralysosomal mitochondria and, eventually, intralysosomal hemosiderin.

The other main pathway for iron acts as an overload

in both ringed sideroblasts and normal sideroblasts and involves formation of ferritin, which is sequestered within lysosomes again by autophagocytosis. This iron is not thought to be utilized, at least over short periods, for heme synthesis. The magnitude of the lysosomal loading decreases to nil with iron deficiency.

Recently, the availability and applicability of X-ray microanalysis has offered great promise for correlating structural and clinical composition of cells and tissues. It is but one example of a set of methods which are based on measurements of characteristic radiation resulting from interactions of electrons with matter and thus far, is the best developed. The technique of X-ray microanalysis permits, in many cases, complete non-destructive analysis by detection of characteristic X-rays which can produce information about the distribution, quantity and chemical form of an element at the cellular and subcellular level in biological samples with a sensitivity of 10^{-18}g. Areas as small as 1000 Å can be measured which permit qualitative and potentially quantitative measurements to be made on organelles or even parts of organelles. This is obviously of value in studying many varieties of cellular inclusions such as those found in iron-overload (Fig. 2), asbestosis, silicosis, etc., and also in the measurements of diffusible ions of physiologic importance such as Na, Mg, K, Ca etc., all of which have been impossible to localize at this level with any other chemical technique because of difficulties of maintaining ion composition when organelles are isolated.

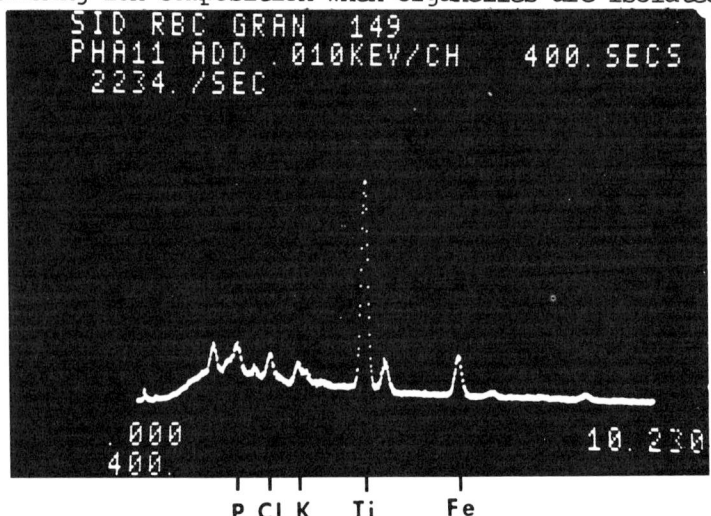

Fig. 2. *X-ray spectrum taken over mitochondrial deposits in an erythroblast from a sideroblastic anemia patient. Note the distinct peaks for P, K and Fe. The Ti peaks are from the grid. Tissues were fixed in a mixture of 4% formalin/1% glutaraldehyde in phosphate buffer, post-fixed in 1% osmium tetroxide and processed for electron microscopy. Unstained Epon sections were examined with an AMR 1000 SEM fitted with a Kevex 30 mm SiLi detector connected to a Tracor Northern NS 880 collecting system. Analyses were done in the STEM mode at an accelerating voltage of 10 KV with 75 μA beam current, a beam diameter of 1000 Å and counting time of 400 seconds.*

The application of this technique to the study of iron-loaded cells allows one not only to measure concentrations of iron in various cell compartments such as lysosomes, mitochondria and cytosome, but also to estimate the concentration ratios between the various cell compartments (Fig. 3).

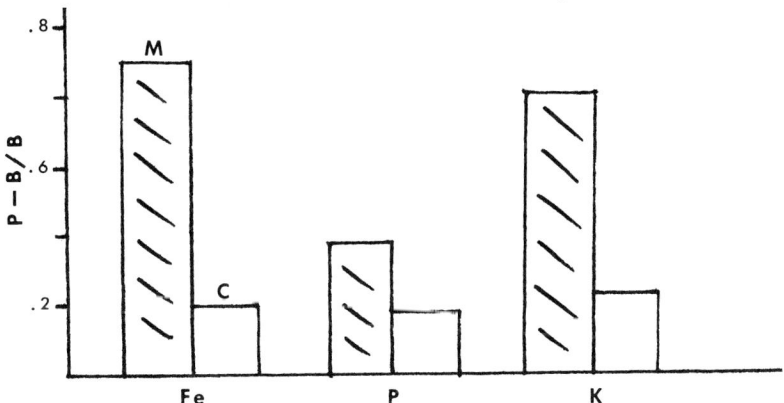

Fig. 3. *Bar graph illustrating P-B/B ratios for iron, phosphorus and potassium taken over mitochondrial deposits (M) and adjacent cytoplasm (C) in a red cell precursor from a patient with sideroblastic anemia. Peak to background ratios were computed by fitting control specimen spectra to experimental specimen spectra using the "Hall X/Super ML" program from Tracor Northern.*

Work is currently in progress in this laboratory analyzing the extensive deposits seen in liver, spleen and bone marrow from patients with thalassemia in order to make more detailed comparisons with results found in sideroblastic cases. Thus far, we characteristically find phosphate to be associated with iron both in sideroblastic and thalassemic cases, while potassium is found in increased amounts in association with iron and phosphorus <u>only</u> in the sideroblastic mitochondrial and lysosomal deposits.

In conclusion, it is now possible to begin to synthesize the limited information available on iron metabolism with newly acquired results and formulate working hypotheses regarding intracellular iron pathways. Among unanswered questions are the nature of the iron in the exchange phase and cell sap that is the definition of some type of carrier, the pathways of entry into the autophagic vacuoles by ferritin, chemical details of the conversion of the hemosiderin, and also the question of whether or not the cisternae of the endoplasmic reticulum even contain either apoferritin or ferritin. The answers to all of these questions are subject to future experimentation.

REFERENCES

1. Trump, B.F., Barrett, L.A., Valigorsky, J.M. and Jiji, H.Y., Ultrastructural Studies of Sideroblastic Anemia, in "Iron Metabolism and Its Disorders" (H. Kief, Ed.), Vol. 3, p. 251, Excerpta Medica, Amsterdam, 1976.

2. Trump, B.F., Valigorsky, J.M., Arstila, A.U. and Mergner, W.J., A Concept of Cellular Iron Metabolism and Iron Overload, in "Iron Metabolism and Its Disorders" (H. Kief, Ed.), Vol. 3, p. 97, Excerpta Medica, Amsterdam, 1976.

3. Trump, B.F., Arstila, A.U., Valigorsky, J.M. and Barrett, L.A., Subcellular Aspects of Ferritin Metabolism, in "Proteins of Iron Storage and Transport in Biochemistry and Medicine" (R.R. Crighton, Ed.), p. 343, North-Holland Publishing Co., Amsterdam, 1975.

4. Trump, B.F., Valigorsky, J.M., Arstila, A.U., Mergner, W.J. and Kinney, T.D., The Relationship of Intracellular Pathways of Iron Metabolism to Cellular Overload and the Iron Storage Diseases. Am. J. Path. 72, 295 (1973).

5. Trump, B.F., in "Ciba Symposium on Iron Metabolism," (D. Fitzsimmons, Ed.), Elsevier/Excerpta Medica/North-Holland Publishing Co., Amsterdam, 1976.

MEMBRANE RECEPTORS FOR MICROBIAL IRON TRANSPORT COMPOUNDS

(SIDEROPHORES)

J. B. Neilands
R. R. Wayne

*Department of Biochemistry, University of California
Berkeley, California*

I. INTRODUCTION

Siderophores, defined as *high affinity microbial ferric ion transport agents*, are widely produced by aerobic and facultative anaerobic species. In general, the ligands of siderophores can be classed as either hydroxamic acids or catechols; ferrichrome and enterobactin are the respective prototypes of the two classes. The former is commonly produced by fungi, such as *Aspergillus niger*, while the latter is the siderophore indigenous to *Escherichia coli* K-12, *Salmonella typhimurium* LT-2 and other enteric bacteria.

The enteric bacteria, even though unable to synthesize ferrichrome, nonetheless possess a transport system for this siderophore. The outer membrane receptor for ferrichrome in *E. coli* is shared by albomycin, phages T1, T5, Φ80, Φ80h and colicin M (1); in *S. typhimurium* the ferrichrome receptor is the common binding site for albomycin, as well as for phages ES18 and ES18h (2).

The ferric enterobactin receptor is shared by colicin B. Phages have not yet been described for this receptor but they very likely exist.

Citrate can also be viewed as a siderophore of high affinity. It supplies iron to *E. coli* via a membrane receptor, but not enough ligand can be incorporated via this route to satisfy the carbon requirements of the organism. *S. typhi-*

murium, in contrast, can grow on citrate as carbon source but cannot transport the ferric chelate of the tricarboxylic acid.

The wealth of genetic information which exists for the enteric bacteria renders these organisms the species of choice for studies of siderophore transport. In regard to the specific siderophore, there are advantages and disadvantages to working with either ferrichrome or ferric enterobactin. Both can be prepared in good yield by culture of an appropriate organism at low levels of iron.

Ferrichrome comes with an impressive array of structural analogs substituted in the cyclohexapeptide and/or hydroxamate moieties - ferrichrysin, ferricrocin, ferrichrome C, ferrichrome A, albomycin, ferrirhodin, ferrirubin. The ligand is rugged and can be converted to the kinetically inert, isostructural chromium-III complex thus affording a convenient means of defeating the rapid exchangeability of the ferric ion. Affinity columns can readily be prepared from ferrichrome compounds and the molecule stands up reasonably well to tritiation by microwave discharge. The phages and colicins which exploit the biochemical function of the ferrichrome receptor are invaluable tools for probing the structure and mechanism of action of the latter. The two major problems with the ferrichrome receptor are, first, that it does not appear to be derepressed by growth of the organism at low levels of iron and, second, that it releases its iron rapidly to enterobactin. Hence all work on ferrichrome transport in enteric bacteria must be performed in mutants defective in the synthesis of enterobactin.

The main advantage of working with enterobactin lies in the fact that it is the siderophore native to enteric bacteria and, like all siderophores, it is strongly overproduced by culture of the organisms at low absolute or available levels of iron. This affords an opportunity for study of its regulation of synthesis. On the other hand the ligand is notoriously unstable since it may suffer both hydrolysis and oxidation.

The current knowledge of the high affinity iron transport systems has been reviewed elsewhere (3). Siderophores interact with specific, outer membrane receptors in *E. coli*. The molecular weights of these receptors by PAGE analysis in SDS are in the 75K-90K range and the one for ferric enterobactin has been shown to be derepressed in its biosynthesis at low intracellular concentrations of iron. The ferric citrate receptor is apparently induced by the presence of citrate in the medium.

II. THE FERRICHROME TRANSPORT SYSTEM

Evidence has been presented for a common binding site for ferrichrome, colicin M, phages T1, T5, Φ80, Φ80h and albomycin in the outer membrane of *E. coli* K-12 (1). This receptor site is the product of the *tonA* gene. In *S. typhimurium*, a similar receptor for ferrichrome, albomycin and phages ES18 and ES18h was identified (2). Working *in vitro* with the isolated T5 receptor from *E. coli*, Luckey *et al.* (4) showed that the ferrichrome receptor is specific for this siderophore and its close relatives, including isostructural metal ion analogs such as the chromium III complex of deferriferrichrome.

Enteric bacteria appear to transport ferrichrome by two mechanisms (5). In the first the iron is either selectively and rapidly removed at the surface or else the siderophore dissociates to allow simultaneous permeation of the parts, followed by fast expulsion of the ligand. In the second mechanism, which proceeds at a slower pace, the intact ferric complex penetrates the cell.

III. THE FERRIC ENTEROBACTIN TRANSPORT SYSTEM

As reviewed elsewhere (3), earlier work by Wang and Newton (6) and Guterman and Dann (7) on the role of catechols and colicin B in iron transport could be interpreted as a competition between ferric enterobactin and colicin B for an outer membrane receptor in *E. coli* (1). This proposal was strengthened by an examination of the nutritional response to siderophores observed in mutants insensitive to members of the B group colicins. Thus mutants lacking colicin B receptor were unable to utilize ferric enterobactin (8). The mechanism of siderophore protection against colicin I and V remains unexplained but it is known to require utilization of siderophore iron (9).

The regulation of the siderophore transport system is of profound biological significance since control of iron uptake is known to be actuated by a repressive type of mechanism throughout the living world. If the process can be understood in a prokaryotic cell then there is hope of extending this knowledge to mitochondria and to eukaryotic systems in general.

Strains capable of excretion of high levels of enterobactin are of obvious interest for studies of regulation of biosynthesis of this siderophore. The *exbB* phenotype hyperexcretes enterobactin and requires methionine. Since the mutation maps near the "polyamine operon" it was decided to examine the effect of polyamines on the sensitivity of exbB strains to colicin B. Strain LD60 plated with zero or a low level of spermidine was fully resistant to colicin B but at 1.0 mM spermidine the organism was as sensitive as the parent. Table I shows the effect of several polyamines on accumulation of catechols by a tonB mutant of E. coli. Those poly-

TABLE I   The Effect of Polyamines and Diamines on Catechol Accumulation by a tonB Mutant of Escherichia coli K-12.

Compound	Concentration	Catechol Production (percent of control)
No addition	----	100
Ethylene diamine	1 mM	12 ± 1
	10 mM	6 ± 1
1,3-propanediamine	1 mM	44 ± 10
	10 mM	51 ± 15
1,4-diaminobutane (putrescine)	1 mM	>90 ± 10
	10 mM	80 ± 20
1,5-diaminopentane (cadaverine)	1 mM	>90 ± 10
Spermidine	1 mM	80 ± 10
	10 mM	24 ± 2
N-(3-aminopropyl) 1,7-diaminoheptane	1 mM	>90 ± 10
2-hydroxyputrescine	1 mM	>90 ± 10
iron citrate	50 µM	13 ± 1

amines deemed most active in metal binding, namely ethylene diamine and spermidine, were the most active in suppressing catechol synthesis. The data in Table I refer to the arithmetic mean ± standard deviation, calculated from eight trials for each compound and concentration tested.

The integrity of S-adenosylmethionine decarboxylase and propylamine transferase, in respect to both Km and V, were found to be maintained in LD60. Similarly, exbB mutants appear to be normal in regard to their ability to effect the iron-dependent thiomethylation of tyrosine tRNA. Transduction of strain RW1934 ($exbB$) to met^+ with P1 donor phage from Eth^r ($metK$) gave transductants all of which were insensitive to ethionine, thus suggesting that the $exbB$ and $metK$ genes, the latter specifying S-adenosylmethionine synthetase, are tightly linked, if not identical. The methionine requirement in $exbB$ was traced to a block in the conversion of homoserine to cystathionine.

REFERENCES

1. Wayne, R., Neilands, J. B., J. Bacteriol. 121, 497 (1975).

2. Luckey, M., Neilands, J. B., J. Bacteriol. 127, 1036 (1976).

3. Neilands, J. B., in "Bioinorganic Chemistry-II" (Raymond, K. N. ed.), Advances in Chemistry Series, American Chemical Society, Washington, D.C., 1977.

4. Luckey, M., Wayne, R., Neilands, J. B., Biochem. Biophys. Res. Commun. 64, 687 (1975).

5. Leong, J., Neilands, J. B., J. Bacteriol 126, 823 (1976).

6. Wang, C. C., Newton, A., J. Biol. Chem. 246, 2147 (1971).

7. Guterman, S. K., Dann, L., J. Bacteriol. 114, 1225 (1973).

8. Wayne, R., Neilands, J. B., Fed. Proc. 35, 1453 (1976).

9. Wayne, R., Frick, K., Neilands, J. B., J. Bacteriol. 126, 7 (1976).

SIDEROPHORE TRANSPORT IN *BACILLUS MEGATERIUM*

B. R. BYERS, J. E. L. ARCENEAUX,
A. H. HAYDON AND J. E. ASWELL

Department of Microbiology
University of Mississippi Medical Center
Jackson, Mississippi 39216 USA

I. THE PROBLEM IN IRON TRANSPORT

Iron is an essential but sometimes elusive metal for life systems. The Ksp for ferric hydroxide is low, which has required many microbes to release siderophores for chelation of iron, making it available for transport (1). Microbial siderophores are either secondary hydroxamic acids or phenolic acids; siderophores have high affinity for ferric iron but lesser affinity for ferrous iron (2). *Bacillus megaterium* produces the citrate-hydroxamate siderophore deferrischizokinen; however, this organism also can use some structurally different siderophores produced by other microorganisms (3).

II. THE TRANSPORT PROCESS

Mutants of *B. megaterium* unable to produce deferrischizokinen have been useful in defining the siderophore transport process. In one of these (strain SK11) ferrischizokinen and the ferrioxamine-type hydroxamate, ferriferrioxamine B, stimulated iron uptake whereas another citrate-hydroxamate, ferriaerobactin, withheld iron (4, 5). Determination of uptake kinetics over a range of concentrations revealed that rates increased sharply as the concentration of ferrischizokinen was increased. High concentrations of ferriferrioxamine B did not generate the rapid rates of radioiron uptake observed with ferrischizokinen, and ferriferrioxamine B delivered only 1/5 of the maximal level of iron transported from ferrischizokinen. Thus, *B. megaterium* SK11 has a recognition capacity for utilization of siderophores based on the structure of the siderophore.

Although siderophores may function as agents only for delivery of iron to the cell surface where iron may be released by special systems (6), in *B. megaterium* the major route of siderophore utilization involves transport of the chelate (7). Uptake of both labels of [^{3}H,^{59}Fe]ferrischizokinen showed that at low concentrations the pmoles of each accumulated were approximately equivalent. At higher concentrations, release of [^{3}H]deferrischizokinen after 30 seconds of uptake was apparent; however, release of [^{3}H]deferrischizokinen was not complete. These results are interpreted to mean that the chelate crossed the membrane and that the iron was rapidly released for metabolic use. The resulting deferrischizokinen was discharged from the cell, probably to function again.

III. FACILITATED DIFFUSION TYPE TRANSPORT

Maximal uptake of ferrischizokinen or ferriferrioxamine B was blocked in energy-starved or sodium azide-poisoned cells (4). Recent work (8) indicates that siderophore transport occurs by a facilitated diffusion process and that energy-requiring release of iron from the chelate drives siderophore transport. *B. megaterium* Ard1, a mutant (isolated from strain SK11) whose growth is inhibited by ferriferrioxamine B, showed only low-level, energy and temperature independent uptake of [^{55}Fe]ferriferrioxamine B. To assay for possible transport of the chelate, protoplasts were exposed to [^{55}Fe]ferrischizokinen or [^{55}Fe]ferriferrioxamine B for 30 seconds and then quickly ruptured. After removal of membranes by centrifugation, the cytoplasm was fractionated by Sephadex G-200 chromatography. Cytoplasm from protoplasts labeled with [^{55}Fe]ferrischizokinen contained 22% of the ^{55}Fe associated with high molecular weight material; the remainder appeared in a low molecular weight fraction (Figure 1A). In protoplasts labeled with [^{55}Fe]ferriferrioxamine B nearly all of the cytoplasmic label appeared in low molecular weight material (Figure 1B) which was identified as ferriferrioxamine B. *B. megaterium* Ard1 transported ferriferrioxamine B, but was unable to release iron from the chelate; such transport appeared to be facilitated diffusion. These results also revealed that a pool of cytoplasmic ferrischizokinen was established from which iron could be quickly removed. Uptake of [^{55}Fe]ferriferrioxamine B did not prevent transport of [^{59}Fe]ferrischizokinen and assimilation of its iron. The early-labeling high molecular weight fraction seen with [^{55}Fe]ferrischizokinen may be an acceptor of iron following its release from this chelate.

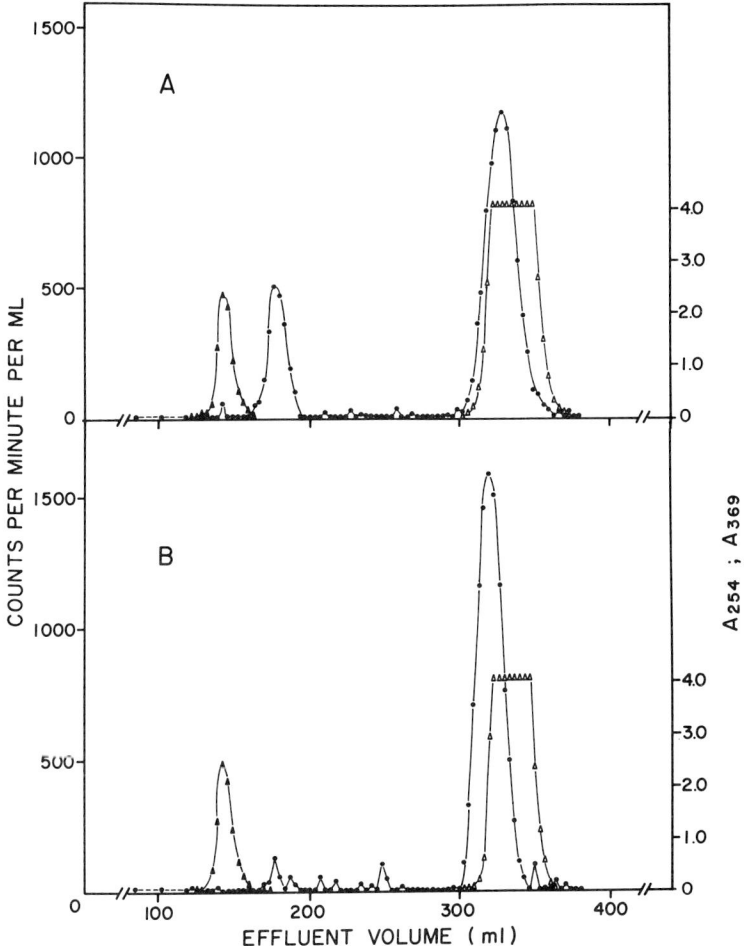

Fig. 1 Cytoplasmic appearance of [^{55}Fe]ferriferrioxamine B without metabolic assimilation of its ^{55}Fe in B. megaterium strain Ard1. Sephadex G-200 chromatographic profiles shown were prepared from cytoplasmic fractions of protoplasts labeled for 30 seconds with [^{55}Fe]ferrischizokinen (Panel A, note ^{55}Fe associated with high molecular weight material) or with [^{55}Fe]ferriferrioxamine B (Panel B, note lack of ^{55}Fe in high molecular weight fractions). Symbols: Closed circle, ^{55}Fe; Closed triangle, Blue Dextran 2000 marker absorbancy at 254 nm (A_{254}); Open triangle, potassium dichromate marker absorbancy at 369 nm (A_{369}). From Ref. 8.

IV. MEMBRANE RECEPTORS FOR SIDEROPHORES

The studies summarized above indicate that *B. megaterium* can discriminate between siderophores of different structure and that separate transport systems may exist for the siderophores which are utilized. Because association with a membrane receptor likely is an early step in siderophore transport, independent and specific receptors for ferrischizokinen and ferriferrioxamine B should occur on the membrane. These have been identified in *B. megaterium* strains SK11 and Ard1 (9). Membrane vesicles (prepared by osmotic rupture of protoplasts) of both strains rapidly accumulated both [^{59}Fe]-ferrischizokinen and [^{59}Fe]ferriferrioxamine B, reaching equilibrium between 2 and 5 minutes of assay. Equimolar accumulation of both labels of [^{3}H,^{59}Fe]ferrischizokinen suggested uptake of the intact chelate. Siderophore accumulation was not influenced by addition of several energy sources and identical assays were noted at both 0° and 37°C. Moreover, dilution of membrane vesicles previously exposed to radioactive siderophores did not cause rapid efflux of the label. These results suggested that the siderophores were not transported (either actively or by facilitated diffusion) by the vesicles and that siderophore accumulation represented only a binding reaction.

When binding of [^{59}Fe]ferrischizokinen was allowed to reach equilibrium (5 minutes) and unlabeled ferrischizokinen added at 100 times the initial concentration of this chelate, more than 90% of the bound [^{59}Fe]ferrischizokinen was released by 60-120 seconds (Table 1). Ferriferrioxamine B did not exchange with bound [^{59}Fe]ferrischizokinen, and of several other structurally different siderophores tested only ferriferrichrome A caused low-level release of [^{59}Fe]ferrischizokinen (Table 1). The ferrischizokinen binding site displayed marked specificity for this chelate. Similarly, membrane-bound [^{59}Fe]ferriferrioxamine B was dissociated rapidly (92%) upon addition of unlabeled ferriferrioxamine B (Table 1); a structurally related chelate, ferriA22765, also exchanged (67%) with [^{59}Fe]ferriferrioxamine B. Several other chelates tested caused lesser release of the labeled siderophore, suggesting that although the ferriferrioxamine B binding site has highest affinity for this chelate, it may be able to bind certain other siderophores.

Scatchard analyses of specific binding at a range of chelate concentrations were used to estimate maximal binding capacities and affinity constants of the membrane vesicles for ferrischizokinen and ferriferrioxamine B. Non-specific

TABLE 1   Dissociation of Bound [^{59}Fe] Siderophores from *Bacillus megaterium* Ard1 Membrane Vesicles By Unlabeled Siderophores. From Ref. 9.

Unlabeled Siderophore Added[a]	Per Cent Dissociation of[b]	
	[^{59}Fe] Ferrischizokinen	[^{59}Fe] Ferriferrioxamine B
Ferrischizokinen	91	6
Ferriaerobactin	0	0
Ferriferrioxamine B	0	92
FerriA22765	0	67
Ferriferrichrome A	11	25
Ferrirhodotorulic Acid	0	29

[a]Added at 100 times the initial concentration (102 pmoles/ml) of [^{59}Fe] siderophore. Assay contained 200 µg protein/ml.

[b]Dissociation of [^{59}Fe] siderophore determined 120 seconds after addition of unlabeled siderophore.

binding was defined as the amount of radioactive siderophore bound by the membranes in the presence of an excess (100-fold) concentration of unlabeled siderophore; specific binding was determined by subtracting non-specific from total siderophore bound. In strain Ard1 binding affinity constants were $1.4 \times 10^7 M^{-1}$ for ferrischizokinen and $11.0 \times 10^7 M^{-1}$ for ferriferrioxamine B (Table 2). Maximal binding capacities were about 186 pmoles ferrischizokinen/mg protein and 23 pmoles ferriferrioxamine B/mg protein (Table 2). These data support the existence of separate specific binding sites (receptors) for the two siderophores examined. The ferrischizokinen binding system had less affinity for its chelate than the ferriferrioxamine B binding system showed for its chelate; however, the membrane sites binding ferriferrioxamine B were fewer in number. These binding sites may be receptors for separate transport processes.

Binding of the two siderophores by membrane vesicles was consistent with the physiological mechanism operative in whole cells. For example, at high concentrations of the chelates the rate of uptake of ferrischizokinen exceeds the

TABLE 2 Affinity Constants and Maximal Binding Capacities[a] of Membrane Vesicles of *Bacillus megaterium* Ard1 From Ref. 9.

[^{59}Fe] Siderophore	Ka	Maximal Binding[b]
[^{59}Fe]ferrischizokinen	$1.4 \times 10^7 M^{-1}$	186
[^{59}Fe]ferriferrioxamine B	$11.0 \times 10^7 M^{-1}$	23

[a]Affinity constants (Ka) and maximal binding capacities calculated from Scatchard plots of specific binding at 8 to 330 pmoles [^{59}Fe] siderophore/ml and 200 μg protein/ml.

[b]pmoles [^{59}Fe] siderophore bound per mg protein.

rate of uptake of ferriferrioxamine B (5). This can be predicted from the greater number of ferrischizokinen receptors present on the membranes. It is also known that high levels of deferrischizokinen will impede transport of ferrischizokinen, possibly through competition for the transport system in whole cells (5). Addition of deferrischizokinen markedly lowers uptake of ferrischizokinen by membrane vesicles (9). Similar results were obtained in both whole cells and membrane vesicles with deferriferrioxamine B and ferriferrioxamine B (5, 9).

V. SUMMARY: MODEL OF SIDEROPHORE TRANSPORT

Because of the tendency of iron to form complexes which make it unavailable for transport, microorganisms have evolved special high affinity systems for acquisition of the metal. The special chelating molecules of the system are called siderophores. In *B. megaterium* the siderophore deferrischizokinen is released from the cells. Externally formed ferrischizokinen first associates with a specific membrane receptor and is thereafter transported across the membrane by a facilitated diffusion process. Cytoplasmic ferrischizokinen forms a pool of available iron for metabolism. Although the precise events associated with release of iron from the chelate are unknown, the low affinity of the ligand for ferrous iron suggests reductive release of the iron, possibly to a ferrous iron acceptor. Such an energy-requiring step would drive high level transport of the chelate. These events are summarized in Figure 2.

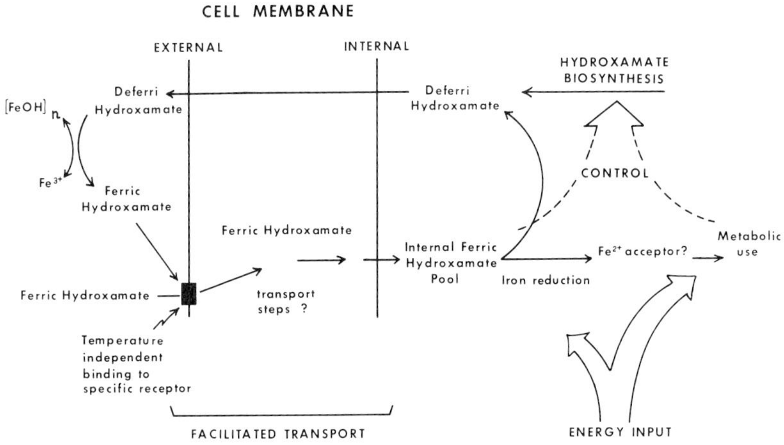

Fig. 2 Model of siderophore transport in B. megaterium. From Ref. 10.

B. megaterium maintains the capacity to transport and utilize not only its own siderophore but siderophores produced by other organisms. This has required synthesis of a number of separate transport systems. At what point in the iron assimilation process these multiple transport systems might converge is not known. The capacity to use exogenous siderophores must be of survival advantage, particularly in an environment where other organisms are releasing their siderophores. This is a curious situation in which several structurally different forms of an essential cofactor exist, and multiple utilization systems have developed for the various forms of the cofactor.

ACKNOWLEDGEMENTS

This research was supported by Research Grant CA 11886 from the National Cancer Institute and by Research Career Development Award GM 29366 from the National Institute of General Medical Sciences (to B.R.B.).

REFERENCES

1. Lankford, C. E., Crit. Rev. Microbiol. 2, 273 (1973).

2. Neilands, J. B., in "Microbial Iron Metabolism" (Neilands, J. B. ed.), pp. 4-31. Academic Press, New York, 1974.

3. Byers, B. R., in "Microbial Iron Metabolism" (Neilands, J. B. ed.), pp. 83-104. Academic Press, New York, 1974.

4. Davis, W. B., Byers, B. R., J. Bacteriol. 107, 491, (1971).

5. Haydon, A. H., Davis, W. B., Arceneaux, J. E. L. and Byers, B. R., J. Bacteriol. 115, 912 (1973).

6. Leong, J. and Neilands, J. B., J. Bacteriol. 126, 823 (1976).

7. Arceneaux, J. E. L., Davis, W. B., Downer, D. N., Haydon, A. H. and Byers, B. R., J. Bacteriol. 115, 919 (1973).

8. Arceneaux, J. E. L. and Byers, B. R., J. Bacteriol. 127, 1324 (1976).

9. Aswell, J. E., Haydon, A. H., Turner, H. R., Dawkins, C. A., Arceneaux, J. E. L. and Byers, B. R., J. Bacteriol., in press (1977).

10. Byers, B. R. and Arceneaux, J. E. L., in "Microorganisms and Minerals" (Weinberg, E. D. ed.) in press. Marcel Dekker Press, New York, 1977.

Section VII

OTHER ASPECTS OF IRON METABOLISM

BRAIN IRON IN THE RAT: A POSSIBLE BASIS FOR IRREVERSIBLE DEPLETION OF BRAIN NON-HEME IRON FOLLOWING A BRIEF PERIOD OF IRON DEFICIENCY

Peter R. Dallman
Robert A. Spirito

University of California, San Francisco

INTRODUCTION

In the rat as well as in man, the amount of non-heme iron in the brain increases rapidly during early postnatal growth and development, a time when the nutritional supply of iron is marginal. We recently reported that total non-heme and ferritin iron became deficient in the brain between 21 and 28 days of age in rats weaned at 21 days but exposed to an iron-deficient regimen from 10 days of age (1). Unlike other manifestations of iron deficiency, the depletion of brain iron could not be reversed even 45 days after treatment with intramuscular iron (5 mg of iron as iron dextran) and initiation of normal diet. The values before treatment, at 28 days, and those after 45 days of treatment, at 73 days of age, are shown below:

	Age, days	Hematocrit, %	Non-Heme Iron		Ferritin Iron
			Liver µg/g	Brain µg/g	Brain µg/g
Deficient	28	13*	17*	6.8*	0.7*
Control	28	38	97	9.3	1.4
Treated	73	41	144	7.1*	1.5*
Control	73	42	126	9.8	2.5

* $p < 0.01$, groups of 8 rats; all other differences not significant.

Iron administration had corrected the hematocrit and liver iron within 14 days, but the deficiency in brain non-heme iron and ferritin iron remained significant even after 45 days. This 45 day period starts shortly after weaning, includes sexual maturation, and continues well into adult life.

Our next experiments were designed to determine the basis for this unusual persistence of a tissue deficiency. We previously found that the rates of repair of tissue cytochrome deficiency after treatment of iron (2,3) or copper (4) deficiency were related to the rate of replacement of the tissue compound or the cell or subcellular particle which contains that compound. We therefore wished to test the hypothesis that an extremely slow turnover of some brain iron compounds might be related to the failure of iron treatment to reverse iron deficiency in the brain. We reasoned that if certain brain iron compounds are characterized by an extremely slow or negligible rate of turnover after their synthesis in the young animals, this could be a basis for the persistence of a deficiency. The results of our study, using rats fed a normal diet, support this hypothesis.

MATERIALS AND METHODS

Fifteen day old, male, Sprague-Dawley rats were injected intraperitoneally with 100,000 cpm/g of body weight [^{59}Fe] ferric chloride (10.3 mCi/mg Fe) in a total volume of 0.1 ml. The isotope (Amersham/Searle Corporation, Arlington Heights, Illinois) was obtained in 0.1 M HCl and was diluted with isotonic saline just before injection. Rats were weaned to a normal stock diet at 21 days of age. In order to rapidly reduce the specific activity of the major pools of iron in red cells and in the iron storage organs, rats were bled about 1 ml from the tail and were given 5 mg of intramuscular iron as iron dextran on 21, 25, and 28 days of age. At 21, 50, 100, and 150 days of age, groups of 8 rats were thoroughly perfused with warm, heparinized, isotonic saline via a blunt tipped needle placed through the wall of the left ventricle of the heart into the ascending aorta, in order to remove most of the blood from the organs. After decapitation, the brain and liver were carefully removed. Thorough perfusion is important because of the low concentration of brain iron, 5 to 7 μg of non-heme iron/g of brain tissue, in comparison to about 1 mg of iron/ml of packed red cells. Efficacy of perfusion was checked in preliminary experiments after intravenous administration of red blood cells labelled with ^{51}Cr. From the number of counts in blood compared to the number in perfused brain we calculated that hemoglobin iron accounted for a maximum of only 0.17 µg of total iron/g of brain. This is further diminished as a potential source of error because only 0.4% of contaminating hemoglobin iron was found to be detected in the assay for non-heme iron.

Total non-heme iron was measured in brain and in liver by the method of Weinfeld (5). A ferritin fraction was isolated under the conditions of heat-treatment and centrifugation described by Linder and Munro (6). An average of 60% of the counts in 0.5 ml samples of the ferritin fraction could be precipitated by the addition of 0.5 ml antiserum against rat liver ferritin and 100 µg carrier rat liver ferritin. Radioactivity of non-heme iron and ferritin fractions was measured in an automatic gamma counter to a 2 sigma statistical counting error of 3% or less. The statistical significance of differences between groups was calculated using Student's t test.

RESULTS AND DISCUSSION

The uptake of label and its retention by the brain in rats injected with ^{59}Fe at 15 days of age was sufficient to allow this group to be followed from 21 to 150 days of age (Fig 1) despite the limitations imposed by the 45-day half-life of ^{59}Fe. Between 21 and 50 days of age, there was a 33% decrease in the total number of counts per brain when corrected for decay ($p < 0.01$). However, between 50 and 100 days of age, there was no further significant change. In the non-heme iron fraction of the brain the total number of counts did not decrease significantly between 21 and 50 days but did fall 34% between 21 and 150 days of age ($p < 0.001$). The number of counts in the ferritin fraction did not change significantly throughout the entire period.

The specific activity of the brain iron fractions remained relatively high despite a profound decrease in specific activity of iron in the red cells (Fig 2). The specific activity of non-heme iron in the brain was almost double that of red cell iron at 21 days of age. Subsequently, the discrepancy increased to a 10 to 20-fold difference between the high specific activity of brain iron and the much lower specific activities of the larger red cell and liver iron compartments. The low specific activities of these major iron pools minimized errors due to the further incorporation of label by the brain following the initial pulse dose. The fall in specific activity of non-heme iron

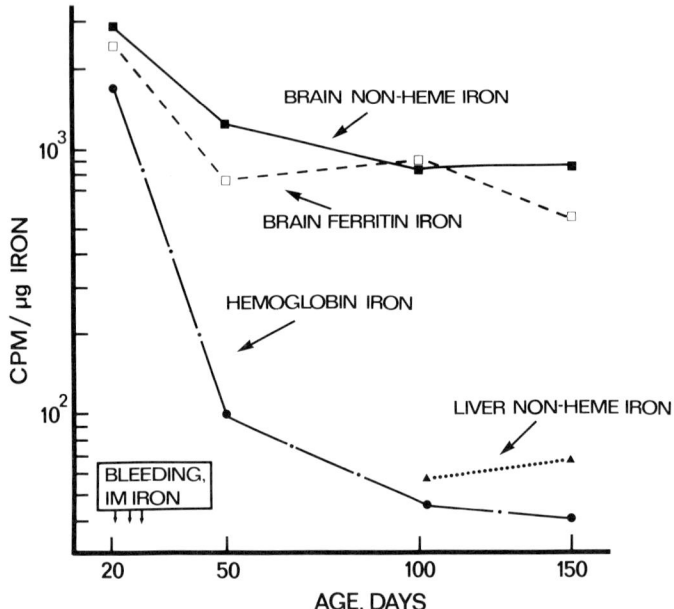

Fig. 1. Counts per minute per brain after intraperitoneal injection of [^{59}Fe] ferric chloride at 15 days of age. Standard errors of the means are indicated.

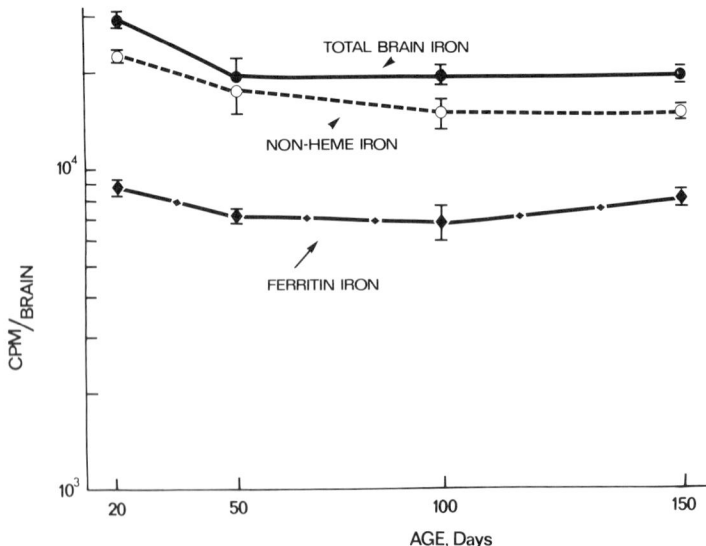

Fig. 2. Counts per minute per μg iron after intraperitoneal injection of [^{59}Fe] ferric chloride at 15 days of age.

in the brain between 21 and 150 days of age could be attributed primarily to dilution by the developmental increase of total non-heme iron in this organ. The maintenance of high total counts and a relatively high specific activity of brain iron indicated that there was little mixing of brain iron with the much larger systemic iron pools. We therefore concluded that the turnover of the brain iron compounds was exceedingly slow.

An alternate but less likely explanation for the maintenance of a high specific activity of non-heme iron only in brain is based on local reutilization of iron. As brain iron compounds are catabolized, the iron could be locally reutilized for synthesis, being prevented from mixing with systemic iron pools by a blood-brain barrier. This possibility seems remote because of the pattern of uptake of ^{59}Fe by the brain. Rats of various ages between 3 and 60 days were injected intraperitoneally or intravenously with [^{59}Fe] ferric chloride, 200,000 cpm/g body weight. Uptake of label by the brain after 24 hours was greatest before 21 days of age when the increase in brain iron was also most rapid. However, the considerable uptake of label even after that age seemed inconsistent with a blood-brain barrier to iron as an explanation for our turnover data.

CONCLUSIONS

The stability of brain iron in the rat is in contrast to the dynamic characteristics of iron compounds elsewhere in the body. Liver ferritin, for example, has a half-life of 2 to 3 days (7,8). Other iron compounds also have relatively rapid rates of turnover that allow the repair of hemoglobin and tissue iron deficiencies (2) as new synthesis of these compounds takes place. In the brain, most ferritin iron and total non-heme iron is deposited during the period of most rapid organ growth (1). Once these compounds are synthesized their iron is not readily mobilized. However, if this iron deposition is decreased because of iron deficiency during a brief critical period of early development, the abnormality is certainly long lasting and perhaps permanent.

REFERENCES

1. Dallman, P.R., Siimes, M.A., Manies, E.C., Brit. J. Haematol. 31, 209 (1975).

2. Dallman, P.R., Schwartz, H.C., J. Clin. Invest. 44, 1631 (1965).

3. Dallman, P.R., Goodman, J.R., J. Cell Biol. 48, 79 (1971).

4. Dallman, P.R., Loskutoff, D., J. Clin. Invest. 46, 1819 (1967).

5. Weinfeld, A., Acta Med. Scand. Suppl. 427, 1 (1964).

6. Linder, M.C., Munro, H.N., Analyt. Biochem. 48, 266 (1972).

7. Ove, P., Obenrader, M., Lansing, A., Biochim. Biophys. Acta 277, 211 (1972).

8. Munro, H.N., Drysdale, J.W., Fed. Proc. 29, 1469 (1970).

IRON STORES, SERUM FERRITIN AND IRON ABSORPTION

R.W. Charlton, D. Derman, B. Skikne, S.R. Lynch, M.H. Sayers, J.D. Torrance and T.H. Bothwell.

MRC Iron and Red Cell Metabolism Unit, University of the Witwatersrand, Johannesburg

I   INTRODUCTION

There is today a good deal of evidence that the serum ferritin concentration normally reflects the quantity of iron stored in the tissues (1, 2, 3). An opportunity to confirm this was afforded by an experiment in which for other purposes, three of us were repeatedly venesected until iron deficiency anemia had been produced. Serial serum ferritin estimations were performed, and could be correlated with the amount of storage iron, which was calculated from the quantity of hemoglobin iron removed.

The body iron stores are one of the determinants of the amount of iron absorbed from the diet. They are generally agreed to be the most important of the factors which act at the level of the intestinal mucosa, as distinct from the intestinal lumen (4). There should therefore be an inverse relationship between iron absorption and the serum ferritin concentration, and this has indeed been demonstrated (5, 6). The relationship should be especially close if the influence of the luminal factors which affect absorption are minimized, as they are when fasting subjects ingest a small dose of a ferrous salt together with ascorbic acid. When otherwise normal individuals with varying iron nutrition have been investigated in this way, a high degree of statistical significance has been found (5, 6), but the correlation coefficients were nevertheless not as high as might perhaps have been expected. Further investigation was accordingly thought to be worth while, and the opportunity afforded by the performance for other purposes of absorption studies on a large number of subjects was accordingly taken.

II MATERIAL AND METHODS

A. Serum Ferritin Concentration and Iron Stores

After an initial baseline period, approximately 600 ml blood was removed at weekly intervals from three of us (RWC, BS, and DD), the accurate volumes and hemoglobin concentrations being recorded and the serum ferritin concentrations (7) being determined. The depletion of the body iron stores was accompanied by a fall in the plasma iron concentration (8), the transferrin saturation, the mean cell volume and mean cell hemoglobin, and a rise in the total iron binding capacity (9). When the hemoglobin concentration had dropped below 11 g/dl the amount of blood removed each week was reduced to 60 ml, and the fact that the stores had been exhausted was demonstrated by the failure of the hemoglobin concentration to rise, while the biochemical and morphologic manifestations of iron deficiency anemia became more pronounced. The size of the initial iron store was then calculated by subtracting the difference between the initial and the final amounts of circulating hemoglobin iron from the total amount of iron removed. The maximum net amount of dietary iron which could have been absorbed was estimated to be 3 mg/day, and this was also subtracted. The residual iron store at the time of each venesection could then be derived, and correlated with the appropriate serum ferritin concentration.

B. Serum Ferritin Concentration and Iron Absorption

The serum ferritin concentration was correlated with the absorption of iron in 340 Durban housewives of Indian descent. They were aged 35 - 50 years, had each had at least three children and belonged to a low socio-economic group. Iron deficiency has previously been shown to occur commonly among such subjects (10). Each individual fully understood the nature, purpose and risk of all the procedures used. The study was approved by the Committee for Research on Human Subjects of the Faculty of Medicine of the University of the Witwatersrand, Johannesburg.

After fasting overnight the volunteers were weighed, and blood samples were taken for serum iron, unsaturated iron binding capacity, hemoglobin and serum ferritin estimations. They then drank 150 ml water containing

30 mg ascorbic acid and 3 mg iron as $FeSO_4 \cdot 7H_2O$ labeled with 2.5 μCi ^{59}Fe. After 14 days blood was collected for determination of the ^{59}Fe content using a scintillation spectrometer (Packard Auto-Gamma Tri-Carb Model 3001). The percentage absorbed was calculated on the assumption that all the radioactivity was in circulation, and that the blood volume was 65 ml/kg body mass.

III   RESULTS

A.   Serum Ferritin Concentration and Iron Stores

The calculated initial iron stores of RCW, BS and DD were 1700 mg, 700 mg and 1600 mg and their mean initial serum ferritin concentrations were 185 ng/ml, 51 ng/ml and 92 ng/ml respectively. In each case the serum ferritin concentrations fell progressively during the venesection program, reaching zero at the point at which the stores were exhausted. As part of a different experiment, after a number of weeks each subject took large doses (at least 400 mg iron/day) of a ferrous salt, and the venesections were resumed as frequently as proved necessary to prevent the hemoglobin concentration from rising. The iron administration was stopped after 3 weeks, but the venesections had to be continued for some time thereafter. The reappearance of measurable amounts of ferritin in the sera of all three subjects during the iron administration supported the conclusion that iron had been absorbed in excess of the amount which the erythroid marrow could utilise immediately. Levels as high as 25 ng/ml were observed, and they returned to zero at the point at which the venesections were found to be no longer necessary, and all the previous manifestations of iron deficiency reappeared.

B.   Serum Ferritin Concentration and Iron Absorption

The inverse relationship between the logarithms of the serum ferritin concentrations and the percentage absorptions which had been previously described (5,6) was confirmed by the findings in the 340 housewives. The coefficient of correlation was not particularly high ($r = 0.42$, P 0.001), however, and in particular there were many individuals with little or no serum ferritin who absorbed very low percentages of the iron salt. As many as

23 of the 97 subjects who absorbed less than 20% of the dose of iron had serum ferritin concentrations of 10 ng/ml or less. Significant correlations between the absorptions and the percentage transferrin saturations ($r = -0.38$, $P < 0.001$), and between the absorptions and the total iron binding capacities ($r = 0.28$, $P < 0.001$), were also present.

IV DISCUSSION

The inverse correlation between the log serum ferritin concentrations and the absorptions ($r = -0.42$, $p < 0.001$) found in this investigation was lower than that reported by Cook et al (5) ($r = -0.74$, $P < 0.001$) in 30 women, and also lower than we (6) had noted in a previous study of this population ($r = -0.67$, $P < 0.001$). In that study there was, however, a smaller proportion of individuals with very low serum ferritin concentrations, transferrin saturations and hemoglobin concentrations.

Several reasons for the relatively poor correlation must be considered. Firstly, the serum ferritin estimations may not be reliable. The procedure is certainly technically difficult, and in our laboratory, as reputedly in others, problems arise from time to time for reasons which are not always apparent. We use a solution of purified human liver ferritin to construct a standard curve for each batch of samples, and an aliquot of the stored serum of one of us (THB) is included as a control. Dr.Cook's estimate of the ferritin concentration in this control serum is similar to ours. The initial experiment with the three subjects provided evidence that our estimates of the serum ferritin concentration are related to the quantity of storage iron, since there was a progressive fall in the values obtained as the stores were reduced by the repeated venesections, and their exhaustion occurred as the ferritin concentrations reached 0-10 ng/ml. The ratios of storage iron to serum ferritin concentration in BS (13.7 mg/ng/ml) and in DD (17.4 mg/ng/ml) were, however, higher than the approximately 8 mg/ng/ml which others (2, 3) have reported. That for RWC was closer (9.2 mg/ng/ml). All the estimations were done in one batch, and were similar when repeated. Conceivably there are antigenic differences between individuals which could account for such discrepancies, or else the ratio may not in fact be exactly the same in each case. Part of the explanation for the imprecision of the correlation between the serum ferritin concentration and iron absorption may therefore lie here.

A second possibility is that some of the women may have ignored the request not to eat or drink anything before coming to the clinic, and this would have reduced their percentage absorptions very considerably. The women seemed highly motivated, however, and the Indian social worker in the team should have eliminated problems of communication. We therefore do not believe that more than a few of the discrepant findings can be explained in this way. Of greater moment may be the malabsorption of iron which was demonstrated in a recent study (11). The majority of the 12 Indian housewives with established iron deficiency who were investigated on that occasion absorbed very little of the nonheme iron in a standard meal, although their ability to absorb the heme iron seemed largely unimpaired. This, together with the efficient absorption of ferrous iron given with ascorbic acid on a second occasion when they were fasting, appeared to indicate that the malabsorption was due to some defect at the luminal level, perhaps achlorhydria. At least one of them, however, also absorbed a low percentage of the ferrous ascorbate so that mucosal malfunction may also occur in this group. Subclinical tropical sprue has been shown to be not uncommon among blacks in the Durban area (12), and may possibly have been present in some of the 23 individuals in the present study whose serum ferritin concentrations indicated little or no stored iron, and who yet absorbed less than 20% of the iron salt. Further investigation is planned.

A final point relates to the possibility of a direct connection between the serum ferritin concentration and the absorptive process. We (unpublished) and others (13) have not been able to alter the absorption of iron by injecting ferritin into experimental animals, and if there is a humoral 'messenger' which alters the mucosal cell function to match the body's need for iron, this does not seem to be it. Even if the iron stores are the major determinants of iron absorption, therefore, and the serum ferritin concentration mirrors them with reasonable accuracy, it may be unreasonable to expect a very close correlation between absorption and the serum ferritin if the latter is not actually the messenger.

V   REFERENCES

1. Jacobs, A., Miller, F., Worwood, M., Beamish, M.R., Wardrop, C.A., Brit. Med. J. 4, 206 (1972).
2. Walters, G.O., Miller, F.M., Worwood, M., J. Clin. Path. 26, 770 (1973).
3. Lipschitz, D.A., Cook, J.D., Finch, C.A., New Eng. J. Med. 290, 1213 (1974).
4. Finch, C.A., in "Iron Deficiency, Pathogenesis, Clinical Aspects, Therapy" (Hallberg, L., Harweth, H-G. Vanotti, A., eds.), p.409, Academic Press, New York, 1970.
5. Cook, J.D., Lipschitz, D.A., Miles, L.E.M., Finch, C.A., Amer. J. Clin. Nutr. 27, 681 (1974).
6. Disler, P.B., Lynch, S.R., Charlton, R.W., Torrance, J.D., Bothwell, T.H., Walker, R.B., Mayet, R., Gut 16, 193 (1975).
7. Miles, L.E.M., Lipschitz, D.A., Bieber, C.P., Cook, J.D., Anal. Biochem. 61, 209 (1974).
8. International Committee for Standardization in Hematology, Blood, 37, 587 (1971).
9. Herbert, V., Gottlieb, C.W., Lau, K.S., Gevirtz, N.R., Sharney, L., Wasserman, L.R., J. Nucl. Med. 8, 529 (1967).
10. Mayet, F.G.H., Adams, E.B., Moodley, T., Kleber, E.E., Cooper, S.K., S. Afr. Med. J. 46, 1427 (1972).
11. Bezwoda, W.R., Disler, P.B., Lynch, S.R., Charlton, R.W., Torrance, J.D., Derman, D., Bothwell, T.H., Walker, R.B., Mayet, F., Brit. J. Haemat. 33, 265 (1976).
12. Moshal, M.G., Hift, W., Kallichurum, S., Pillay, K., J. Trop. Med. Hyg. 78, 2 (1975)
13. Greenman, J., Jacobs, A., Gut 16, 613 (1975).

A NEW IRON BINDING PROTEIN WITH WIDE TISSUE DISTRIBUTION

Simeon Pollack
Fred D. Lasky

Albert Einstein College of Medicine, Bronx, New York

We have proposed that an iron carrier in the intestinal mucosa could be located by observing the movement of an intraluminal pulse of ^{59}Fe through various fractions of intestinal mucosa. A fraction containing a carrier should show prompt uptake of iron introduced into the intestinal lumen followed by rapid transfer of iron into the carcass; when absorption is increased, movement of iron across the mucosa by this postulated carrier should take place at an accelerated rate.

We identified a mucosal fraction ("Fraction II") with these properties (using Sephadex chromatography)(1) by studying the distribution of ^{59}Fe immediately after an intraluminal pulse of ^{59}Fe and comparing it to the distribution of ^{59}Fe fifteen minutes after the pulse. (Fig. 1).

The iron binding protein contained in Fraction II was purified and studied. The new protein has a MW of 78,000 (as estimated with gel filtration) and consists of two subunits of MW 43,000 (as estimated on SDS gel electrophoresis). It binds iron in a ratio of 2 gram atoms of iron per mole of protein. The apparent formation constant for the iron is $10^{19} M^{-1}$ at pH 7. The iron is in the ferric oxidation state as demonstrated by electron spin resonance spectroscopy and by the rapid loss of iron from the protein on addition of reducing agents. It has an absorption peak at 336nm in the UV. On electron spin resonance spectroscopy it shows a major absorption peak near $g'=4.22$, characteristic of iron with almost complete rhombic symmetry, and a minor absorption peak near $g'=5.55$. On isoelectric focusing electrophoresis, it separates into two peaks with pI=6.16 and pI=6.23. The new protein was distinguished from transferrin by these properties and also by amino acid analysis and by lack of immunologic cross-reactivity (2)(3)(4).

The new protein, called GIBP in earlier publications (an acronym for Gut Iron Binding Protein), elutes from DEAE-Sephadex at a buffer conductivity of 3 mmho; it was quantified in guinea pig brain, heart, spleen, gut, kidney and liver by labelling homogenates of these organs with ^{59}Fe, loading the labelled homogenates onto DEAE-Sephadex and counting the radioactive peak eluting in a linear gradient of .01M HEPES*, .01M NaCl pH 8, .01M HEPES, .075M NaCl pH 8.

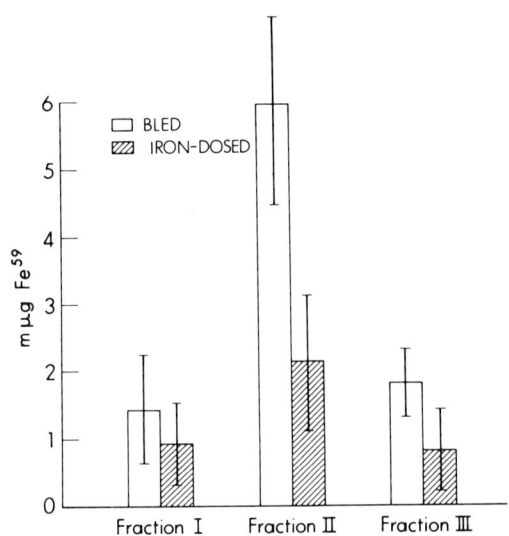

Fig. 1. Fe^{59} transfer-to-carcass of duodenal supernatant fractions following an intraluminal pulse of ^{59}Fe. Animals were sacrificed either immediately after the pulse or 15' after the pulse. 0' and 15' groups of animals were paired and the difference between the Fe^{59} content of the designated fration at each time was obtained by subtraction. The bars represent the mean $\pm$ S.E.M. Twelve pairs of Bled and Iron-dosed groups were studied. The transfer of Fe^{59} by Fraction II in the Bled group was significantly greater than the transfer of Fe^{59} by Fraction II of the Iron-dosed group ($p < 0.05$). From Reference 1 by permission from C. V. Mosby.

Tissues were minced, homogenized in 0.25M Mannitol, .05M Tris, .03M NaCl, pH 7.4 and subjected to centrifugation at 140,000xg for 1 hour. An equal volume of saturated $(NH_4)_2SO_4$ was added to the 140,000xg supernatant and the precipitate discarded. Six additional volumes of saturated $(NH_4)_2SO_4$ were then added, the precipitate collected and dissolved in .1M NaCl, .01M HEPES, pH 7. ^{59}Fe was added to a final iron concentration of 5 micromoles with citrate in 120-

---

* HEPES: hydroxyethylpiperazine ethanesulfonic acid.

fold molar excess. This mixture was equilibrated with .01M HEPES, .01M NaCl pH 8 in Bio-Rad Hollow Fiber Beakers, applied to DEAE-Sephadex and eluted with the previously described gradient.

GIBP yield was proportional to the amount of tissue in the range of 0-1.5gm wet weight for liver, kidney and intestinal mucosa (using a 6 gm DEAE-Sephadex column). One gram or less of tissue was assayed for all the organs studied. Bleeding and iron loading produced a 6-fold difference in iron absorption between the two groups of guinea pigs (5).

We considered that, if GIBP were regulatory in iron transport, the amount in intestinal mucosa might vary with changing iron absorption. However, there was 0.47 nanomoles of GIBP-iron per gram of intestinal mucosa in bled guinea pigs and 0.49 in iron-loaded guinea pigs; this difference was not significant.

TABLE 1
GIBP Iron Binding Protein Content of Guinea Pig Tissues

Organ	Bled	Iron-loaded
	(nanomoles GIBP iron per gram wet weight)	
Brain	.03(1)	.04(1)
Heart	.12(1)	.10(1)
Spleen	.12(1)	.17(1)
Intestinal Mucosa	.47 ± .05(4)	.49 ± .3(5)
Kidney	1.02 ± .16(4)	.89 ± .16(4)
Liver	1.60 ± .4(4)	1.66 ± .16(4)

GIBP is expressed as mean ± S.E.M. The number of determinations is shown in parentheses. There was no significant difference in GIBP content of bled and iron-loaded organs.

GIBP was detected in liver, kidney, gut, spleen, heart and brain (Table 1). The amounts of GIBP in tissues obtained from bled and iron-loaded guinea pigs were not significantly different. GIBP varied from 0.035 nanomoles of GIBP-iron per gram of brain to 1.63 nanomoles of GIBP-iron per gram of liver.

Although the amount of GIBP-iron does not vary significantly with either phlebotomy or iron loading, GIBP may nevertheless have a role in iron transport in the intestine and in intracellular iron exchange in other tissues. The assay used detects GIBP by its iron-binding properties; the iron binding sites of GIBP might be unvarying while another portion of the molecule, involved in transport or intra-cellular iron metabolism, might be modified. The wide distribution of

GIBP suggests that it may be linked to heme or ferritin metabolism.

In conclusion, a new iron-binding protein has been isolated from an intestinal mucosal fraction which appears to participate in iron transport. However, there is not more of the new protein present when iron absorption is increased; also, it has a wide tissue distribution. The function of this new iron-binding protein, therefore, remains to be elucidated.

REFERENCES

1. Pollack, S., Campana, T. and Arcario, A., J. Lab. and Clin. Med. 80, 322 (1972)

2. Pollack, S. and Lasky, F., J. Lab. and Clin. Med. 87, 670 (1976).

3. Pollack, S. and Lasky, F., Biochem. Biophys. Res. Comm. 70, 533 (1976).

4. Pollack, S., Lasky, F., in "Proteins of Iron Storage and Transport in Biochemistry and Medicine"(Ed. Crichton, R. R.) p. 389. North Holland Publishing Co., Amsterdam, Netherlands, 1975.

5. Pollack, S. and Campana, T.C., Scand. J. Haemat. 7, 208 (1970).

IRON BINDING COMPOUNDS IN PARTICULATE FRACTION
OF INTESTINAL MUCOSA

Yoshio Yoshino
Sanae Yamakawa
Yukihiko Hirai
Nippon Medical School

I. INTRODUCTION

Studies of several investigators indicated the presence of ferritin and transferrin-like compounds in supernatant fraction from mucosal homogenate of upper small intestine during iron absorption (1 - 5). The authors drew attention to the fact that almost 50 % of Fe59 in mucosal homogenate remained in the particulate fraction, at the time when the homogenate was prepared from rats 1 h after Fe59 administration. Accordingly the analysis of iron compounds in particulate fraction was carried out by the authors to elucidate the mechanism of iron transport by mucosal cells. A large molecular Fe59 containing compound which showed different attitude in Fe59 binding activity from ferritin derived from supernatant fraction at time course was detected previously from SDS extract of insoluble particles (6,7).

In this paper the properties of Fe59 binding compound from particulate fraction will be described further on molecular weight, density, isoelectric point, immuno precipitin reaction for antiferritin and antitransferrin, and attitude of its iron binding activity at time course.

II. MATERIALS AND METHODS

Male albino rats of Wister strain (200 - 250 g of body weight, fed on regular diet) were used for *in vivo* Fe59 absorption. A dose of material was composed of 4 μci Fe59Cl3, 2 μmole /100g FeSO4, and 2 μmole/100g Sodium ascorbate dissolved in 5 % fructose, and given the stomach by

intubation. The scraped mucosa was prepared from upper small intestine (20 cm length from pylorus to jejunum) and homogenized with 40 volumes of 0.25 M sucrose by Dounce homogenizer.
The homogenate was separated by centrifugation at 20,000 g for 30 min. The particulate fraction so obtained was washed twice by the same solution, and treated with 1 % Triton X-100 solution dissolved 10 mM Tris-HCl (pH 8.2). Then Triton X-100 extract (Triton extract) was separated by the same centrifugal force. Gel filtration on Sepharose 6B column of 25 x 450 mm was performed immediately to soluble supernatant and Triton extract by vehicles of 10 mM Tris-HCl and 0.02 % Triton X-100 in Tris-HCl (pH 8.2) respectively. All procedures mentioned above were carried out under chilled condition. Fe59 activity, protein content and in some cases iron content were determined in each fraction, using Aloka autowell system for gamma ray, Lowry method for protein determination (8), and atomic absorption spectrometer (Shimazu AA-650). Fe59 containing fractions detected by these procedures were measured on following properties: immuno precipitin reaction for anit-rat-liver-ferritin serum and anti-rat-transferrin serum, molecular weight by gel filtration, density gradient centrifugation in a 10 - 40 % sucrose dissolved in 0.02 % Triton X-100 at 39,000 rev/min for 3 h in RPS 40A roter (Hitachi Ultracentrifuge 55P-2) (9), and isoelectric point using isoelectric focussing column of Ampholine at pH 3 - 10 and 4 - 6 in 0.1 % Triton X-100.

III. RESULTS AND DISCUSSION

20,000 g supernatant from mucosal homogenate (soluble supernatant) and Triton extract were fractionated by gel filtration on Sepharose 6B column. The elution pattern of soluble supernatant could show two main Fe59 peaks. The one appeared immediately after void volume Ve/Vo 1.08, and the other was at Ve/Vo 1.75. They were named peak 1 and 2 of soluble supernatant with the respect of their appearance (Fig. 1). The elution pattern of Triton extract also could show two Fe59 peaks, which located similar positions with those of soluble supernatant, namely Ve/Vo 1.08 and 1.78. These peaks were named tentatively peak 1 and 2 of Triton extract in the same order (Fig. 2).

Fe59 distribution into subfractions 1 h after the administration was summarized as percentages of mucosal homogenate (TABLE I). The results were shown by average of 5 experiments composed of 4 rats in each group. On soluble supernatant, Fe59 in peak 1 contained few ferritin and

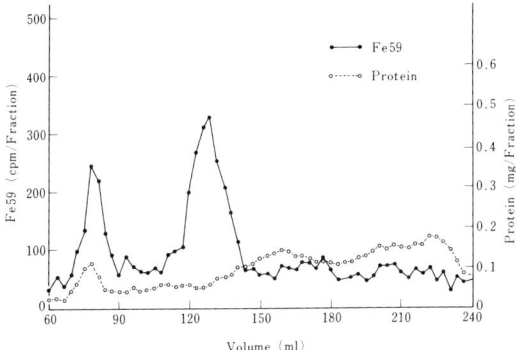

Fig. 1 Typical elution pattern of soluble supernatant obtained from mucosal homogenate 1 h after Fe59 administration on Sepharose 6B.

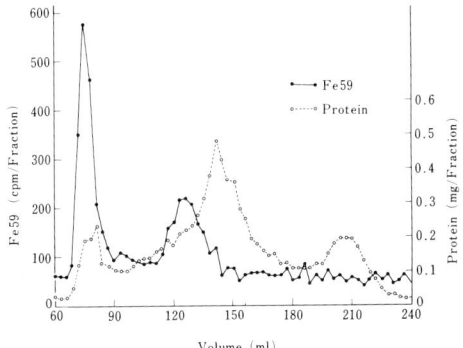

Fig. 2 Typical elution pattern of Triton X-100 extract obtained from particulate fraction of mucosal homogenate 1 h after Fe59 administration on Sepharose 6B.

transferrin, but 62 % of Fe59 in peak 2 was recognized as ferritin by precipitin reaction. On Triton extract, the major portion of peak 1 did not belong to ferritin or transferrin, but 74 % of Fe59 in peak 2 was precipitated by antiferritin. These results suggest the presence of two kinds of iron compounds in soluble supernatant and Triton extract. The one will be large molecular iron binding compound other than monomeric ferritin, and the other will be ferritin, which is soluble form in cytoplasma and fixed form attached to particulate fraction.

The properties of Fe59 binding compounds from both sources were summarized in TABLE II. Peak 2 compounds from both sources were quite similar in molecular weight, density, isoelectric point and attitudes to antiserum, and corresponded to rat liver ferritin contained in the reaction

TABLE I Fe59 Distribution in Intestinal Mucosa 1 h after Fe59 Administration to Normal Rats

	Fe59 % of homogenate	% of each peak	No. of experiment
Homogenate	100		5
soluble Supernatant	55.7 ± 6.3*		5
Peak 1	6.6	-	3
Ppt by antiferritin	0.3	4.6	3
Ppt by antitransferrin	0	0	3
Peak 2	17.6	-	3
Ppt by antiferritin	10.6	62.2	3
Ppt by antitransferrin	0	0	3
Triton extract	12.9 ± 1.6		5
Peak 1	1.9 ± 0.8	-	5
Ppt by antiferritin	0.5 ± 0.3	26.3	5
Ppt by antitransferrin	0.2 ± 0.2	10.5	5
Peak 2	2.7 ± 1.0	-	5
Ppt by antiferritin	2.0 ± 0.7	74.0	5
Ppt by antitransferrin	0.1 ± 0.1	3.7	5

* average ± 1SD

listed in TABLE II. Accordingly both peak 2 were recognized as ferritin. However peak 1 from Triton extract had larger molecular weight, heavier density, and lower isoelectric point than rat liver ferritin contained in the same materials. The iron binding compound in peak 1 will be supposed as either ferritin oligomers or nonferritin iron compound surrounded by acidic membrane fragments, because cell membrane shows the property of polyanion and its isoelectric point is measured in the vicinity of pH 2 (10). Owing to the poor reaction exhibited between peak 1 and antiferritin, it may be presumed as nonferritin iron compound. The comparison of Fe59 turnover time between peak 1 and 2 of each extract will contribute effectively to the solution of this problem. The time course experiment of iron binding activity was tested at time points of 1, 3 and 5 h after Fe59 administration to rats, which were fed on iron deficient diet for 7 - 10 days and fasted overnight before the experiment (Fig. 3). All results were expressed by the percentages for Fe59 activity in mucosal homogenate obtained from 1 h after the administration. The amounts of radio-

TABLE II Properties of Fe59 Binding Compounds Separated from Mucosal Homogenate 1 h after Fe59 Administration

	Molecular weight	Density	Isoelectric point
Triton extract			
Peak 1	over 10^6	1.15-1.16	2.5
Peak 2	4.8×10^5	1.07-1.09	5.2
Soluble sup.			
Peak 2	4.8×10^5	1.08	5.2
Rat liver			
ferritin	4.8×10^5	1.08, 1.05	5.2

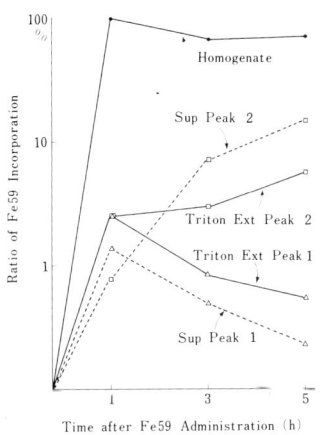

Fig. 3 Time course of Fe59 incorporation into mucosal homogenate, peak 1 and 2 obtained from soluble supernatant and Triton extract of particulate fraction.

activity in the homogenate had a tendency to be slightly decreasing with time from 1 to 5 h. At 1 h, Fe59 content in all four subfractions were almost in similar levels, but remarkable alternations were recognized by the time course. Fe59 in peak 2 from both fractions clearly increased with time. The rate of magnification at 5 h was 19 and 2.3 times as much as that at 1 h in soluble supernatant and Triton extract respectively. These results may indicate the increase of ferritin molecules at time course. On the contrary Fe59 content in peak 1 exhibited decreasing tencency. Radioactivity at 5 h became almost 1/6 and 1/5 of that at 1 h respectively in the same order. These results suggest the metabolic role of peak 1 as a labile iron acceptor, which attached to the particulate fraction and could work at iron transport in its early stage. Differentiation of each iron compound derived from both sources was not successful so far as the properties shown in this experiment. Especially peak 1 from soluble supernatant could not be clarified on the origin. There may be two possibilities about it. One is cytoplasmic and the other is from decomposed product of insoluble particles. Because there is observations which exhibit the contamination of decomposed materials into soluble supernatant produced mechanically by share forces of the homogenizer (11, 12). The findings of peak 1 iron compound and fixed ferritin from particulate fraction could bring some contribution for understanding of iron transport.

REFERENCES

1. Worwood, M., Jacobs, A., Life Sciences 10 (1), 1363 (1971).
2. Huebers, H., in "Proteins of Iron Storage and Transport in Biochemistry and Medicine" (Crichton, R. R. ed.), p.381. North Holland Publishing Co., Amsterdam, 1975.
3. Linder, M. C., Munro, H. N., ibid., p.395.
4. Halliday, J. W., Powell, L. W., Mack, U., ibid. p.405.
5. Pollack, S., Lasky, F. D., J. Lab. Clin. Med. 87, 670 (1976).
6. Yoshino, Y., Manis, J., Am. J. Physiol. 225, 1276 (1973).
7. Yoshino, Y., Hiramatsu, Y., J. Biochem. 75, 221 (1974).
8. Lowry, O. H., Rosebrough, N. J., Farr, A. L., Randall, R. J., J. Biol. Chem. 193, 265 (1951)
9. Gabuzda, T. G., Pearson, J., Biochim. Biophys. Acta 194, 50 (1969).
10. Dowben, R. M., in "Biological Membrane" (Dowben, R. M. ed.), p. 1. Little, Brown and Co., Boston, 1969.
11. Kowarski, S., Schachter, D., J. Clin. Invest. 52, 2765 (1973).
12. Porteous, J. W., in "Subcellular Components" (Birnie, G. D. ed.), p. 157. Butterworths, London, 1972.

REGULATION OF IRON ABSORPTION BY CONTROL OF HEME BIOSYNTHESIS IN THE INTESTINAL MUCOSA

Jack Hegenauer
Larry Ripley
Paul Saltman

Department of Biology
University of California San Diego

I. INTRODUCTION

Our present understanding of intestinal iron absorption and its various control mechanisms is still very primitive. The literature of iron, however, is rich in the phenomenology of iron absorption, and there are many clues about the nature of transport and regulation. One persistent observation is the finding that iron absorption increases in bled, anemic, or iron-deficient animals and decreases in iron-overloaded animals with ample stores. The interpretation that body stores regulate intestinal iron absorption by erecting a "mucosal block" has been modified over the years, and recent attention has focused on how various non-heme iron fractions within the mucosal cell act as the body's "signal" to the absorptive cells (1). These theories, however, do not account particularly well for the observations that iron absorption is influenced by dietary iron concentration (2) or that body iron stores appear to "regulate" the absorption of inorganic cobalt (3). In recent experiments, we have tried to unify the last two seemingly disparate observations by searching for their common biochemical denominator.

We report here some new data that clearly show the influence of dietary iron on the absorption of both iron and cobalt. We are developing an hypothesis that previous dietary iron content "regulates" subsequent iron and cobalt absorption not by affecting the level of an iron transport system, but by affecting the biosynthesis of certain heme-containing microsomal monoxygenases in the intestinal mucosa. We present recent experiments to determine the concentration and turnover of intestinal cytochrome P-450 *in situ* by measurement of the EPR signal of low-spin ferric heme. We will relate the concentration of duodenal cytochrome P-450 to intestinal iron absorption, as well as classify experimental treatments that affect both iron absorption and heme/hemoprotein biosynthesis.

II. RESULTS AND DISCUSSION

A. Regulation of Iron Absorption by Dietary Iron

Mice with uniform iron status acclimate quickly to a change in dietary iron concentration, so that their ability to absorb iron is determined not by ferritin iron stores but by the concentration of iron in the diet immediately preceding the diagnostic test dose (Fig. 1). The efficiency of iron ab-

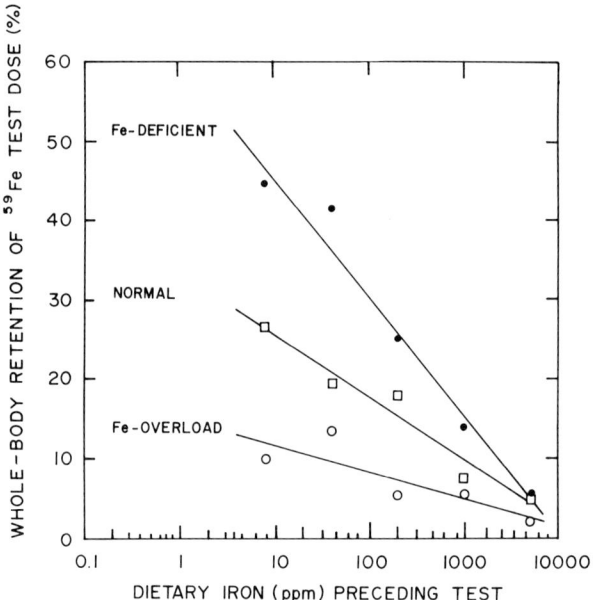

FIG. 1. *Effect of dietary iron concentration on capacity of iron-deficient, normal, and iron-overloaded mice to absorb iron. Iron-deficient mice ate a low-iron diet after weaning. Normal mice ate a low-iron diet and drank 5 mM Fe (as ferric fructose). Iron-overloaded mice ate a low-iron diet but received 3 intraperitoneal injections of iron-dextran at 3.6 mmoles Fe/kg body weight at weeks 1, 2, and 3. After 5 weeks of supplementation, all mice ate a low-iron diet and drank distilled water for another 2 weeks to allow assimilation of iron. Groups of 25 mice were fed solid diets containing varying amounts of $FeSO_4$ for 42 hours and starved 6 hours before administration of a diagnostic iron absorption test (4).*

sorption is also definitely influenced by general iron status. Increased ferritin stores and increased dietary iron act together to decrease iron absorption. Regulation of absorption as a function of dietary iron is clearly discernible even in iron-deficient animals. Absorptive capacity responds immediately to a change in dietary iron concentration and reaches its new level within 24 hr (Fig. 2). We are evidently observing a readily inducible biosynthetic system.

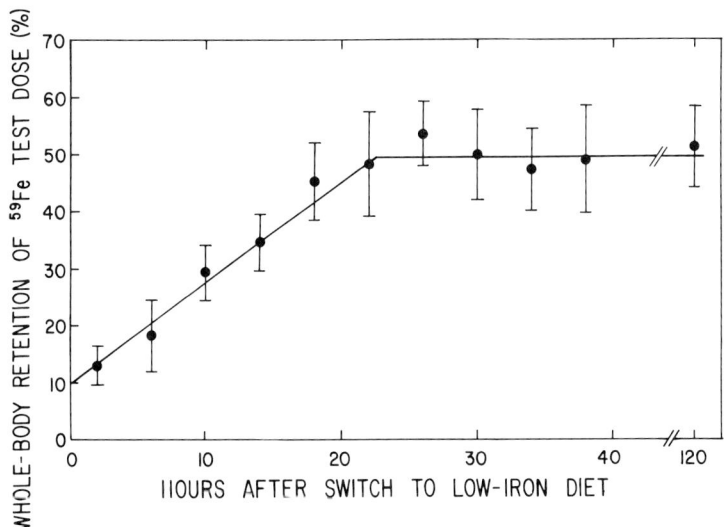

FIG. 2. *Response of iron absorption to a change in dietary iron concentration. Groups of 25 mice, maintained for 2 weeks on a liquid cow's milk diet (5) containing 5 mM $FeSO_4$, were switched to a milk diet containing 10 uM $FeSO_4$. A diagnostic iron absorption test (4) was administered after a 2-hr starvation.*

B. Regulation of Cobalt Absorption by Dietary Iron

Theory and experiment have suggested that cobalt and iron share a common absorptive pathway in the intestinal mucosa (6); this assumption forms the basis for a diagnostic cobalt excretion test to estimate iron stores (3). Unlike iron and cobalamin, however, inorganic cobalt has no storage mechanism. Since absorbed cobalt is promptly excreted in the urine, the elimination of a test dose of radiocobalt serves as a convenient measure of its absorption. When radiocobalt was administered to hematologically normal mice in a diagnostic absorption test, dietary iron rapidly influenced the efficiency of absorption (Fig. 3). The effect of dietary iron on the absorption of both cobalt and iron is thus strikingly similar.

C. Regulation of Hemoprotein Synthesis in Duodenal Cells

1. *Effect of Dietary Iron*

In the rat, the microsomal monoxygenase, cytochrome P-450, is concentrated in the villous tip cells of the upper duodenum, and its synthesis responds to the iron content of the diet (7). Using a sensitive EPR technique (8), we have been able to measure the very low concentration of cytochrome P-450 in unfractionated duodenal cell scrapings. The concen-

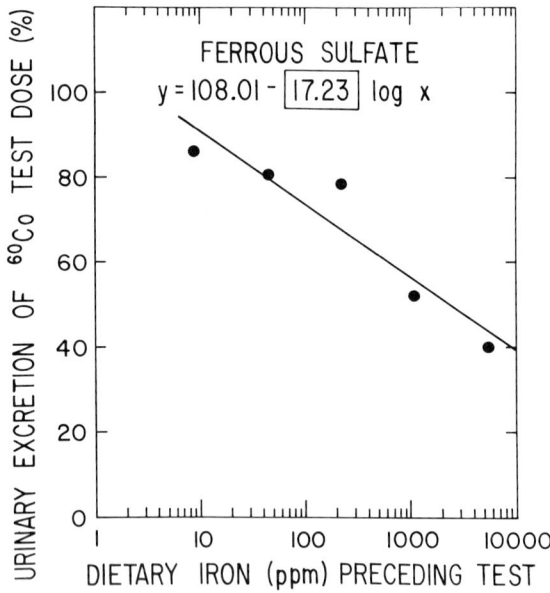

FIG. 3. *Effect of dietary iron concentration on capacity of normal mice to absorb cobalt. Groups of 25 mice were fed solid diets containing varying amounts of $FeSO_4$ for 42 hr and starved 6 hr before administration of a diagnostic cobalt absorption test. After receiving an oral dose of $60\text{-}CoSO_4$ at 0.25 umoles Co/kg body weight, mice were housed in metabolic cages for 24 hr to separate urinary radioactivity (absorbed Co) from fecal radioactivity (unabsorbed Co).*

tration of this hemoprotein not only increases with increasing dietary iron concentration, but can be increased, within 24 hr, to a new level as the dietary iron concentration is increased (Fig. 4).

2. *Effect of Inflammation*

Bacterial endotoxins and turpentine are potent inflammatory agents that cause rapid derangements in iron metabolism,

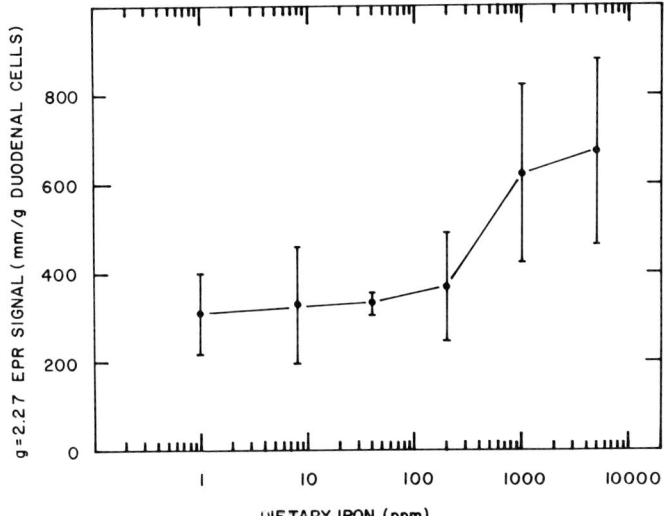

FIG. 4. *Effect of dietary iron concentration on cytochrome P-450 content of mouse duodenal cells. Groups of 5 iron-deficient mice were fed solid diets containing varying amounts of $FeSO_4$ for 24 hr. After a 6-hr starvation, the first 12 cm of small intestine was flushed with cold 0.154M KCl/0.05M Tris-HCl (pH 7.8), and duodenal cells from individual mice were collected by scraping the villi into cold KCl/ Tris buffer containing glycerol, heparin, and soybean trypsin inhibitor (14) to prevent formation of cytochrome P-420. Cells were washed once by centrifugation; 50-100 mg of cell paste was packed into an open-ended EPR tube and stored in liquid N_2. EPR spectra were obtained at $77°$ K and 9.07 GHz with the Varian E-3 spectrometer. The peak-to-peak height of the g=2.27 first-derivative signal was taken as a measure of cytochrome P-450 (8). Instrument settings: microwave power 160 mW; modulation amplitude 40G; receiver gain 10^5; field scan time 1000 G/8 min at recorder time constant 1 sec.*

including reduction in plasma iron and plasma iron turnover (9, 10). We find that administration of turpentine to mice leads to equally rapid changes in the cytochrome P-450 concentration of duodenal cells. About 6 hr after onset of inflammation, duodenal cytochrome P-450 reaches a minimum value (Fig. 5); this decrease roughly parallels the decrease in plasma iron concentration (not shown). If a diagnostic iron or cobalt absorption test is administered to mice 6 hr after injection of turpentine, iron and cobalt absorption are actually enhanced (Table 1). Turpentine inflammation has generally been thought to *decrease* iron absorption (10, 11).

D. Relationship between Duodenal Cytochrome P-450 and Iron Absorption

Rapid induction and turnover of cytochrome P-450 in response to administration of xenobiotics are well-recognized phenomena. Both iron absorption and duodenal cytochrome P-450 concentration respond to alterations in dietary iron content with a 24-hr induction period (Figs. 2 & 4); decreased iron absorption is associated with increased duodenal cytochrome

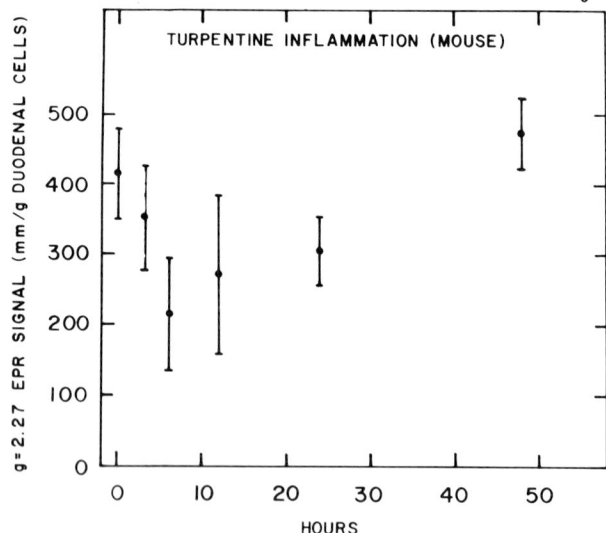

FIG. 5. *Reduction of cytochrome P-450 content of mouse duodenal cells during the course of turpentine inflammation. Groups of 5 normal mice were injected with 25 ul of turpene (distilled from turpentine) into the thigh muscle. Intestinal cells from individual mice were prepared and EPR spectra were obtained as described in the legend of Fig. 4.*

TABLE I
Effect of turpentine inflammation on
iron and cobalt absorption in mice

Treatment	Absorption of Test Dose (%)	
	$^{59}FeSO_4$	$^{60}CoSO_4$
Control	28.1	50.9
Turpentine	36.7	72.3

Inflammation was produced in groups of 15 normal mice by injection of turpentine, as described in Fig. 5; groups of 15 control mice were injected with saline. After 6 hr of starvation, a diagnostic radioiron (or radiocobalt) absorption test was administered. Mice were housed in metabolic cages for 48 hr. Whole-body (WB), fecal (F), and urinary (U) radioactivity was measured in a whole-body gamma counter. Fe absorption = 100 x WB/(WB+F+U). Co absorption = 100 x (WB+U)/(WB+F+U). Individual mice were not separated, so the standard deviation cannot be estimated.

P-450 (Figs. 1 & 4). During turpentine inflammation, decreased duodenal cytochrome P-450 (Fig. 5) is associated with increased absorption of iron and cobalt (Table 1). That both iron and cobalt absorption share a common "regulation" by dietary iron concentration may be related to the fact that both metals form stable complexes with protoporphyrin IX and that both Fe-heme and Co-heme are degraded by the same heme oxygenase (12). Since turnover of heme from cytochrome P-450 is known to exceed the turnover of the protein moiety (13), previous induction of cytochrome P-450 in duodenal cells by the availability of dietary iron may divert substantial amounts of iron, after initial uptake, toward heme synthesis and thus deplete transportable iron within the intestinal mucosa. In this tentative mechanism, cobalt absorption would be affected because the higher level of protoporphyrin synthesis induced by dietary iron would deplete cobalt by diversion toward Co-heme during transport.

ACKNOWLEDGEMENTS

The work of the laboratory is supported by research grant AM 12386 from the National Institute of Arthritis, Metabolic, and Digestive Diseases; by a research grant from the National Dairy Council; and by a contract from the Dairy Council of California.

REFERENCES

1. Crosby, W. H., in "Regulation of Hematopoiesis" (Gordon, A. S., ed.), p. 519. Appleton-Century-Crofts, New York 1970.

2. Pollack, S., Kaufman, R. M., Crosby, W. H., Science 144, 1015 (1964).

3. Wahner-Roedler, D. L., Fairbanks, V. F., Linman, J. W., J. Lab. Clin. Med. 85, 253 (1975).

4. Carmichael, D., Christopher, J., Hegenauer, J., Saltman, P., Am. J. Clin. Nutr. 28, 487 (1975).

5. Carmichael, D., Hegenauer, J., Lem, M., Ripley, L., Saltman, P., J. Nutr. (in press).

6. Shade, S. G., Felsher, B. F., Bernier, G. M., Conrad, M. E., J. Lab. Clin. Med. 75, 435 (1970).

7. Hoensch, H., Woo, C. H., Raffin, S. B., Schmid, R., Gastroenterol. 70, 1063 (1976).

8. Gabrielle, L., Leterrier, F., Cristau, P., Laverdant, C., Clin. Chim. Acta 60, 147 (1975).

9. Cortell, S., Conrad, M. E., Am. J. Physiol. 213, 43 (1967).

10. Hershko, C., Cook, J. D., Finch, C. A., Brit. J. Haematol. 28, 67 (1974).

11. Shade, S. G., Proc. Soc. Exp. Biol. Med. 139, 620 (1972).

12. Maines, M. D., Kappas, A., Biochem. 16, 419 (1977).

13. O'Carra, P., in "Porphyrins and Metalloporphyrins" (Smith, K. M., ed.), p. 123. Elsevier, Amsterdam, 1975.

14. Stohs, S. J., Grafstrom, R. C., Burke, M. D., Moldeus, P. W., Orrenius, S. G., Arch. Biochem. Biophys. 177, 105 (1976).

THE INDUCTION OF DIABETIC CHANGES IN RATS

BY INTRAPERITONEAL INJECTION OF Fe^{3+}-NTA

MICHIYASU AWAI
MIKIO NARASAKI
SATIMARU SENO

Department of Pathology
Okayama University Medical School
Okayama, Japan

I. Introduction

Hemochromatosis is characterized clinically by cirrhosis of the liver, diabetes mellitus and skin pigmentation. The latter two conditions constitute bronzed diabetes[1,2,3]. An experimental model of hemochromatosis or bronzed diabetes has been difficult to create. Long term parenteral administration of large dosages of iron to experimental animals has not produced specific changes of hemochromatosis, i.e., no liver cirrhosis, pancreatic fibrosis or diabetes[4].

In the present study, we introduced iron to rats in the form of iron nitrilotriacetate chelate complex(Fe^{3+}-NTA) which is known in vitro as an efficient iron transporter to serum transferrin[5]. The iron solution was injected intraperitoneally daily for over 130 days. At about 60 days of injection, these rats showed glycosuria, as well as ketone bodies in both blood and urine. Later histological examinations showed iron depositions that were similar to those in hemochromatosis.

II. Materials and Methods

Inbred Wistar rats three weeks of age were used(N=120).

The animals were fed Oriental rat chow and water was available ad libitum. The animals were divided into four groups of 30 animals each. Animals in Group A and C received a daily intraperitoneal injection of 0.2 to 1.0 mg Fe/100g B.W. daily for five months. The iron was in the form of Fe^{3+}-NTA(NTA, Eastman Kodak) in Group A and chondroitinsulfate ferric iron colloid(Blutal, Dainippon Pharmaceutical Co.) in Group C. Each animal received 0.2 mg Fe/100g B.W. daily for 3 weeks, 0.6 mg Fe/100g B.W. daily for three weeks and 1.0 mg Fe/100g B.W. daily for the remaining period. The total amount of iron administered to each animal was about 200 mg. Group B animals were treated with pure NTA at the same dose and schedule as the NTA used in Group A animal. The animals in Group D were the non-treated controls. The animal body weight and urinary glucose concentrations were measured twice a week in the morning. Blood sugar, blood acetone bodies, and urinary acetone bodies were measured in the morning at about day 80 of injection(100 days of age). Urinary glucose was tested quantitatively(Ames Dexter/Dexterstick) with a drop of urine obtained by pressure on the abdomen. Measures on blood sugar were obtained by the Ames Dexter/Dexterstick and occasionally by Hoffman's method[7] with blood collected from the tail. Acetone bodies in both blood and urine were measured by the Ames Ketostick[8]. The animals were sacrificed at 160 days of age by severing the carotid artery. The organs were fixed with phosphate buffered 10% formol or Bouin's solution, embedded in paraffin, and sectioned. The sections were stained with H.E., Perls' method, Van Gieson's stain and Mallory's stain. Fe^{3+}-NTA was prepared by the procedure of Bates and Schlabach[5]. The final solution contained 0.5 mg of iron/ml, and the NTA/Fe ratio was 5 : 1.

III. Results

The body weight increased progressively in all groups to about 60 days of age(40 injection days). At about this period Group A animals showed a growth stoppage and manifested glycosuria. The body weight of Group B and D animals continued to increase to about 70 days(50 injection days) when a stationary phase was reached(Fig. 1). During glycosuria manifestation Group A rats showed a striking elevation in blood glucose. An overnight fast lowered the blood sugar to nearly the control level but the acetone body level was remarkably high both in blood and urine(Table 1). In Group B, C and D an overnight fast produced no effect on blood

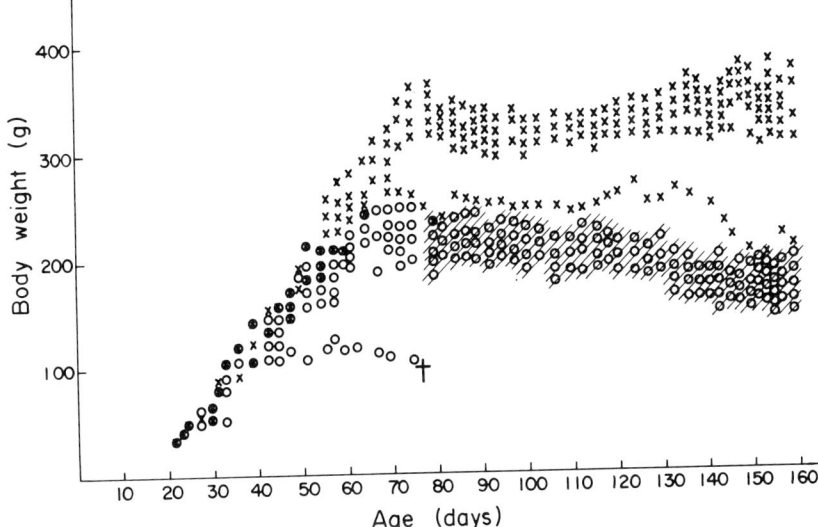

Fig. 1 Body weight of rats treated with Fe^{3+}-NTA(o) and NTA(x). Shaded areas indicate glucosuria.

TABLE 1 Effects of Fasting on Diabetic Rats

Diabetic rats	Blood glucose M ± SD (mg/dl)	Blood acetone bodies (mg/dl)	Urine acetone bodies (mg/dl)
Without fasting N=8	173.3 ± 16.5	40	10
After fasting overnight N=8	84.1 ± 12.1	80 - 100	40

sugar or acetone body level. The urine remained free of
sugar and acetone bodies. H.E.-stained sections revealed no
recognizable histologic changes after 130 injection days,
except for deposition of brown pigments in some organs of
Group A and C animals. Perl's stain disclosed that the brown pigments were deposits of iron-containing substances. In
Group A rats remarkable iron depositions were found in the
parenchymal cells of the liver and pancreas and some cells of
the spleen. Few iron granules were found in Kupffer cells.
Moderate iron reactions were also found in the parietal
cells of the glandular stomach and the epithelial cells of
adrenal glands and sweat glands. Iron depositions were
hardly detectable in the kidneys and heart muscles. No iron
reaction was found in the pancreatic islet cells. In Group
C animals iron depositions were found in the liver and
spleen, but mainly in Kupffer cells and macrophages or RE
cells and much less in the parenchymal cells compared to
Group A animals. In the pancreas iron granules were found
within the interstitial macrophages, but no iron depositions
were found in acinar cells. The Van Gieson and Mallory-
Azan stained sections revealed no fibrotic changes in the
liver, pancreas, or other organs in any group.

IV. Discussion

The present study has succeeded in producing diabetes in
rats by intraperitoneal of iron as Fe^{3+}-NTA after about 60
injection days (80 days of age). Iron deposition was remarkable in parenchymal cells and sparse in macrophages or RE
cells. The histological picture was similar to hemochromatosis, although the pancreatic islets were free of visible
iron depositions and no fibrotic changes were observed in
the organs. Brown et al.[4] loaded iron to dogs over a period
of 7 years but did not produce liver cirrhosis or diabetes.
The iron was deposited mainly in the macrophages or RE cells
and only slightly in parenchymal cell. In the present study
colloidal iron which was deposited in macrophages but not in
parenchymal cells did not induce diabetes, even though the
iron administration was for a long term period at a high total dose. Recently, Brown et al.[9,10] demonstrated in rats
in vivo that transferrin binding iron, especially that of
high saturation, was largely taken up by liver parenchymal
cells. After isolation by Rappaport's method[11] parenchymal
cells showed a high quantity of iron, but Kupffer cells isolated by Mills and Zucker-Franklin's method[12] contained only
traces of iron. This is the same condition as in human hemo-

chromatosis in which heavy iron deposition is manifested in parenchymal cells rather than in macrophages. Bates and his co-worker[5] have demonstrated that iron bound to NTA transferred efficiently to transferrin *in vitro* without forming spurious iron hydroxide complex at neutral pH. In preliminary studies iron introduced intraperitoneally in the form of Fe^{3+}-NTA(1.0 mg Fe/100g B.W.) produced a highly saturated state of serum transferrin that continued for nearly 12 hours after introduction(unpublished data). In the present study iron was deposited in parenchymal cells of the liver, pancreas, and other secretory glands resulting in a distinct diabetic state similar to hemochromatosis or bronzed diabetes of human. No fibrotic change, however, was observed in specimens taken after about four months of iron injections. Cirrhosis and hemochromatosis may be quite separate entities, as suggested by Mac Donald and Pechet[13]. Fe^{3+}-NTA is very toxic and usually kills rats treated by a single large dose, but a gradual increase starting from 0.2 mg/100g B.W. creates tolerance in rats later administered a dosage as large as 1.0 mg/100g. Experiments with adult rats weighing 150g revealed that it took longer to attain a diabetic state, even though the same quantity of iron per body weight was administered. The administration of iron during rapid growth, when efficient assimilation of iron may occur, seems to be effective producing experimental diabetes.

In conclusion the present experiment clearly shows that diabetes can be induced by excessive iron in the form of Fe^{3+}-NTA or iron easily transferable to transferrin, though the induction mechanism of hemochromatosis is still unknown.

REFERENCES

1. Dubin, I.N., Am. J. Clin. Path. 25, 514 (1955).

2. Finch, S.C., Finch, C.A., Medicine 34, 381 (1955).

3. Jacobs, A., Semin. Hematol. 14, 89 (1977).

4. Brown, E.B., Dubach, R., Smith, D.E., Reynaferje, C., Moore, C.V., J. Lab. Clin. Med. 50, 862 (1957).

5. Bates, C.W., Schlabach, M.R., J. Biol. Chem. 248, 3228 (1973).

6. Seno, S., Awai, M., Kobayashi, J., Ose, S., Kimoto, T., Tohoku J. Exp. Med. 76, 179 (1962).

7. Hoffman, W.S., J. Biol. Chem. 120, 51 (1957).

8. Fraser, J., Fetter, M.C., Mast, R.L., Free, A.H., Clin. Chim. Acta 11, 373 (1965).

9. Brown, E.B., Awai, M., Chipman, B., Proceeding of the 14th Congress of the International Society of Hematology, July (1972).

10. Brown, E.B., Okada, S., Awai, M., Chipman, B.J., Lab. Clin. Med. 86, 576 (1975).

11. Rappaport, C., Hanze, C.B., Proc. Soc. Exp. Biol. Med. 121, 1010 (1966).

12. Mills, D.M., Zucker-Franklin, D., Am. J. Pathol. 54, 147 (1969).

13. Mac Donald, R.A., Pechet, G.S., Am. J. Pathol. 46, 85 (1965).

PATHOGENESIS OF IMPAIRED IRON RELEASE IN
INFLAMMATION

C. Hershko and A.M. Konijn

Department of Hematology, Hadassah University Hospital and
Department of Nutrition, Hebrew University-
Hadassah Medical School

I. INTRODUCTION

One of the most common disorders of iron metabolism is the abnormality associated with inflammation. It is characterized by a blockade of iron release from tissues resulting in reduced transferrin saturation, and iron deficient erythropoiesis manifested in hypochromic anemia and increased red cell protoporphyrin concentrations (1). Impaired iron release from reticuloendothelial (RE) cells has been emphasized in several studies (2-4), and more recently a synchronous effect on iron transport from hepatocytes and intestinal mucosa as well as R.E. cells has been demonstrated (5).

According to prevailing concepts of tissue iron exchange, iron entering the cell is partitioned between a rapidly exchanging labile or pre-release iron pool, and a more inert storage iron pool represented by ferritin and hemosiderin (4-6). Kinetic studies in animals and in man have shown, that in inflammation iron is shifted into the compartment of slowly released storage iron at the expense of the rapid early phase of iron release (4,5). Such a change could be explained by one or more of the following mechanisms:

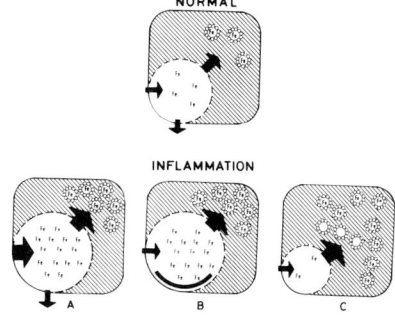

Fig. 1 Alternative mechanisms for explaining impaired iron release in inflammation.

A. Increased cellular uptake of iron in the form of nonviable erythrocytes by RE cells or transferrin iron by parenchymal cells would expand the pre-release iron pool which in turn would stimulate ferritin synthesis and divert incoming iron into ferritin stores. Accordingly, reduced release of a radioiron label would reflect a relative reduction in flow from the pre-release pool but not necessarily a reduction in the absolute amounts of iron released.
B. Interference with the migration of iron through the cell membrane would increase the pre-release iron pool leading to stimulation of ferritin synthesis and deposition of iron in ferritin stores. Unlike the previous mechanism, this would involve an absolute reduction in iron release. C. Inflammation may stimulate ferritin synthesis nonspecifically, as part of a general stimulus for the synthesis of acute phase reacting proteins. Increased ferritin synthesis, in turn, would result in the diversion of iron from the pre-release pool into ferritin stores. In contrast to the previous two mechanisms in which increased ferritin synthesis follows expansion of the pre-release iron pool, the last one assumes a primary enhancement of ferritin synthesis leading to a reduction in the size of the pre-release pool.

In the present study, an attempt was made to define the role of ferritin synthesis in the pathogenesis of impaired tissue iron release by in vivo correlations between iron uptake, iron release and rates of amino-acid incorporation into newly synthesised ferritin at various times following the induction of inflammation in rats by a single injection of turpentine.

II. METHODS

Female rats (Hadassah strain) weighing 125 to 135 grams were used. They were fed ad.lib. on a standard rat stock diet (AmRod 931, Ambar, Emek-Hefer, Israel), and were non-fasting at the time of the experiments. Animals were killed under light ether anesthesia by exsanguination into heparinized syringes through the abdominal aorta. Inflammation was produced by the intramuscular injection of 0.25 ml turpentine into each hind leg. All procedures involving iron analysis were performed in iron-free glassware. Serum iron was determined by the ICSH Iron Panel method (7). Ferritin iron was determined by the dipyridyl color reaction as described by

Drysdale & Munro (8).

Counting techniques. Radioactivity in ferritin protein was measured after its chemical purification followed by immunoprecipitation employing anti rat ferritin serum raised in goats. The immunoprecipitate was washed twice in 0.02M sodium phosphate buffer pH 7.2 containing 0.9% NaCl, 1% Triton X-100 (Koch-Light Labs.) and 1% sodium deoxycholate (Schwartz-Mann, Orangeburg, New York, N.Y.) followed by one washing in 0.9% NaCl. 200 ug of bovine serum albumin (Sigma Chemical Co. St.Louis, Mo.) was added, mixed with perchloric acid (final concentration 0.22M) and heated to $65°C$ for 5 minutes (9). The denatured apoferritin and albumin carrier was centrifuged and washed twice in 5% TCA and once in ether. The washed protein precipitate was dissolved and counted as described for total liver protein.

Uptake of L-(4,5,^{3}H)-Leucine was measured 2 hours after its injection, since it has been shown that a plateau of leucine uptake into liver ferritin is reached at 2 hours and that of total liver protein at 0.5 hour (10).

S.D.S. acrylamide electrophoresis was carried out on 100 ug samples of purified ferritin as described by Linder et.al. (11). 3mm slices were incubated at $37°C$ in a toluene based medium containing 12% protosol and counted for ^{3}H-leucine.

Ferritin in liver homogenates was determined by the method of Linder & Munro, (12) with a slight modification in which the ammonium sulfate precipitate was filtered through Sepharose 4B instead of Sephadex G-200. In some cases, parallel measurements employing an electroimmunoassay (Carmel and Konijn, in preparation) were performed in which ferritin protein was determined directly in the heat supernatant. Ferritin measurements obtained by both methods were essentially identical.

The protein content of liver homogenate, purified ferritin, or ferritin-immunoprecipitate was estimated by Lowry's method using bovine serum albumin-fraction V (Sigma Chemical Co.) as standard reference. The colour value of apoferritin and bovine serum albumin using Lowry's method was identical.

Plasma iron turnover (P.I.T.) was measured as described in detail in a previous study (5).

III. RESULTS

Incorporation of ^{3}H-leucine into ferritin and total protein has been examined in the spleen as well as the liver 6 hours after turpentine injection.

TABLE I Incorporation of ^{3}H-leucine into ferritin and total protein at 6 hours of inflammation (cpm/mg protein)*

	(n)	Liver Ferritin	Liver Total protein	Spleen Ferritin	Spleen Total protein
Normal	(3)	14,356 ±2,424	5,897 ±816	17,204 ±789	7,932 ±742
Inflamed	(3)	35,848 ±4,105	8,183 ±563	30,757 ±4,523	7,966 ±973

*intravenous dose of 100 uCi/100 gm body weight.

Ferritin synthesis in the liver, reflecting mainly parenchymal cell activity, and in the spleen representing RE cells, was roughly twice as much as in the normal controls. Total protein radioactivity was slightly higher than normal in the liver of animals with inflammation, but no such difference in total protein synthesis could be demonstrated in the spleen.

In order to examine whether the observed increase in ferritin synthesis in the spleen and liver of animals with inflammation was preceded by an increased input of iron into these organs, an attempt was made to estimate the rate of transferrin iron uptake by hepatocytes and of iron in non-viable erythrocytes by RE cells. The rate of transferrin ^{59}Fe clearance from the plasma, the absolute rates of plasma iron turnover, and ^{59}Fe uptake by the liver were measured in groups of animals at various times following turpentine injection.

TABLE II Transferrin ^{59}Fe turnover and hepatic uptake at 2 hours

Time after injection (hr)	0	2	4	8	12	24
T½ in plasma (min.)	88 ±4	80 ±3	90 ±5	86 ±9	79 ±6	73 ±5
P.I.T. (µg/day)	212 ±20	240 ±11	181 ±23	129 ±14	134 ±18	116 ±6
Hepatic uptake (%)	12.1 ±0.5	12.9 ±0.7	12.3 ±1.1	12.9 ±0.9	11.2 ±0.8	11.8 ±0.8

Throughout the 24 h study period, the T½ in plasma of transferrin ^{59}Fe was remarkably constant, and only at 24 h could a slight but significant (p < 0.05) reduction in clearance time be demonstrated. P.I.T. was unchanged within the first 4 hours after turpentine injection, but thereafter P.I.T. was reduced to about one half normal (p < 0.001). Since iron uptake by the liver is a product of P.I.T. and the percent uptake of radioiron per unit time, and since the percent uptake of transferrin ^{59}Fe by the liver remained constant, these findings indicate a reduction in hepatocellular iron uptake. RE cell uptake of iron was estimated by measuring the survival of ^{51}Cr-labelled autologous erythrocytes in 6 normal rats and 7 animals with inflammation. The half life of labelled erythrocytes was 16.8 ± 1.3 days (S.E.M.) in normal controls and 14.7 ± 1.3 days in rats with inflammation. The difference between the two groups was not statistically significant. Thus, no evidence in support of increased iron uptake by tissues could be found in these studies, and the reduction in P.I.T. implicated reduced iron release as the underlying anomaly.

A reduction in iron release from tissues could be produced by suppression of iron transport through cell membranes or, alternatively, by increased ferritin synthesis resulting in increased diversion of iron from the pre-release iron pool into cellular ferritin stores. In an attempt to distinguish between these two alternative mechanisms of iron retention, rates of hepatic ferritin synthesis were correlated in vivo with simultaneous sequential measurements of serum iron in the same group of animals.

TABLE III Effect of inflammation on serum iron and hepatic protein synthesis

Time after injection (hr)	S.I. (μg/dl)	^{3}H-leucine uptake Total protein	(DPM/mg)* Ferritin
0	208±11	1592±109	3548±369
2	228±15	2167±74	4306±703
4	222±26	1931±93	6217±985
6	157±16	2179±181	6713±661
8	175±22	2130±112	6783±246
12	77±9	1937±114	3486±186
16	86±7	1838±90	2251±242
24	72±9	2293±70	3512±581

*intraperitoneal dose of 15 uCi/100 gm body weight

A significant reduction in serum iron was first observed at 6 hours of inflammation, and the lowest levels corresponding to about one-third of normal were found between 12 and 24 hours. The reduction in TIBC was less pronounced than the changes in serum iron, and significantly lower levels were only found at 16 and 24 hours. A slight increase in ^{3}H-leucine incorporation into ferritin was already seen at 2 h of inflammation, and maximal rates of incorporation corresponding to about twice the baseline values were found at 4, 6 and 8 hours ($p < 0.001$). Thereafter, ferritin synthesis declined rapidly, and at 12, 16 and 24 hours ^{3}H-leucine incorporation into ferritin was equal to, or less than the corresponding normal values. In contrast to ferritin synthesis, no consistent changes in total protein synthesis could be demonstrated throughout the study. Comparison of the sequential changes in serum iron and rates of hepatic ferritin synthesis following turpentine injection shows that alterations in ferritin synthesis preceded the changes in serum iron throughout the study. Thus, at 4 hours ferritin synthesis was twice-normal whereas serum iron, and P.I.T. were still unchanged. Conversely, maximal reduction of serum iron took place at 12 hours of inflammation, at a time when ferritin synthesis has already declined to 0 time rates. These correlations indicate that ferritin synthesis in inflammation is not the result of a preceding block in iron release, and seem to support the alternative mechanism in which increased ferritin synthesis is the primary event responsible for iron retention.

In an additional study, ferritin synthesis in inflammation was compared with ferritin synthesis following stimulation by the parenteral injection of 400 µg iron in the form of ferric ammonium citrate.

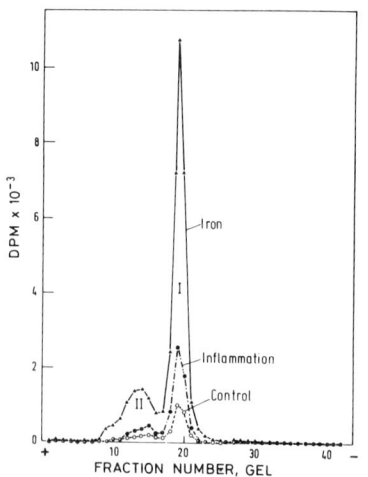

Fig. 2 SDS acrylamide electrophoresis of ferritin synthetised in normal controls, and following turpentine, or iron injection. I large subunit. II small subunits.

Total ferritin synthesis at 6 hours of inflammation was 2.4 times normal, and 5 hours after iron injection 9.6 times normal. The ratio of small to large subunit synthesis (Fig. 2) was 0.320 in normal controls, 0.275 in inflammation and 0.321 following iron injection.

IV. DISCUSSION

In the first part of the present study, iron uptake by parenchymal cells was estimated by measurements of P.I.T. and percent hepatic uptake of transferrin iron. R.E. cell uptake of iron was assessed by comparing the survival of ^{51}Cr-labeled autologous erythrocytes in normal, and turpentine-treated animals. These studies showed no increase in the uptake of iron by R.E. or parenchymal cells in inflammation, thereby excluding mechanism A in which increased iron uptake is assumed to stimulate ferritin synthesis and to divert incoming iron into ferritin stores.

The reduction in P.I.T. observed at 8 to 24 hours of inflammation indicated impaired cellular iron release. This could be caused by a primary interference with iron transport through the cell membrane (mechanism B), or a primary stimulation of ferritin synthesis resulting in the diversion of iron from the pre-release pool into ferritin stores (mechanism C). A distinction between the two mechanisms could be made by comparing the rates of ferritin synthesis, plasma iron turnover and plasma iron levels in the same group of animals at various times following the induction of inflammation by turpentine injection. As shown by Drysdale and Munro (10), the increase in ferritin synthesis induced in vivo by iron administration to rats, occurs at about 5 hours following injection. Thus, in case of a primary membrane effect on iron release, a delay of several hours would be expected between reduction of serum iron or P.I.T. and an increase in ferritin synthesis caused by the expansion of the pre-release iron pool. Conversely, primary enhancement of ferritin synthesis would be manifested by increased ferritin synthesis preceding, or simultaneous with the first manifestations of impaired iron release from cells. Results of the present study provide strong evidence supporting the last mechanism. Increased ferritin synthesis in inflammation preceded any reduction in serum iron or P.I.T., and can only be explained as a direct effect of inflammation which is independent of any prior changes in the size of the pre-release iron pool.

Increased serum levels of fibrinogen, and various other proteins are often found in clinical conditions characterized by inflammation, and can be elicited within minutes to a few hours following the introduction of phlogistins to experimental animals (14). Most of these proteins, often referred to as acute phase reactants, are synthesized in the liver, and their increased production in inflammation is probably a nonspecific response to a variety of substances released from the site of inflammation (14-17). In view of the present findings, increased ferritin synthesis in inflammation is most probably mediated by stimuli which are analogous to those responsible for the increased production of other acute phase reacting proteins. Rather than a result of a block in tissue iron release, increased ferritin synthesis in inflammation appears to be the primary event responsible for increased retention of labile iron and interference with the early phase of iron release. Whether this anomaly in tissue iron release, and the associated reduction in serum iron have any useful role to play in the systemic response to infective agents, is at present unknown.

REFERENCES

1) Cartwright, G.E., Lee, G.R., Brit. J. Haemat. 21, 147 (1971).
2) Freireich, E.J., Miller, A., Emerson, C.P., Ross, J.F., Blood. 12, 972 (1957).
3) Noyes, W.D., Bothwell, T.H., Finch, C.A., Brit. J. Haemat. 6, 43 (1960).
4) Fillet, G., Cook, J.D., Finch, C.A., J. Clin. Invest. 53, 1527 (1974).
5) Hershko, C., Cook, J.D., Finch, C.A., Brit. J. Haemat. 28, 67 (1974).
6) Dresch, C., Najean, Y., Rev. Eur. Etud. Clin. Biol. 17, 930 (1972).
7) I.C.S.H. Expert Panel on Iron, Brit. J. Haemat. 20, 451 (1971).
8) Drysdale, J.W., Munro, H.N., Biochem. J. 95, 851 (1965).
9) Drysdale, J.W., Shafritz, D.A., Biochim. Biophys. Acta. 383, 97 (1975).
10) Drysdale, J.W., Munro, H.N., J. Biol. Chem. 241, 3630 (1966).
11) Linder, M.C., Moor, J.R., Munro, H.N., J. Biol. Chem. 249, 7707 (1974).
12) Linder, M.C., Munro, H.N., Anal. Biochem. 48, 266 (1972).
13) Finch, C.A., Deubelbeiss, K., Cook, J.D. et.al., Medicine. 49, 17 (1970).

14) Glenn, E.M., Bowman, B.J., Koslowske, T.C., in "Chemical Biology of Inflammation" (Houck, J.C. & Forscher, B.K. eds.), p. 27. Pergamon Press, Oxford, 1968.
15) Benjamin, D.C., Weimer, D.E., Nature 209, 1032 (1966).
16) Jamieson, J.C., Ashton, F.E., Canad. J. Biochem. 51, 1281 (1973).
17) Bratcher, S.C., Shetlar, M.R., Amer. J. Physiol. 227, 1394 (1974).

Supported in part by Grant 015.0171 of the Hadassah-Hebrew University Research Foundation, and Grant No. 750 of the U.S.-Israel Binational Science Foundation.

IRON STATUS AND HOST DEFENSE

A. M. Ganzoni and M. Puschmann

*Division of Transfusion Medicine,
University of Ulm, D-7900 Ulm*

I. INTRODUCTION

A direct relationship between transferrin iron saturation and bacterial growth in serum has been demonstrated for a variety of microorganisms (1). The high affinity of transferrin for iron appears to underly this observation; it creates an extreme discrepancy between free iron present and bacterial iron requirements. The host's iron metabolism may therefore interact with bacterial disease. We examined salmonella infection in mice whose iron status had been altered in three ways.

II. INFECTION STUDIES

A. Acute Ferritin Overload

Transferrin body distribution allows to assume that all body fluids may act bacteriostatically through iron deprivation of bacteria. In view of the wide spread occurrence of lactoferrin this might be extended to body secretions. Bacterial multiplication, however, may take place within cells. Inoculation of mice with virulent *Salmonella typhimurium* is followed by a phase of bacterial multiplication in liver and spleen. The rate of this process appears to determine the fate of the host (2). The availability of iron from ferritin may therefore be of importance.

We have compared the effect of 100 µg of iron given as $NH_4Fe(SO_4)_2 \cdot 12\ H_2O$ (FAS) or as horse spleen ferritin on salmonellosis of male NMRJ mice. The iron preparation and the bacterial suspension were administered in 0.5 ml of saline 2 hrs apart by the intraperitoneal route. Fig. 1 shows that most animals which were pretreated with the iron salt died during the first 48 hrs of the infection. A different pattern was seen after an equivalent load of ferritin iron. The first death was recorded on day 4, and after 2 weeks 70 % of the animals were alive. A protective effect of the iron protein may be assumed from Fig. 1. But in further experiments the ferritin treated mice died at the same rate as the controls.

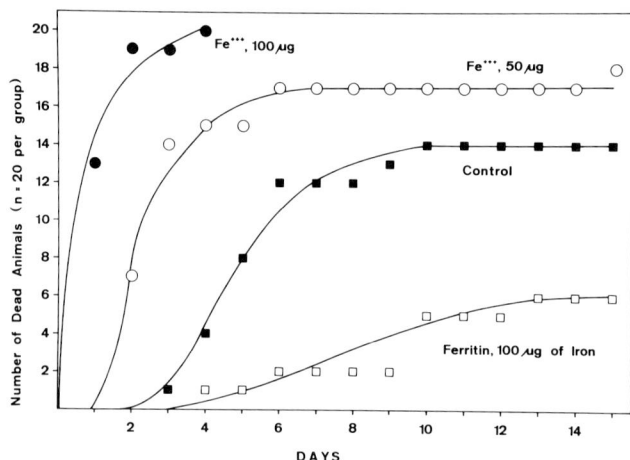

Fig. 1. Cumulative mortality of mice intraperitoneally infected with 1.8×10^5 organisms of S. typhimurium L15403. 2 hrs prior to infection iron was given by the same route.

For iron acquisition microorganisms depend on specific iron transport systems. The capacity to synthesize powerful iron chelators appears to represent an important factor of virulence (3). We believe that transformation of this relatively benign pathogen into a highly virulent form followed the removal of iron restriction at the site of infection. Interestingly, 100 µg of iron (FAS) had no effect on a more virulent type (2386) of S. typhimurium (see C.). Most relevant, however, appears that ferritin iron was relatively unavailable for type L15403. An additional experiment supported this assumption. Thus, S. typhimurium L15403 infection in mice loaded with 25 mg of iron 2 months previously took the same course as in untreated controls.

B. Infection Induced Hyposideremia

Bacterial infection leads to a rapid fall in serum iron. This type of iron redistribution has been considered as a protective mechanism. If true, a sublethal infection with pathogen A inducing hyposideremia should attenuate the infection by pathogen B. This model was tested by subsequent intraperitoneal infections with E. coli (untyped) and S. typhimurium L15403. E. coli infection (10^5 microorganisms) led to a serum iron reduction from 353 $\pm$ 65 (S. D.) to 38 $\pm$ 62 µg/100 ml within 10 hrs. As shown in Fig. 2 salmonellosis set 10 hrs following coli infection indeed took a milder course. This result is compatible with the above idea. But future models must exclude variables besides serum iron changes possibly altering host resistance.

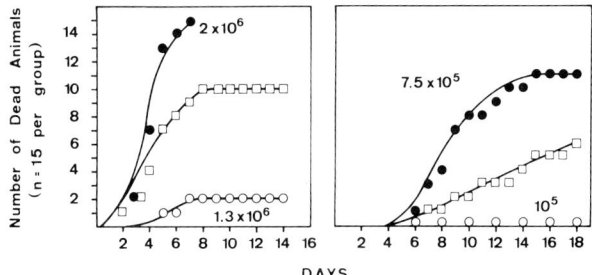

Fig. 2. Two experiments evaluating cumulative mortality of mice intraperitoneally infected with E. coli (o), S. typhimurium (●), or with both organisms at an 10 hrs interval (□). The figures stand for the number of organisms injected. For the double-infection the same doses were applied.

C. Nutritional Iron Deficiency

An increased susceptibility to certain types of bacterial and fungal disease in states of iron overload seems fairly well documented. Concerning iron deficiency more controversial information exists. A decreased resistance of iron deficient rats has been reported against S. typhimurium (4) and Staphylococcus pneumoniae (5) respectively. But African investigators observed much less bacterial complications in anemic patients lacking stainable marrow iron as compared to those with abundant iron stores (6). Attempts to characterize cellular defense mechanisms in iron deficiency anemia again produced conflicting results (7). The fate of a bacterial infection in an iron deficient host might be determined by a balance between at least three variables: a̲ a resistance enhancing "milieu interne" of the host; b harmful effects of the deficiency state on the host, and c the pathogen's intrinsic ability to acquire iron.

Recently, we reported on the increased survival of iron deficient mice infected with S. typhimurium (2386, virulent type!) as compared to iron replete controls[1]. The average hemoglobin concentration of the anemic animals was 10 g/100 ml. It was speculated that the moderate degree of the deficiency state was an important factor for the result observed. There are now data available from one experiment where several degrees of iron deficiency were evaluated.

Animals prepared as indicated in the legend to Fig. 1 were infected at the age of 84 days. Whole body iron measurements reflected the substituted amounts of iron (Table 1). Although the differences between groups were not impressive increasing amounts of body iron were associated with increasing mortality rates. In addition, as body iron increased the animals appeared to die earlier in the course of the disease (Fig. 3).

If one accepts a that the differences between iron deficient and iron replete animals are reality, and b that the

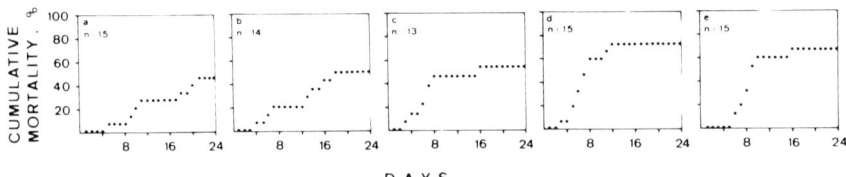

Fig. 3. Cumulative mortality of mice with various degrees of iron deficiency intraperitoneally infected with 160 organisms of S. typhimurium 2386. Groups a - d were fed an iron poor diet after weanling. 18 days prior to infection iron (as iron dextrin) was given parenterally: a: none; b: 0.3 mg, c: 0.6 mg; d: 2.0 mg. Group e was fed a normal stock diet.

---

[1] M. Puschmann and A. M. Ganzoni. Submitted to publication.

TABLE 1

Hematological Data and Average Mortality in the 5 Mice Groups Studied

Group	Iron dextrin 18 d prior to infection (mg [iron])	Whole body[2] iron (μg/animal)	Hemoglobin (g/100 ml)	Serum iron (μg/100 ml)	Total iron binding capacity (μg/100 ml)	Dead animals at d 26 (%)
a (15)[1]	–	795 ± 176	4.6 ± 0.7	78 ± 36	484 ± 31	47
b (14)	0.3	1252 ± 151	*			50
c (13)	0.6	1622 ± 125	*			54
d (15)	2.0	2315 ± 549	*			73
e (15)	(stock diet)	2047 ± 139	14.6 ± 0.9	261 ± 28	450 ± 55	67

*Hemoglobin concentration normalized within 12 to 14 days after 1 mg of iron as iron dextrin.

[1]Number of animals infected.

[2]Animals were dissolved in 100 ml of concentrated nitric acid and the iron was quantitated on a 1 : 10 dilution with distilled water following the recommendations of the ICSH.

improved survival of severely anemic animals is a consequence of iron deficiency, a number of questions arises. As pointed out above salmonella type 2386 did not undergo transformation into a more virulent form in the acutely iron loaded animal. Presumably, this pathogen remains unrestrained in terms of iron acquisition within a normal host. But, the same organism appears to encounter increasing difficulties to meet its iron needs as the host's body iron is progressively reduced. *In vitro* growth of *E. coli* has been found inhibited by specific antibodies and transferrin, but in presence of both components inhibition was more pronounced (8). Lack of iron may optimize some effects of specific immunity. For *S. typhimurium* infection in mice a decisive role of humoral antibodies for disease control was postulated (9). Furthermore, specific humoral response has usually been found intact in iron deficiency. Examination of other host-pathogen combinations is certainly urgent. Future work also needs to be closely monitored by *in vitro* studies of bacterial behaviour.

III. ACKNOWLEDGMENTS

We thank Mrs Tellervo Kathke-Nieminen for secretarial assistance. This work was supported by a grant from the Deutsche Forschungsgemeinschaft.

IV. REFERENCES

1. King, R. D., Khan, H. A., Foye, J. C., Greenberg, J. H., and Jones, H. E., *J. Lab. Clin. Med.* 86, 204 (1975).
2. Collins, F. M., *J. Reticuloendothel. Soc.* 10, 58 (1971).
3. Payne, S. M., and Finkelstein, R. A., *Infect. Immun.* 12, 1313 (1975).
4. Baggs, R. B., and Miller, S. A., *J. Nutr.* 103, 1554 (1973).
5. Chu, S.-h. W., Welch, K. J., Murray, E. S., and Hegsted, D. M., *Nutr. Rep. Int.* 14, 605 (1976).
6. Masawe, A. E. J., and Muindi, J. M., *Lancet* 2, 314 (1974).
7. Suskind, R. M., and Adeniji-Jones, S., *Pediatrics* 88, 696 (1976).
8. Bullen, J. J., Rogers, H. J., and Leigh, L., *Br. Med. J.* 1, 69 (1972).
9. Marecki, N. M., Hsu, H. S., and Mayo, D. R., *Br. J. Exp. Pathol.* 56, 231 (1975).

THE INHIBITORY EFFECT OF SERUM ON THE IMMUNORADIOMETRIC ASSAY OF FERRITIN

DAVID LIPSCHITZ
JAMES COOK

University of Kansas Medical Center
Kansas City, Kansas

I. INTRODUCTION

Measurements of serum ferritin by immunoradiometric assays (IRMA) are now widely used for clinical and nutritional evaluation of iron status. Epidemiologic studies in particular require careful standardization and validation of these measurements to permit longitudinal studies and to allow comparisons of iron status in different geographic areas of the world. An important technical problem with the two-site IRMA assay commonly used to measure serum ferritin is the inhibitory effect of serum. The present study was undertaken to evaluate this problem.

II. METHOD

Serum ferritin was measured by a two-site IRMA which has been fully described in a previous report (1). Measurements are performed on 1-10μl of a serum unknown diluted to 200μl in Veronal buffer containing bovine serum albumin (BSA-buffer) and added to a polystyrene tube previously coated with rabbit antiserum raised against recrystallized human ferritin. Following 24 hours incubation and rinsing of the tube, ^{125}I-labeled antiferritin antibody is added to the tube and incubated for 24 hours to allow maximum binding to the insolubilized antigen. The tube is washed to remove unbound labeled antibody and the residual radioactivity measured in a well-type scintillation counter.

Seven standards containing 0.25-10ng ferritin protein are prepared by addding recrystallized human liver ferritin to BSA-buffer containing a final concentration of 5% normal rabbit serum (5% NRS) unless otherwise stated. Ferritin protein was determined by the Lowry method (2) using bovine serum albumin as a standard.

III. RESULTS

A. Dilution Studies

Measurements of ferritin by the two-site IRMA are inhibited by the presence of serum in the measured sample. For example, when human serum is diluted with BSA-buffer to obtain final serum concentrations of 5, 2 and 1% a progressive rise in the measured value is observed. These gross disparities can be minimized by introducing sufficient rabbit serum at each dilution to maintain a constant serum concentration of 5% (Fig. 1).

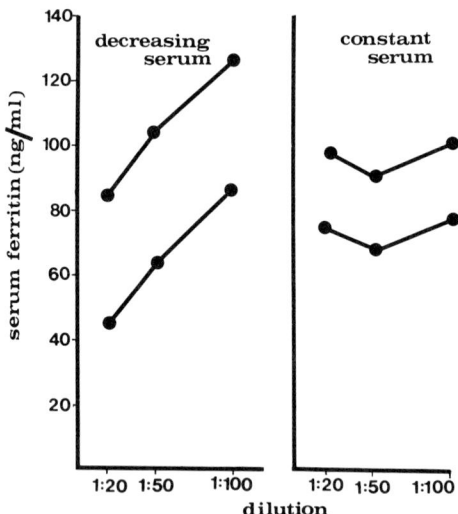

Fig. 1. The effect of serum on serum ferritin measurements. When human serum was diluted in BSA-buffer alone (left), a progressive rise in values was observed due to a reduced serum concentration. When the same serum was diluted with sufficient NRS to maintain a serum concentration of 5% at all dilutions, constant values were obtained (right). All samples were read against standards containing 5% NRS.

Because the inhibitory effects of rabbit and human serum are equivalent systematic errors due to serum inhibition can be avoided by preparing standards which contain the same quantity of NRS as the serum unknown. Sera containing 0-200ng ferritin/ml are conveniently measured by diluting 1:20 with BSA-buffer. Error due to serum inhibition can be eliminated by preparing standards in 5% NRS. At serum concentrations above 200ng/ml requiring greater dilution of the serum unknown, standards are prepared with proportionately smaller amounts of NRS. Because this serum effect disappears at concentrations below 1%, dilutions greater than 1:100 can be measured against standards prepared in BSA-buffer alone.

B. Recovery Studies

Recovery studies provide an important means of validating laboratory measurements. However, when adding known quantities of ferritin to serum and measuring recovery with the two-site IRMA a second type of inhibitory serum effect is observed which is unrelated to the final concentration of serum in the assayed sample. Low recovery of added ferritin was pronounced when added to undilute human serum. The results of a typical study are shown in Table 1. Recrystallized human ferritin in a volume of 10µl containing 112ng (as measured in 5% NRS) was added to 8 samples of human sera with basal ferritin concentrations between 1-97ng/ml (average 35ng/ml). When diluted 1:20 with BSA-buffer and read against standards containing 5% NRS, recovery of the added ferritin ranged from 58-80% (average 68%) and was not related to the basal ferritin concentration.

A study was then performed to determine whether or not this incomplete recovery was an artefact related to the use of recrystallized ferritin. Serum from a patient with untreated hemochromatosis in a volume of 30µl containing 108ng ferritin (as measured in 5% NRS) was added to the same sera used for recovery studies with recrystallized ferritin (Table 1). Recoveries were again incomplete ranging from 67-84% (mean 69%) and did not differ from the recoveries using recrystallized ferritin. Percentage recovery of human ferritin added to whole NRS was of the same order as observed with human serum ferritin.

This second type of serum effect could be eliminated by performing recovery studies with dilute rather than whole serum as shown by the following study. Recrystallized ferritin (10µl) was added to varying concentrations of rabbit serum which were then further diluted in the assay to obtain

Table 1. Serum ferritin concentration (ng/ml) before and following the addition of recrystallized and native circulating ferritin to human serum.

Sample	Basal Serum Ferritin	With Added Recrystallized Ferritin[a]	With Added Native Ferritin[b]
A	1	74(65)[c]	70(64)
B	3	73(63)	75(67)
C	12	92(71)	83(66)
D	15	91(68)	87(66)
E	39	121(73)	116(71)
F	51	125(66)	125(68)
G	61	151(80)	133(67)
H	97	162(58)	188(84)
Mean	35	111(68)	110(69)

[a] 0.01ml containing 112ng recrystallized human ferritin as assayed in 5% NRS was added to 1ml whole serum.
[b] 0.03ml serum from a patient with untreated hemochromatosis (ferritin concentration 3624ng/ml) containing 108ng ferritin as measured in 5% NRS was added to 1ml whole serum.
[c] percentage recovery shown in parenthesis.

a final concentration of 5% NRS in all samples. Serum ferritin values of 121 and 118ng/ml were obtained when added to 5% and 25% NRS respectively as compared with values of 102, 106 and 102ng/ml when added to 50%, 75% and undilute NRS respectively. Thus incomplete recovery was only observed when ferritin was added to serum more concentrated than 25%. Similar results were observed in studies with human serum.

IV. DISCUSSION

Two distinct types of inhibitory serum effects were observed in the present study. The first, identified by dilution studies, was proportional to the amount of serum at concentrations greater than 1%. This type of inhibition has been observed with previous two-site IRMA (3) but its cause has not been adequately explained. It could be due to an exchange of IgG in the serum unknown for specific IgG bound to the wall of the tube which would reduce the number of sites at which ferritin is insolubilized during the first stage of the assay. Alternatively, serum might interfere with the insolubilization of antigen or may inhibit the binding of ^{125}I - labeled IgG to antigen during the second stage of the assay. The specific factor in serum is not known but high molecular weight proteins such as IgM or alpha-2-macroglobulins have been incriminated (4).

Prior attempts to eliminate this type of serum inhibition have been disappointing. One approach has been to use immunologic spacer arms so that the antibody to which the unknown antigen reacts is at some distance from the wall. Other maneuvers which have decreased the serum effect in some assays have involved the addition of dextran solutions or Tween 20 to samples prior to assay (3). None of these measures were found to decrease serum inhibition in the ferritin assay. Because of the unique ability of ferritin protein to resist denaturation by heating, we also attempted to eliminate the serum effect by prior heating of the serum unknown to precipitate the bulk of non-ferritin protein. No improvement in serum inhibition was observed.

Although it has not been possible to eliminate this first type of serum inhibitory effect, systematic errors can be avoided by preparing standards with the same concentration of rabbit serum as the diluted serum unknown. While rabbit and human serum appear roughly equivalent in their inhibitory effect, a slightly higher ferritin value is often obtained when serum is assayed at a 1:100 dilution against standards containing 1% NRS as compared with the 1:20 dilution against standards containing 5% NRS. This residual difference is small.

The second type of serum inhibitory effect not previously described was observed in recovery studies when purified or native ferritin was added to whole rabbit or human serum prior to dilution in the assay. From the aspect of recovery studies this can be eliminated by initially diluting the serum to at least 25% with BSA-buffer. However this error may also apply to endogenous ferritin which circulates in undilute plasma. The best approach for standardization with other assays is therefore to prepare standards which are identical with the serum unknowns by adding ferritin to undilute rabbit serum or human serum which has been rendered ferritin-free. Both standards and serum unknowns can then be diluted in parallel with BSA-buffer alone. We have also found with this approach that human sera assayed at 1:20 and 1:100 give identical ferritin values.

ACKNOWLEDGEMENTS

This work was supported by USPHS grant 1 R01 AM19011 and Contract 223-76-2112 from the Food and Drug Administration.

REFERENCES

1. Miles, L.E.M., Lipschitz, D.A., Bieber, C.P., Cook, J.D., Anal. Biochem. 61, 209 (1974).

2. Lowry, O.H., Rosebrough, W.J., Farr, A.L., Randall, R.J., J. Biol. Chem. 193, 265 (1951).

3. Miles, L.E.M., Bieber, C.P., Eng, L.F., Lipschitz, D.A., in "Symposium on Radioimmunoassay and Related Procedures in Clinical Medicine and Research" pp. 149-164. International Atomic Energy Agency, Vienna, 1973.

4. Reuter, A.M., Hendrick, J.-C., Sulon, J., Franchimont, P., Acta. Endocr. (Kbh.) 72, 235 (1973).

THE DEVELOPMENT OF NEW IRON CHELATING DRUGS FOR THE
TREATMENT OF PATIENTS WITH THALASSEMIA

Anthony Cerami, Robert W. Grady, Charles M. Peterson,
Robert L. Jones, and Joseph H. Graziano*
The Rockefeller University and
New York Hospital-Cornell Medical Center*

Transfusion therapy on a continual basis is the only available treatment for the severe anemia associated with the genetic disease, β-thalassemia major. Because the body lacks an effective means of excreting iron, virtually all the iron administered as transfused erythrocytes is retained within the body and stored in various tissues as ferritin or hemosiderin. This accumulation of iron, particularly in the heart, liver and pancreas, leads to fibrotic changes in these tissues resulting in organ failure and early death (1). Most thalassemia patients on a regular transfusion regimen die by the end of their second decade. This transfusion-induced secondary hemochromatosis is not unique to thalassemia, but also occurs in aplastic anemia, sideroblastic anemia and sickle cell anemia (2). Although one might correct the defect in thalassemia by preventing the switch from fetal hemoglobin (Hb F) to adult hemoglobin (Hb A) (3), this is not possible at the present time. Accordingly, it appears that one of the most effective ways of dealing with this disease is to remove the excess iron that accumulates in these patients.

Several years ago we undertook a program to design and evaluate new iron-chelating drugs with emphasis on those which might be orally effective. At that time most attempts at preventing excess iron accumulation centered on the use of desferrioxamine (4-11), an iron chelator produced by *Streptomyces pilosis*. Initially our investigation focused on compounds which were derived from specific iron chelators produced by microbes. Since the pioneering work on desferrioxamine, many iron chelators have been isolated and characterized from bacteria, yeast, fungi and plants (12). The existence of these compounds stems from the extreme insolubility of ferric oxides ($Fe(OH)_3$: $K_{sp} = 10^{-38}$). During the course of evolution the requirement for iron by microbes has led to the development of specific mechanisms for sequestering environmental iron. The selection pressure for such mechanisms is great since iron is often the rate limiting nutrient for microbial, plant, and even human growth. The chelating agents involved in these systems primarily are either hydroxamic acids (13), of which desferrioxamine is an example, or derivatives of 2,3-dihydroxybenzoic acid (2,3-DHB) (14-17). Many naturally occurring iron chelators are large complex multidentate compounds difficult to obtain in large quantities (12). Accordingly we chose first to look at smaller molecules which incorporate the ligands selected for by the microbes.

To identify new agents we developed an animal model of iron overload by chronically hypertransfusing rats (18). Drugs under investigation were administered to these animals and the amount of iron excreted in the urine and feces determined by atomic absorption.

During the course of our studies three potentially useful agents have been identified and are currently undergoing more extensive evaluation. The first agent that we identified was 2,3-dihydroxybenzoic acid (18). When 2,3-DHB was administered orally to iron-overloaded rats, a significant increase in the amount of iron found in the urine was observed. This effect was specific for iron. No changes were observed in the excretion of copper, zinc, magnesium, calcium or potassium. 2,3-DHB was evaluated in experimental animals and found to be non-toxic. The LD_{50} is in excess of 6 g/kg. The drug was administered chronically to mice, rats, and dogs for periods up to one year and found to be without side effects. A clinical evaluation of the compound was thus initiated.

Administration of 2,3-DHB at 25 mg/kg for eight days to five patients with β-thalassemia major caused an average increase in iron excretion of 4.5 mg/dy (19). When the drug was administered four times per day at the same dose to eight patients, the average excretion was found to be 6.5 mg/dy with a range of 1.4-19 mg/dy. The chelation was highly specific for iron with no increased excretion of copper, zinc, magnesium, calcium, or potassium. It is of interest that the route of excretion in man is different from that in the rat. In man the excretion was primarily via the fecal route. The reason for this change in the mode of excretion remains unclear as does the variability of iron excretion from one patient to another. It is of interest that White et al. (20) have found that the ability of 2,3-DHB to chelate iron in vitro using a Chang liver cell system was potentiated by prior incubation of the drug with a liver homogenate. The possibility of 2,3-DHB being biotransformed into a more effective chelator in the liver remains a possibility. The extent to which this change takes place could explain the variability in iron excretion as well as the altered route of excretion in man. Further work is needed in this area.

Having obtained a measure of success with 2,3-DHB, we undertook a systematic screening program with the hope of discovering an even more efficacious analog. Of the 26 benzoic acid derivatives studied none appeared to be more effective than 2,3-dihydroxybenzoic acid (21). Moreover, toxicity increased in the case of the 3,4 analogs.

A one-year double-blind evaluation of this drug was initiated in collaboration with the New York Hospital Transfusion Clinic. The drug was administered at a dose of 25 mg/kg $q.i.d.$ and a number of clinical parameters were assessed. At the present time it appears that the drug was well tolerated during the course of this study and that there may be a retardation, but not a complete inhibition of iron accumulation in these patients. It is hoped that an evaluation of 2,3-DHB in combination with desferrioxamine will be undertaken in order to determine if there is an additive effect justifying the use of both drugs in these patients.

The second compound that is undergoing evaluation at the present time is rhodotorulic acid (RA). This compound is produced by and isolated from cultures of *Rhodotorula pilimanae*. When administered parenterally, RA induced iron excretion via both the urinary and the fecal routes and was more than twice as potent on a weight basis as desferrioxamine (22). However, like desferrioxamine, RA is not effective when administered orally. The relative insolubility of RA in water should make it possible to inject the drug as a suspension which will dissolve slowly thereby achieving maximal chelation. This is found to be the case in experimental animals. After intramuscular injection of suspensions of RA, significant serum levels are maintained over 6-12 hours whereas an intramuscular injection of desferrioxamine is completely cleared in less than 2 hours. The slow infusion of desferrioxamine over 12-24 hours leads to a significant increase in the efficiency of chelation (23-24). It is believed that this is due to the fact that desferrioxamine removes iron from a chelatable pool which is replenished slowly from less accessible pools in the body. Clinical evaluation of RA is currently underway.

In order to overcome the problem of absorption encountered with hydroxamic acids such as RA and desferrioxamine we have synthesized a number of hydroxamic acid molecules of differing lipophilicities. Unfortunately most of the derivatives that we have synthesized are either not orally absorbed or ineffective in removing iron from the hypertransfused rat (21). We have, however, found one orally active hydroxamic acid, cholylhydroxamic acid (CHA). This compound is orally absorbed and is as effective as parenterally injected desferrioxamine in the iron-overloaded rat model. Presumably CHA is transported into the liver via the enterohepatic circulation where it chelates iron after which the chelate is excreted through the bile. It is hoped that a non-absorbable iron chelator in the diet, such as the tannins of tea, will then remove iron from the chelate so as to allow the CHA to be recycled. In addition such compounds would prevent orally absorbed iron from contributing to the inexorable iron accumulation seen in these patients. Further pharmacological and toxicological studies of CHA are in progress. In addition we are synthesizing bile acids containing multidentate centers in the hope of obtaining a compound that will have a higher affinity for iron and thus be more efficacious than CHA.

Another interesting group of compounds having a high affinity for iron and being biologically active *in vitro* (25) and *in vivo* are the tropolones. These molecules have a very high affinity for iron which is not as dependent on pH as the hydroxamic acid moiety. Unfortunately most of the tropolones that we have studied have significant toxicity thus precluding their use in patients. We hope to design a tropolone that has decreased toxicity while retaining its ability to chelate iron *in vivo*. These experiments are currently underway.

All in all, the development of agents to remove iron from patients with β-thalassemia major remains a challenge to the medicinal chemist. Many structures are now known which are specific for chelating iron yet the task remains to design an agent which has the capability to remove sufficient iron from these patients to keep them in iron balance. This is not a simple task, however, when one realizes that these patients are receiving the equivalent of about 20 mg of iron per day as transfused erythrocytes. This means that a significant amount of drug must be administered in order to remove this large quantity of iron. The obvious need for a non-toxic highly efficient chelator is clear. Hopefully, in the years ahead, a series of drugs will be found such that therapeutic programs utilizing several of these agents will alow the physician to maintain these patients in negative iron balance.

REFERENCES

1. Ellis, J.T., Schulman, I., and Smith, C.H., Am. J. Pathol. 30, 287 (1954).
2. Walker, R.J., and Williams, R., in "Iron in Biochemistry and Medicine" (A. Jacobs and M. Worwood, Eds.), p. 596. Academic Press, New York, 1974.
3. Conley, C.L., Weatherall, D.J., Richardson, S.N., Shepard, M.K., and Charache, S., Blood 21, 261 (1963).
4. Wohler, F., Acta Haematol. 30, 65 (1963).
5. Erlandson, M.E., Golubow, J., Wehman, J., and Smith, C.H., Ann. N.Y. Acad. Sci. 119, 769 (1964).
6. Smith, R.S., Ann. N.Y. Acad. Sci. 119, 776 (1964).
7. Gevirtz, N.R., Tendler, D., Lurinsky, G., and Wasserman, L.R., N. Engl. J. Med. 273, 95 (1965).
8. Hwang, Y.-F., and Brown, E.B., Arch. Intern. Med. 114, 741 (1964).
9. Hedenberg, L., Scand. J. Haematol., Suppl. 6 (1969).
10. Barry M., Flynn, D.M., Letsky, E.A., and Risdon, R.A., Br. Med. J. 1, 16 (1974).
11. Letsky, E.A., Miller, F., Worwood, M., and Flynn, D.M., J. Clin. Pathol. 27, 652 (1974).
12. Neilands, J.B., Struct. Bonding 1, 59 (1966).
13. Emery, T., Adv. Enzymol. 35, 135 (1971).
14. Ito, T., and Neilands, J.B., J. Am. Chem. Soc. 80, 4645 (1958).
15. O'Brien, I.G., Cox, G.B., and Gibson, F., Biochim. Biophys. Acta 177, 321 (1969).
16. Corbin, J.L., and Beulen, W.A., Biochemistry 8, 757 (1969).
17. Tait, G.H., Biochem. J. 146, 191 (1975).
18. Graziano, J.H., Grady, R.W., and Cerami, A., J. Pharmacol. Exp. Therap. 190, 570 (1974).
19. Peterson, C.M., Graziano, J.H., Grady, R.W., Jones, R.L., Vlassara, H.V., Canale, V.C., Miller, D.R., and Cerami, A., Br. J. Haematol. 33, 477 (1976).
20. White, G.P., Jacobs, A., Grady, R.W., and Cerami, A., Br. J. Haematol. 33, 487 (1976).
21. Grady, R.W., Graziano, J.H., Akers, H.A., and Cerami, A., J. Pharmacol. Exp. Therap. 196, 478 (1976).
22. Grady, R.W., Graziano, J.H., Akers, H.A., and Cerami, A., Blood 44, 911 (1974).
23. Propper, R.D., Shurin, S.B., and Nathan, D.G., N. Engl. J. Med. 294, 144 (1976).

24. Hussain, M.A.M., Flynn, D.M., Green, N., Hassein, S., and Hoffbrand, A.V., Lancet 2, 1278 (1976).
25. White, G.P., Jacobs, A., Grady, R.W., and Cerami, A., Blood 48, 923 (1976).